Man's Anatomy, Physiology, Health and Environment

NANCY ROPER MPhil, SRN, RSCN, RNT

British Commonwealth Fellow 1970

FIFTH EDITION

CHURCHILL LIVINGSTONE EDINBURGH LONDON AND NEW YORK
1976

CHURCHILL LIVINGSTONE
Medical Division of Longman Group Limited

Distributed in the United States of America by
Longman Inc., 19 West 44th Street, New York,
N.Y. 10036 and by associated companies,
branches and representatives throughout
the world.

First Edition	1963
Second Edition	1965
Reprinted	1967
Third Edition	1969
Reprinted	1970
Fourth Edition	1973
Fifth Edition	1976

ISBN 0 443 01497 3

Library of Congress Cataloging in Publication Data

Roper, Nancy
  Man's anatomy, physiology, health, and environment.

  Includes bibliographical references and index.
  1. Hygiene. 2. Human physiology. 3. Anatomy, Human.
I. Title. [DNLM: 1. Anatomy—Nursing texts.
2. Hygiene—Nursing texts. 3. Physiology—Nursing texts.
QT200 R784m]
RT69.R63 1976   613   76–25811

Preface to the Fifth Edition

The reorganization of the Health Services in April 1974 has necessitated a fifth edition of this book. Although all the countries in the U.K. have a similar service, now that I am domiciled in Scotland, I have described its service. For the sake of consistency and also because it is likely that there will be change in the examinations set by the General Nursing Council for England and Wales, the questions set by the Council have been omitted.

My thanks are hereby given to many people who have wittingly and unwittingly contributed to the book, but I accord a special 'thank you' to Win Logan who was always willing to help with this edition.

I hope that many people will continue to find the book interesting and useful.

NANCY ROPER

EDINBURGH, 1976

There has been a major revision in the fourth edition of this book. It now starts with a discussion of the origin of life, followed by modern cell theory, with an introduction to genetics and inheritance. Concepts of health are discussed, together with the development and organization of societies, and national expenditure to maintain society's health. This is put in a global context with discussion of the World Health Organization. Man's responsibility to himself, to others and to his environment is stressed throughout the book. The structure and function of each system is discussed together with health education concepts of achieving maximal functioning of that system; mentioning the varying levels of achievement in the developing countries. The development of infants through to maturity and old age is discussed with consideration for the essential contribution of leisure, recreation, work, rest and sleep. Addiction to, and dependence on tobacco, alcohol, marijuana/pot/cannabis, amphetamines, barbiturates, LSD, heroin, cocaine and morphia are discussed and set in a world context. The nutritional problems of obesity, malnutrition, and food shortage are presented, together with the many ways in which these problems are being tackled throughout the world. Provision of milk and water is similarly surveyed in different countries, along with the ever-present danger of pollution of these commodities. Society's problems and solutions for sanitation and the disposal of refuse, and pollution of the atmosphere are mentioned. Research work on smoking is referred to, the effect of smoking on the unborn child, and the problem of children smoking in different countries. Deafness and blindness are discussed in relation to provisions by varying countries for those so afflicted. The varying levels of resources of many countries are mentioned in the presentation of housing, heating and lighting. There is discussion of different society's provisions for, and attitudes to, pregnancy, marriage, the rights of children, illegitimacy, marriage guidance counselling, divorce, abortion, contraception, and the sexually transmitted diseases.

Because my prime concern is to make this book useful to the nursing profession, I have included relevant portions from questions in the Final Papers for the General Register set by the General Nursing Council for England and Wales. However, it is hoped that other health workers and school children will find it useful, especially as they are being increasingly encouraged to develop a responsible attitude to health and their environment.

NANCY ROPER

EDINBURGH, 1973

vi

Contents

The origin of life	1
Modern cell theory	4
Genetics and inheritance	10
The growth of a human being	13
Development of societies	20
Different patterns of living	21
The effect of outside influences on a group	22
Metropolitan planning	22
National expenditure	22
Health education	26
The United Nations Organization	30
The World Health Organization	32
General survey of the Health Service	33
The responsibility of being a citizen	39
The importance of mental and physical health	40
The skeleton	43
posture	57
care of feet in maintenance of health	70
conditions which interfere with health of feet	72
care of hands in maintenance of health	88
conditions which interfere with health of hands	89
The muscular system	92
exercise	108
fatigue	110
Joints or articular system	112
The skin	121
care of skin in maintenance of health	125
conditions which interfere with health of skin	129
external animal parasites	129
external vegetable parasites	131
clothing	133
The circulatory system	141
factors affecting the circulatory system	165
care of circulatory system	167
The lymphatic system	173
care of lymphatic system	177
The spleen	179
The nervous system	181
normal development	195
care of nervous system	204
addiction/dependence	207
conditions which interfere with nervous system	212
Nutrition in relation to health	217
wholesale storage of food	232
hygiene of food shops	233
quality of food	233
food poisoning	238
food-borne infection	240
clean milk supply	246
diseases spread by milk	255
clean water supply	255
diseases spread by water	260
Digestive system	265
care of mouth	270
conditions which interfere with mouth	271
care of digestive system	293
conditions which interfere with digestive system	294
Sanitation and disposal of refuse	303
Air or atmosphere	333
pollution of the atmosphere	337
prevention of pollution	342
Respiratory system	347
care of respiratory system	362
smoking	363
conditions which interfere with respiratory system	365
Urinary system	370
care of urinary system	378
Endocrine system	382
care of endocrine system	398
The eye	399
visual perception	399
care of eyes	406
conditions which interfere with eyes	409
concerning blind people	411
Lighting	414

The ear   420
  care of ears   423
  concerning deaf people   424
  hearing as it affects mental
  health   425
  noise   426
Reproductive system   429
  care of reproductive system   442
  pregnancy   444
  marriage   445
  rights of children   446
  illegitimacy   446
  marriage guidance counselling   447
  divorce   448
  contraception   448
  abortion   452

  sexually transmitted diseases   453
Housing   464
  safety of the home   470
  overcrowding   477
  modification of a home   478
  care of the home   478
  household pests and vermin   479
  relationship of bad housing to
  disease   484
Heating   485
Prevention of infection   491
  control of epidemics   499
Disinfection   501
References   505
Index   513

The Origin of Life

One of the main functions of a nurse is to preserve life. There is no simple definition of the word 'life', yet man has for centuries been fascinated by questions about the origin of life. Like any other individual, a nurse establishes her personal belief about this subject, but she cares for people holding various beliefs, so that it is necessary to survey, albeit briefly, some of the propositions for the origin of life.

Was man 'created' as in Genesis, Chapter 1? Thinking along these lines, Aristotle, a Greek philosopher, over 2,000 years ago assumed that life could be produced in matter by an 'active principle'. He used this assumption to account for what he thought was the sudden and apparently causeless appearance of some animals and plants. It is a *hypothesis of spontaneous generation* of life. It fits in with the *ideology of vitalism* defined about 1,500 years later. This states that living things are intrinsically different from non-living things, and that their study and ultimate explanation is beyond the physical laws of chemistry, physics and mathematics, that govern the rest of the universe.

In the early seventeenth century, an Italian biologist, Francesco Redi, challenged the idea of spontaneous generation and supported the *theory of biogenesis*—that all life comes from pre-existing life. A few years later Anton van Leeuwenhoek, a Dutch biologist, perfected a simple microscope through which teeming bacteria could be seen. People had not previously suspected that these tiny creatures existed. Could they have been 'spontaneously generated'? So people turned back with new interest to the hypothesis of spontaneous generation, and proposed the idea that life arose from time to time from non-living materials.

In the eighteenth century, John Needham of London carried out experiments in support of spontaneous generation of life. He heated vegetable juice, placed it in a test tube, sealed the tube and heated it again. After a few days the fluid was swarming with organisms which he thought proved spontaneous generation. About 25 years later Lazzaro Spallanzani, an Italian priest, repeated Needham's experiments. After sealing the tubes he boiled them for one hour. After several days there was no sign of life in the tubes. Spallanzani suggested that Needham had not heated the test tubes at a high enough temperature to kill all the organisms contained therein. Needham riposted that boiling for one hour would destroy the 'active principle' of the infused substances. He was supported by the popular belief at that time, as opinion is often slow to change. Spontaneous generation and vitalism continued to be more favoured than biogenesis as an explanation of the origin of life.

In the latter half of the nineteenth century Louis Pasteur, a French biologist, performed conclusive experiments that rejected the theory

1

of spontaneous generation and brought about gradually the acceptance of the theory of biogenesis—all life comes from pre-existing life.
Not that this acceptance meant that all questions had been answered,
but a sufficient number were answered to give impetus to the
mechanist ideology which stated that living things are but machines,
exquisitely and delicately complex perhaps, but nevertheless subject

to the general laws of nature and as such, given time, capable of being
understood by human minds and duplicated by human hands. Despite
the tremendous increase in knowledge during this century, not all
the questions about the origin of life have been answered. The
assumption that it takes a living organism to produce a living
organism gives rise to the question: Do all living things have a
common ancestor? A partial answer is the *theory of evolution*; the
theory that all living things on the earth today are modified descend-
ants of protists, plants, and animals that have lived before them.
(Protists are a miscellaneous group—neither plants nor animals.)
The two theories (biogenesis and evolution) have a common content,
but they differ in the key word *modified*.

Returning to biogenesis, if it takes life to produce life, where did
the first life come from? In this day and age it seems reasonable to
ask: Did it come from outer space? This implies that the spores of
organisms were sufficiently strong to withstand the rigours of outer
space. To investigate the possibility, present-day scientists are
investigating space for signs of life.

An organism that is capable of making its own food is called an
autotroph. All green plants and some bacteria are autotrophs. Green
is the key word and refers to chlorophyl which can harness energy
from the sun to produce complex food molecules of carbohydrate, fat
and protein. Photosynthesis is the name given to this process.

The autotroph hypothesis assumes that the first form of life on
earth was a food-maker, i.e. it was a complex organism that origin-
ated in simple surroundings, such as those on primitive earth. This
does not fit the theory of evolution which states that complex
organisms are the result of many minute changes that take place and
add up to significant change only after a very long period. It may be
that future scientists will gain knowledge that will make the autotroph hypothesis a tenable one.

An organism that is not capable of making its own food is called a
heterotroph. It may be capable of making a few compounds, but in
general it has to rely on an outside source for its food. Man, nearly
all animals, many bacteria, and some plants such as moulds and
mushrooms are heterotrophs. Originally all natural food comes from
the process of photosynthesis in autotrophs. Algae are eaten by tiny
fish, which are eaten by bigger fish, which are eaten by mammals.
Plant-eating mammals (herbivores) provide food for flesh-eating
mammals (carnivores). Man is classified as 'omnivore' since he eats
both plants and animals

The heterotroph hypothesis assumes that the earliest form of life
developed from non-life and was a simple organism incapable of
making its own food. The original conception of spontaneous genera-

2

tion was that complex organisms such as animals and plants could arise suddenly and daily from non-living matter, thus differentiating the spontaneous generation from the heterotroph hypothesis. The latter fits in with Darwin's theory of evolution by natural selection. At the same time it would appear to presume a further hypothesis that there was a long, slow evolution of chemical compounds over millions of years preceding the evolution of a simple organism.

Scientists estimate that the earth is about 4,700 million years old, and that its primitive atmosphere probably lacked oxygen, therefore life on earth probably began as life without oxygen. The oldest known fossils are estimated to be 3,100 million years old and are rod-shaped, bacteria-like objects in ancient rocks from South Africa. It is thought that these bacteria-like objects may have lived without oxygen in water or mud.

When trying to describe 'life', one might say that all living things have a boundary separating them from their environment, from which they absorb molecules, from which they obtain chemical energy. They are also able to reproduce themselves. Yet in systems termed 'non-living' some of these properties are found, e.g. if a salt crystal is added to a concentrated salt solution, the crystal will 'grow' and start the formation of another crystal. Viruses form a group between living and non-living things. They can survive outside the living cells of plants and animals, but they can only reproduce themselves while they are parasites inside plant or animal cells.

To support, but not confirm the heterotroph hypothesis, Stanley Miller, University of Chicago, demonstrated in 1952 that amino acids (complex basic units that make up *protein*, the nitrogenous food essential for life) can be produced from simple gases such as are thought to have surrounded the earth. Energy is necessary for this conversion and Miller used a spark-discharge apparatus in the laboratory. We now know that energy is used in each living cell to combine amino acids into proteins. That is all very well but does not explain how amino acids were made into protein before there were living cells to do this. Sidney Fox, Florida State University, demonstrated in the mid 1960s that heating a dry mixture of amino acids results in many of the amino acids bonding together to form larger, more complicated molecules that have some of the properties of proteins. It is thought that the energy to form compounds from the gas molecules of primitive earth's atmosphere was ultraviolet radiation from the sun, electrical energy from lightning, radiation from radioactive elements in the earth's crust, and erupting volcanoes.

The heterotroph hypothesis goes on to say that complex organic molecules became grouped in clusters to form *pre-cells*. Different types of pre-cells may have competed for the organic molecules needed for growth and reproduction in the waters (sometimes likened to a 'thin soup') of primitive earth. The more successful pre-cells may have crowded out the less successful ones, thereby increasing in complexity until they resembled a heterotroph cell, eventually presenting the features of a *living cell* as currently described.

Knowledge of cells began to accumulate in the latter half of the seventeenth century when improvements in the microscope were made by Robert Hooke, an English scientist. After looking at a piece of cork through his microscope he used the word 'cells' to describe what he saw. At that time cells meant little cubicles or rows of small rooms. Early in the nineteenth century, Dutrochet, a French biologist compared plant and animal tissues (tissue=collection of cells) and stated that all the organs (organ=collection of tissues) of animals and plants are really only a 'cellular tissue' diversely modified. About the same time two German biologists, Schwann and Schleiden, further paved the way for the *modern cell theory*. This theory states that the cell is the basic unit of structure and function of a multicellular organism and that cell division brings about genetic continuity in the newly produced cell.

Acceleration of the production of new instruments and techniques in the first half of the twentieth century has so increased knowledge about cell structure and function (cytology), that whole books are now written on this subject; in fact books have been written about one component of a cell, popularly known as the 'double helix'. So what follows must of necessity be a mere introduction for those students who have not studied biology and a brief revision for those who have studied the subject.

MODERN CELL THEORY

A plant or animal cell is the 'unit of life', the 'atom of biology', and can be likened to an organized chemical factory. In a one-celled (unicellular) organism it is of the all-purpose variety, absorbing nourishment from its environment, manufacturing all the products necessary for the continuance of its life, and ridding itself of the waste products of metabolism. In a many-celled (multicellular) organism each cell becomes a specialty shop, doing one specialty job, e.g. muscle cells for movement, nerve cells for communication. Just as there is no typical mammal so there is no typical cell, but for descriptive purposes there are three main regions in a cell—the nucleus, cytoplasm and membranes (Fig. 1).

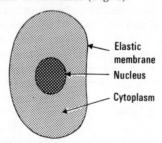

Fig. 1 A cell.

NUCLEUS

The nucleus is sometimes referred to as the 'brain of the cell'. It is generally a rounded body, though it may be flattened into a lens

4

shape, or it may even be lobed. It is capable of movement within the cell. The contained nuclear sap is bounded by a nuclear membrane. The nuclear sap bathes an intricate system of chromatin filaments or threads. In 1944 three American bacteriologists, Avery, McCarty and McLeod, found that the most important components of chromatin are *nucleic acids* which are responsible for the hereditary processes and the production of protein-enzymes that control all the activities essential for the cell's life. In 1953 Watson and Crick, of Cambridge, detailed the double helix structure (likened to a spiral staircase, Fig. 2) of a particular nucleic acid—*deoxyribonucleic acid* (DNA).

Fig. 2 A double helix.

DNA acts like a blueprint set out in the form of a chemical code for exact duplication of the cell, and for the structure of every molecule manufactured by the cell. DNA repeats itself like the patterns on wallpaper. It is thought that breakdown of the control exerted by the nucleus may be responsible for many diseases (particularly cancer) which are as yet poorly understood.

Though a cell without a nucleus can live for some time it cannot reproduce itself and it eventually ceases the activities necessary for its life and thus it dies. An example is the human red blood cell.

CYTOPLASM

Cytoplasm surrounds the nucleus and is bounded by the *cell membrane* which is made of fat molecules (phospholipids) that play a part in the absorption of fat-soluble substances, and protein molecules that are long and complex. The protein molecules can fold and unfold so that the membrane can, by expansion and contraction (molecular spacing) control which molecules can enter from the environment and which molecules can pass to the environment. Thus the membrane is said to be *selectively permeable*, the degree of permeability depending on the state of the protein molecules at any given time. It is not clearly understood why the concentration of sodium inside a cell is low and that of potassium is high. Cytoplasm contains other complex membranous structures: *mitochondria* that have a 'respiratory' function, and *lysosomes* that have a 'digestive' function. The protein-enzymes necessary to bring about these functions are made in the ribosomes (composed of another nucleic acid called *ribonucleic acid*—RNA) scattered throughout the cytoplasm. One form of RNA called Transfer RNA brings the amino acids (which reach the cell from the blood) from the cell membrane to the ribosomes which build them into a particular protein or protein-enzyme according to the 'plan' laid down by the nuclear DNA. Another type of RNA, appropriately called Messenger RNA, passes from the nucleus through the nuclear membrane and transmits the 'plan' from nuclear DNA to the ribosomes, (Fig. 3).

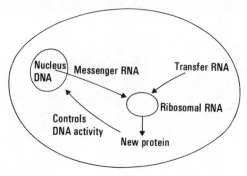

Fig. 3   Synthesis of proteins in a cell.

CELLULAR FLUID

In multicellular organisms, the majority of the cells continue to be bathed in fluid, spoken of as *extracellular* or *tissue fluid*, that acts as the transport medium bringing oxygen and nutrients to the cell wall, and carrying away carbon dioxide, waste products and water. Adult man has approximately 48 litres (80 pints) of water in his body, 12·4 litres (24 pints) of which is in the form of extracellular fluid. Fluid within each cell is termed *intracellular* and there is approximately 24·8 litres (48 pints) in man. Both fluids have to be kept at a constant composition, chemical reaction and temperature to maintain the health and vitality of the cells.

MOVEMENT OF CELLS

Some cells can move. A unicellular organism, the amoeba, moves by pushing out part of its cell membrane and then allowing the contents of the cell to flow into the pushed-out part. The movement is in response to something in the environment such as a food particle. The now projecting part of the cell encloses the food particle and digests it, a process called *phagocytosis* (Fig. 4). In cells other than the amoeba this movement is called *amoeboid action* (Fig. 5). White cells in human blood can move from their containing vessel, enter the tissues and engulf any invading germs, thus acting as a protection

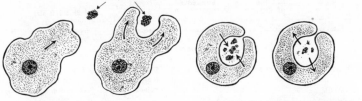

LOCOMOTION IN RESPONSE TO STIMULUS   DIGESTION ABSORPTION AND ASSIMILATION

Fig. 4   Phagocytic action.

against infection (Fig. 83). Some of the cells lining the blood vessels in the liver, spleen and bone marrow are phagocytic, i.e. they remove unwanted particles of solid matter, ranging from bacteria to old and dead red blood cells, from the blood.

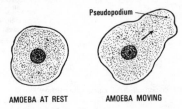

AMOEBA AT REST   AMOEBA MOVING

Fig. 5   Amoeboid action.

CELLULAR ENERGY

The chemical energy released in the cell by the breakdown of large molecules from food (carbohydrate and fats as well as proteins) is changed into one special form which is carried by a molecule known as *adenosine triphosphate* (ATP), sometimes called the 'energy currency' of living cells (Fig. 6). ATP can be likened to electricity (Fig. 7) in that it can be made in many different ways, it can be stored, and it can be used for many different purposes. The majority of ATP in mammals is from breakdown of substances in the presence of

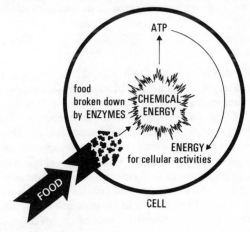

Fig. 6   Cellular energy.

7

	SOURCE OF ENERGY	CAN BE CONVERTED INTO:	CAN BE STORED AS:	CAN BE USED TO:
Plant world	Sun	Protein Fat Carbohydrate	Protein Fat Carbohydrate	Produce growth Reproduce species
Physical world	Falling water Coal Gas Oil The atom	Electricity	Electricity in power stations	Provide heating, lighting, cooking, refrigeration Drive electric motors for washing, drying, cleaning, lifting, pushing, transport, etc
Animal world	Breakdown of larger into smaller molecules, especially in the case of proteins, fats, carbohydrates	ATP	ATP	Build smaller molecules into larger ones, e.g. protein, fat, glycogen, enzymes, hormones and other secretions. Bring about movement through contraction of muscle. Transmit information via a network of nerves. Produce growth. Reproduce species

Fig. 7   Comparison of sources and uses of energy.

oxygen (oxidation or aerobic catabolism). A minority of ATP is from breakdown in the absence of oxygen (anaerobic catabolism). Energy carried by ATP is used to bring about the activities mentioned in the previous paragraph. To sum up: energy is used to build larger molecules from smaller ones (anabolism); to break down larger molecules into smaller ones (catabolism); to carry substances into the cell and arrange them inside the cell; to move unwanted substances out of the cell; to secrete substances if that is the function of the cell; to move; to perform mechanical work such as muscular activity, and to bring about cell division. In October 1971 Dr Hugh Huxley, of Cambridge University, proposed his 'sliding hypothesis', which explains how *chemical* energy can be transformed into *mechanical* energy in living organisms.

CELL DIVISION

One process of cell division is called *mitosis*. The nucleus enlarges, the thread-like filaments of chromatin merge into structures called chromosomes (molecules of DNA), there being a different number characteristic of the cells belonging to each species; for human beings it is 46. The nuclear membrane disappears and the chromosomes arrange themselves along the equator of the cell. Each chromosome divides longitudinally into halves (Figs 8, 9) which travel to opposite poles, there to disentangle into chromatin threads. Meantime the cytoplasm elongates and divides so that there are two cells, in each of which the chromatin threads are collected into the reformed nucleus (Fig. 8). Bacteria are examples of unicellular organisms and they can divide by mitosis every 20 minutes.

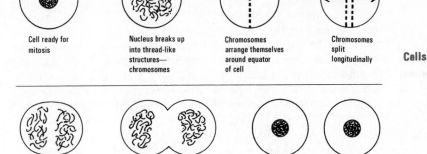

Cell ready for mitosis	Nucleus breaks up into thread-like structures—chromosomes	Chromosomes arrange themselves around equator of cell	Chromosomes split longitudinally

'Split' chromosomes travel to opposite poles. Elongation of cytoplasm	Cytoplasm begins to divide	Chromosomes collected together to form nucleus in each new cell

Fig. 8   Reproduction by mitosis.

In sexually reproducing species a female cell (gamete) is fertilized by a male cell (gamete). If the cells of the offspring resulting from this union are to have the number of chromosomes characteristic of the species, only half that number needs to come from each gamete. This reduction division is accomplished by an even more complicated cell division called *meiosis,* in which there are two nuclear divisions but only one division of chromosomes. The single new cell resulting from the fertilization of a female gamete by a male gamete is called a zygote and has the full complement of

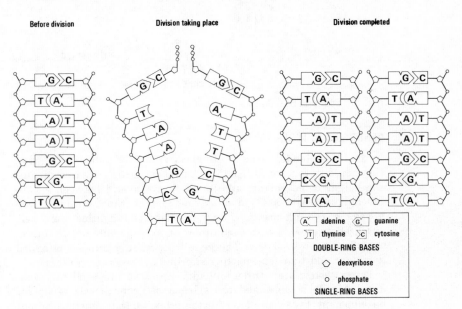

Fig. 9   Division of DNA molecule.

9

chromosomes for that species. The zygote then replicates by mitosis. The 46 human chromosomes are arranged in pairs. In 22 of these pairs, each member of the pair is similar, so these pairs are called *autosomes*. In the twenty-third pair, each member is dissimilar and these pairs are called the *sex chromosomes*. In the female the chromosomes can be written as 44+X+X and at each reduction division they can only contribute 22+X and disperse 22+X. In the male the chromosomes are written as 44+X+Y, and at each reduction division they can contribute 22+X or 22+Y. It is therefore the male cell that determines the sex of the unborn child, (Fig. 10).

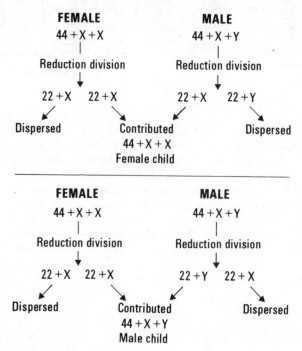

Fig. 10   Diagram of reduction division of gametes.

GENETICS AND INHERITANCE

In the 1860s Mendel worked with garden peas and published his view that heredity was controlled by independent factors. He investigated the height of pea plants and found that a hybrid between a true-breeding tall strain (T) and a true-breeding short strain (t) was always a tall offspring. If two of these tall hybrids were self-fertilized or crossed with each other, on average only one in four of the offspring was short[1]. The gene transmitting tallness is called *dominant* and that transmitting shortness is *recessive* (Fig. 11).

Thirty years later, three biologists working independently validated Mendel's work and the science of genetics was ready for development. Five years later Sutton proposed that *genes* are actual physical units located on *chromosomes*. An English physician,

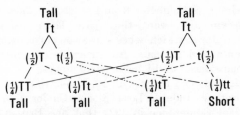

The gene transmitting tallness is called dominant and
the gene transmitting shortness is called recessive.

Fig. 11   Mendelian inheritance.

Garrod, in 1908 described several illnesses in children that were due
to an enzyme deficiency which behaved as if they were caused by a
Mendelian recessive gene. Morgan, around 1910 discovered the
X and Y chromosomes already mentioned, from work that illustrated *sex-linked inheritance*, haemophilia being the classical
example in man. About the same time Bridges published his work,
Non-disjunction as Proof of the Chromosome Theory of Heredity. The
Watson-Crick model of DNA structure described in 1953 suggests
that an exact copy is made in duplication of a cell, but molecular
changes can occur. A variation in the DNA 'message' is called a
mutation, and this is replicated in ensuing cell divisions. One of the
causes of mutation is irradiation as was evident after the dropping of
an atom bomb on Hiroshima. We now know that one gene carries
the instructions for the synthesis of one enzyme that does a particular
job. One such gene can be faulty even though the rest are normal.
The diseases described by Garrod are collectively spoken of as in-
born errors of metabolism. Around 1956 human chromosome study
improved with the discovery of how to separate chromosomes from
the rest of the cell. At that time it was thought that human cells
carried 48 chromosomes, but with this new technique it was discovered that there were 46. Then in 1959 the non-disjunction described by Bridges in 1910 explained the discovery of an extra
chromosome in the twenty-first pair of chromosomes in children with
Down's syndrome, causing characteristic facial features, an affectionate disposition and lowered intelligence. This was the first hint that
chromosomal abnormality could affect behaviour. Several other
variations in the number of human chromosomes are now recognized
some of these involving the X and Y chromosomes. There is some
evidence that an extra Y chromosome leads to diminished responsibility, outside an individual's control[2]. It is now possible to test for
genetic disorder *before* a baby is born, i.e. prenatally.

Research shows that genetic damage is caused by chronic abuse
of alcohol. In this study by de Torok[3], apart from the normal controls,
whose blood tests were normal, chromosome damage of varying
degrees was found in all the alcohol drinkers.

In 1960 an International Conference was held in Denver to establish a standard numbering-system for human chromosomes. The

autosome pairs were numbered from 1 to 22. The 2 sex chromosomes X and Y became the twenty-third pair. There are genetic maps (showing the order in which genes appear on chromosomes) for some protists, plants and animals. Construction of genetic maps for human chromosomes is still in its early stages.

The genetic code can be broken, i.e. a mutation can be produced in the laboratory. Indeed, Lord Ritchie-Calder told the British Pharmaceutical Conference in 1971 that one British scientist invented a new germ by altering the code of an existing type. He died because he had no natural immunity to it. Could this knowledge be applied in biological warfare? Probably with increasing knowledge, interference with the code could be beneficial, e.g. a change of message to plants to produce more food; to cancer cells to make them harmless; a change of message to affected cells in hereditary diseases; a change of message for the production of chemicals in the brain to improve memory and reduce forgetfulness. But by the same token we could have chemicals to repress intelligence, to produce living robots reduced to conformity! Could breaking of the code lead to parents' choice of giving birth to a boy or a girl? These form some of the ethical problems with which succeeding generations are faced.

Even this brief survey shows how the pieces in the jigsaw of knowledge are put into place by each succeeding generation. Robert Hooke's seventeenth century description of cork 'cells' is vastly different from today's description of cells, and today's scientists are aware of gaps in their knowledge about the minute details of the structure and functions of a cell. Man is a collection of millions of cells and since human beings are our main concern we will explore this collection.

The Growth of a Human Being

When a female gamete is fertilized by a male gamete a zygote with its full complement of 46 chromosomes (which carry the plans for the development of the whole body) results. This zygote undergoes mitotic division, which process is repeated until there is a mass of undifferentiated cells. These then differentiate and settle into three layers, the inner one being the endoderm or endoblast, the middle one being the mesoderm or mesoblast, and the outer one being the

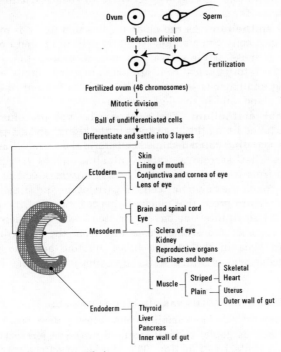

Fig. 12   Growth of a human being.

ectoderm or ectoblast. From each of these layers cells collect into special tissues, and tissues are grouped to form organs (Fig. 12). The study of tissues is known as *histology*. An example is tissue typing which is now done prior to organ transplant operations. An elementary knowledge of tissues is necessary for the nurse to understand the *pathological* processes going on in the patients whom she tends.

The four main groups of tissues are epithelial, muscular, nervous and connective. Muscular and nervous tissue are discussed with these systems on pages 92 and 181 respectively, so that here we can master the rudiments of the epithelial and connective tissues.

13

EPITHELIUM
SIMPLE EPITHELIUM

There is only one layer of cells, cemented together much the same as brickwork, and supported on a fine basement membrane. It is further classified according to the arrangement and shape of the cells it contains.

Squamous or pavement epithelium. The fine cells are laid edge to edge as in a pavement. It is found lining the heart and blood vessels, where it provides a smooth surface which is essential to prevent the blood from clotting. (Later you will learn of conditions which roughen this layer of cells and thus predispose to the clotting of blood—thrombosis.) It is present in the alveoli of the lungs (where it allows diffusion of gases) and in the lymphatic vessels. It forms the serous membranes lining body cavities to reduce friction between opposing surfaces.

Cubical epithelium. As the name implies this tissue is made up of cube-shaped cells. Simple, non-secretory cells are found lining small ducts, e.g. those of the salivary glands, to protect the underlying tissues. Secretory cubical cells are found in many glands, e.g. serous secreting and mucous secreting cells are both present in the mixed salivary gland, and in the digestive tract.

Columnar epithelium. The most simple type of column-shaped cells are found lining the large ducts in the kidney and are protective. Mucous secreting column-shaped cells form the surface lining of the stomach. They secrete a watery, alkaline mucus to protect the stomach from its own acid and digestive enzymes. Goblet cells are found between the columnar cells in parts of the intestine, wherever a thicker, more protective mucus is needed. Some of the columnar cells in the small intestine have a striated border and their function is absorption of foodstuffs. Projecting from the free edge of the columnar cells in the uterine tubes are fine hair-like processes (cilia), movement of which wafts the ovum towards the womb.

PSEUDOSTRATIFIED COLUMNAR CILIATED EPITHELIUM

The columnar cells have smaller cells between their bases, at intervals, as well as goblet cells. It is found in the respiratory passages and has a protective function. The blanket of mucus traps foreign particles, such as dust and soot. The cilia form currents which move these trapped particles towards the mouth.

COMPOUND (TRUE STRATIFIED) EPITHELIUM

There are many layers of cells and two types are recognized.

Transitional epithelium. Found in the urinary passages. The cells are of the columnar type with rounded ends. They are extensible and protective, preventing penetration of urine into underlying tissues. As the deeper cells multiply those at the surface are shed. They are less scale-like than those of stratified squamous epithelium.

Stratified squamous epithelium. This is the most resistant type and is found where there is great wear and tear. The uncornified type is present on internal surfaces such as the mouth and gullet and it protects against friction. The cornified type forms the skin. In its deepest layers the cells possess shape and a nucleus, but rising to the surface they first lose their shape, then their nucleus, and are eventually shed as scales. Dead cell layers are thickest on the soles of the feet and palms of the hands. As well as protecting against wear and tear, cornified epithelium protects against evaporation and extremes of temperature.

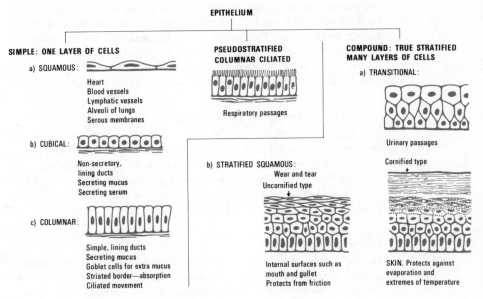

Fig. 13   Pictorial summary of epithelial tissue.

CONNECTIVE TISSUE

The term 'connective tissue' has been wisely chosen for this group, as one of the prime functions of each member is 'connecting', packing or supporting. To do this the basic unit is formed more like a fibre than a cell, though cells are to be seen interspersed throughout the bundles of fibres. Blood is classed as a connective tissue. It is discussed on page 141.

AREOLAR TISSUE

This is a loose connective tissue containing both white, and yellow elastic, fibres. It allows considerable stretching and recovery. The limit of stretch is reached when the bundles of white fibres are straightened out. Recovery to the normal position is due to the elasticity (recoil) of the yellow fibres. It is widely distributed throughout the body, as it fills spaces (interstitial tissue); it supports highly specialized ·structures such as the kidney tubules; it connects

15

mucous membrane to the muscle layer in many tubes (submucosa), where it supports the blood vessels and nerves needed by the mucous membrane and muscle.

ADIPOSE TISSUE

Again this is a loose connective tissue, but it has many fat cells interspersed between the white, and yellow elastic, fibres. It too is used to fill many large and small spaces and is present throughout the subcutaneous tissue. It supports and protects the spherical eyes in their conical sockets, and maintains the normal position of the kidney in the loin. (This perirenal fat from animal kidney is sold as suet.) As well as the function of packing, this tissue acts as a store-house for food which can be utilized in starvation. In addition it acts as an insulator, since fat is a bad conductor of heat.

RETICULAR (ADENOID OR LYMPHOID) TISSUE

The white, and yellow elastic, fibres in this connective tissue are arranged like a net, hence the name *reticular* tissue. It is found in the lymph nodes scattered throughout the body, in the spleen, bone marrow, the tonsil and along the wall of the gut, especially in the appendix. It contributes to the prevention of infection by (*a*) filtration, and (*b*) provision of fresh white blood cells (lymphocytes).

WHITE FIBROUS TISSUE

In this connective tissue white fibres predominate and are densely packed together to produce great strength rather than resilience. The main constituent of white fibrous tissue is an insoluble, albuminoid substance arranged in bundles and called collagen. It is very resistant to the catalytic action of body enzymes. Current research is investigating the role of collagen in wound healing. White fibrous tissue forms the strong covering for many organs, muscles, bones and blood vessels. The tendons which join muscle to bone, the ligaments which join bone to bone, and the aponeuroses which join muscle to muscle are made of white fibrous tissue. (The socket of the hip joint is deepened by a rim of white fibro *cartilage*, but it is called the cotyloid ligament.)

YELLOW ELASTIC TISSUE

In this connective tissue the yellow elastic fibres predominate and are densely packed together to produce resilience rather than sheer strength, as in the lining of large arteries and the lower respiratory tract. The main function of elastic fibres is recoil.

CARTILAGE

This connective tissue is of the consistency of indiarubber, used for erasing pencil marks. There are three types.

Hyaline cartilage. This is bluish-white in colour. It forms the long bones of the skeleton in early fetal life, and remains as a layer between the shaft and each end facilitating growth until maximum height is reached. Hyaline cartilage joins the ribs to the sternum, prevents compression in the air passages and is present on all articulating surfaces to provide smoothness, durability and resilience.

White fibrocartilage. Like the white fibrous tissue, white fibrocartilage is built mainly for strength, but does provide some resilience. White fibrocartilage is found separating the bone surfaces in all slightly movable (cartilaginous) joints. White fibrocartilage is present in the knee joint as two semilunar cartilages to deepen the socket (p. 116); between the vertebrae as discs to act as shock absorbers and to allow forward, backward and sideward movement by compression (p. 55) surrounding the rim of the sockets of the hip (p. 114) and shoulder (p. 118) joints to deepen the sockets.

Yellow elastic cartilage. Like the yellow elastic tissue, this is is built mainly for resilience, being combined with great strength. It is found in the epiglottis and parts of the external ear.

BONE

This is the most dense tissue in the human body and it is opaque to X-ray. In early fetal life long and irregular bones start off as hyaline cartilage, flat bones as sheets of white fibrous tissue (membrane). Specialized cells (osteoblasts) take the soluble calcium salts from the blood and cause them to be deposited in the hyaline cartilage as insoluble (solid) calcium salts. In this way the hyaline cartilage and fibrous tissue are converted into bone and the process is known as ossification. (Osteoblasts build bone.) Other cells, the osteoclasts, are capable of converting the insoluble calcium salts laid down as bone back into soluble salts to be washed away in the blood stream. Using bone as a storehouse for calcium, the blood is able to maintain its normal calcium content which is 9 to 11 milligrams (mg) per 100 millilitres (ml) of blood. The osteoclasts are also active in the formation of grooves, hollows (fossa, sing.; fossae, pl.), holes (foramen, sing.; foramina, pl.) and cavities (sinuses) in bone.

Periosteum. This is the white fibrous tissue (membrane) which continues to surround bone throughout life. Periosteum carries blood vessels to the underlying bone tissue and is essential for regeneration of bone. Hence when a surgeon has to remove a piece of bone he carefully strips away the periosteum, removes the piece of bone and replaces the periosteum so that new bone formation can take place. Bone tissue is further divided into two types: compact and cancellous.

Compact bone tissue. As the name implies this is a very dense substance and it is therefore found as the strengthening layer immediately beneath the periosteum. Compact bone tissue thickness varies according to the strength required.

Cancellous bone tissue. This is enclosed by compact bone tissue and is sponge-like in structure, the spaces being filled with bone marrow, red and yellow.

17

Red bone marrow. This is found in the ends of long bones, and in flat bones. Red blood cells are made in this active tissue which is richly supplied with blood (highly vascular). For the formation of red blood cells protein, iron, folic acid, vitamins C and B_{12} and a trace of copper are required. (Anaemia is the name given to lack of red blood cells.) Platelets and white blood cells (leucocytes) are also made in the red bone marrow.

Yellow bone marrow. This is found in the shafts of long bones. Yellow bone marrow can be converted to red in an emergency such as after excessive bleeding.

Summary of Connective Tissues

Areolar tissue. This is widely distributed as interstitial tissue, supporting delicate structures such as the kidney tubules, blood and lymphatic vessels, nerves, etc.

Adipose tissue. This is also widely distributed and contains fat cells. It forms the subcutaneous tissue. It supports, protects and maintains the position of the eyes and kidneys; it acts as a storehouse for food, and as an insulator, since fat is a bad conductor of heat.

Reticular (adenoid or lymphoid) tissue. It is found in lymph nodes, spleen, tonsil and along the gut wall, especially in the appendix. It protects against infection by filtration and provision of lymphocytes.

White fibrous tissue. This strong tissue forms the tendons and ligaments and the outer capsule of many delicate organs.

Yellow elastic tissue. This is strong and very resilient.

Cartilage. This dense tissue is like indiarubber. *Hyaline cartilage* forms the long bones in fetal life. It is present on all articulating surfaces and in the air passages to prevent compression. *White fibrocartilage* separates the bone surfaces in all slightly movable (cartilaginous) joints. *Yellow elastic cartilage*, combined with strength it has great resilience. Found in the epiglottis and external ear.

Bone. This is the most dense tissue in the body and is opaque to X-ray. It acts as a storehouse for calcium, osteoblasts depositing the calcium and osteoclasts dissolving it, so that the blood level remains at 9 to 11 mg per 100 ml. Throughout life bone is covered with periosteum for nourishment and regeneration. *Compact* bone tissue lies deep to the periosteum, giving tremendous strength, and within that lies *cancellous* bone tissue, the spaces of which are filled with *red bone marrow* in the ends of long bones and with *yellow bone marrow* in the shafts. Red blood cells, platelets and leucocytes are made in the red bone marrow.

Thus far we have discussed the structure and function of a cell, and the growth of a new human being, surveying the epithelial and connective tissues of which he is composed, and leaving the muscular and nervous tissues until later.

18

It is expedient at this point to consider the measures available to ensure the healthy development of this new human being from conception until his emergence into this world. It is necessary to consider the health of the mother during pregnancy, and her safety during confinement. Consideration needs to be given to the family so that there is satisfactory integration of the new member into it. An atmosphere has to be created in which this new child can develop his full potential—intellectual, emotional and physical, while other members of the family continue to develop and adjust in these spheres. He has to learn to participate in groups other than his own family, e.g. school, children's clubs, youth clubs, etc. Choice has to be made about a career for which this growing person is suited and an environment has to be created in which he can maintain his health throughout his working life, enjoy his retirement, maintain his independence and human dignity in the process of senescence, and when his span of life is ended, have the last rite of burial performed according to his faith and the wishes of his family.

There are many definitions of the word health. The one accepted by the World Health Organization and published in 1946 reads: 'Health is a state of complete physical, mental and social well-being and not merely the absence of disease or infirmity.' Many people now think that the word health is only meaningful when defined in personal, functional terms. For instance a man might be said to be healthy when he is functioning at the maximum intellectual, emotional and physical level of which he is capable. *Prospects in Health*, a publication from the Office of Health Economics, states, 'A person should be regarded as healthy provided he can remain socially and economically active, even though he may have to suffer some health disability or discomfort'. The World Health Organization goes on to state: 'The enjoyment of the highest attainable standard of health is one of the fundamental rights of every human being without distinction of race, religion, political belief, economic or social condition. . . .' Informed opinion and active co-operation on the part of the public are of the utmost importance in the improvement of the health of the people. Florence Nightingale with her clarity of thought pin-pointed the important fact that one has to *participate* in the maintenance of health—'Health is not only to be well, but to be able *to use well* every power we have.' Crawford[4] believes that 'There is as much need for a Health Code as there is for a Highway Code', and Wintersgill[5] suggests the name National Health Code. These codes presuppose a well organized and administered national society, but societies throughout the world are at varying stages of development. Nurses in this country are concerned with the health of people from many other lands and indeed a nurse can choose to work in any country. Also, increasing numbers of young people give a year's service with the Voluntary Service Organization (VSO) using their skills in a developing country so that a brief global look at the development of societies and their health and welfare programmes is essential.

DEVELOPMENT OF SOCIETIES

There are still a few primitive groups left in which visiting anthropologists and social psychologists study group structure, patterns of living and the effects of outside influences on the group. Anthropologists recognize two types of families: the extended, and the nuclear; the *extended family* being more common before industrialization. An example of an extended family is the Zulu family which includes the father, mother and their children; the other wives of the father and their children, and the father's brothers and their wives and children. All members live nearby in huts and all the children regard themselves as brothers and sisters. Breast feeding is the rule and this can go on for 18 months to 2 years. There are many adults around to help with the weaning process and there are likely to be several weaners at any one time, so that a weaner is not the centre of attraction as in a nuclear family. In primitive extended families the children either 'grow into' the adult role, having many adult role models around them for the 24 hours of each day, or they attend an 'initiation' ceremony at which they leave the child's world and are accepted into the adult world, thus avoiding the area of marginality that causes many teenagers distress in the west. Though there is inevitable grief, the effect of the death of a parent on a child is likely to be less traumatic where others are readily available to take over parental function, as in an extended family. The term as used by social workers in this country refers to a family of father, mother, and their children living in one house, with grandparents, aunts, uncles and cousins living in houses nearby.

After industrialization, the *nuclear family* is more common, i.e. the father, mother and their children living in one town, and grandparents, uncles, aunts and cousins living in other and various towns, so that visits may be infrequent. This often means a reduction in mutual help and the influence of one generation on another. In each stage of development the child is likely to be the centre of attraction. There are likely to be less adults to talk to the child—an important process in the successful acquisition of language. The Government makes laws about what can and what can not be done to children, e.g. in the United Kingdom no child under 16 can be left alone in a house. Laws state the age at which a child must go to school and the minimum age at which he can leave. The Government provides training for various kinds of work. It also provides homes for orphaned children who have no adults to care for them, so that not only is a child faced with grieving for his parent, but has to adapt to a totally different pattern of living. Punishment comes not from the family but from the State for offences against property and person. It can therefore be seen that in modern complex societies the State takes over many of the tasks of training a child for work and citizenship. The demarcation between school and work can result in feelings of insecurity in teenagers. They no longer belong to the child's world and are unsure of belonging to the adult world of work.

A Zulu extended family has stability at the expense of initiative and creativity. The modern nuclear family has geographical and

20

social mobility, of which many members are able to take full advantage, but some members suffer social isolation which has a detrimental effect on their health (p. 466).

DIFFERENT PATTERNS OF LIVING

The family is the basic unit in which each child learns the pattern of living of that particular unit. Even within a nation, considerable variation in this pattern can be found in the different social classes to which families belong. Throughout the world there are a variety of patterns,[6] the most notable being the differences associated with food, clothes and behaviour.

Patterns of Living

FOOD

The differences with regard to food are about the type of food served, and the method of preparation. Some religions incorporate instruction about the preparation of food and followers adhere to these instructions. The place food is served can be determined by the weather, some families taking all or most of their meals outside; or by the type of dwelling, e.g. whether it has a dining room, or whether food is eaten in the living room or kitchen. The commitments of the family members can determine how many of them are present at meal times. The method of serving food can vary; it may be customary for portions to be served on to individual plates or members may each partake from a communal serving vessel. There is variation in how food is eaten, e.g. with fingers, chopsticks or cutlery. There can be a religious taboo on a particular food, and there can be religious observance of feast days and periods of fasting. For some people, certain foods can only be eaten in specified seasons.

CLOTHES

In the last two decades there has been a much more informal approach to clothes in the western countries. There are still a few groups in the world where the males and females only wear a loin cloth. There are groups where it is traditional for the women to wear trousers and the men to wear flowing robes. A form of head-dress can be worn by both male and female, male only, female only, or by neither male nor female. Works of art and literature abound with examples of fluctuation in the length of hair of males and females over the centuries.

BEHAVIOUR

All gradations of behaviour in sex role are to be found in communities in the world. This may be surprising to those brought up in the west where the boys are encouraged to be 'manly' and the girls are encouraged to be 'gentle and feminine'. In some groups anthro-

21

pologists found sensitive, artistic, child-rearing males and acquisitive, aggressive, outdoor-working females.

THE EFFECT OF OUTSIDE INFLUENCES ON A GROUP

In the primitive groups, tribal members grow what food they can, and for the rest they keep animals, go hunting, shooting and fishing according to the geographical location. Any wealth accumulated is in the form of bartered goods. At some stage there is outside influence on the group, and the question of money enters the arena—the process of *commercialization* and *industrialization* begins. Education is all important in this process and often starts by the young children attending a school on the compound for a few hours daily. As they grow older they are sent away to school, until such time as the community can afford to build and staff its own schools. As literacy is gained, each country is encouraged to register births, marriages and deaths. As a society becomes more sophisticated other records are kept, for example sickness absence from school and work, reasons for this; incidence of accidents—at work, in the home, while travelling; immunization and so on. This allows not only comparison between different parts of the same country, but comparison between countries.

METROPOLITAN PLANNING

Throughout the world increasing numbers of people are attracted from rural to urban areas in search of a living. The United Nations Organization said in 1964 that after the question of keeping world peace, metropolitan planning is probably the most serious single problem faced by man in the second half of the present century. People leaving the land for industry are not prepared for city life; the city's resources are often inadequate, e.g. housing, water supply, sewage disposal, garbage disposal, schools, social services and recreational facilities. Ways of coping with water and waste that may be tolerable in the sparsely populated village result in dangerous conditions in the city. In tropical and subtropical areas, lack of sanitation, overcrowding, low earning capacity and poor nutrition offer a far greater threat of disaster than the industrial revolution in the temperate latitudes a century ago. The World Health Organization states that the disease pattern is changing and one of the threats comes from the mosquito which now flourishes in the mushrooming cities of warm countries. Among measures to combat pollution, the World Health Organization experts recommend the following: satellite towns in which no pollution-producing fuels are used; green belts and wide open spaces to dilute and disperse pollution; central plants to provide heat and hot water to whole districts; less traffic flow in metropolitan areas and devices for car engines to minimize pollution by waste products.

NATIONAL EXPENDITURE

The largest group to which a person belongs is his national group. Money is necessary for the functioning of such a group and it is obtained by levying taxes in various forms. International comparison of health expenditure is extremely difficult, but Table I was prepared by The Office of Health Economics (Information Sheet No. 22, May 1973) and is an estimated comparison of 7 countries.

Table I.   Total expenditures for health services as percentage of the gross national product for seven countries over selected periods 1961 to 1969.

Country Ranked by percent of GNP	WHO[1] estimates		SSA[2] estimates	
	Year	Percent of GNP	Year	Percent of GNP
Canada	1961	6·0	1969	7·3
United States	1961–1962	5·8	1969	6·8
Sweden	1962	5·4	1969	6·7
Netherlands	1963	4·8	1969	5·9
Federal Republic of Germany	1961	4·5	1969	5·7
France	1963	4·4	1969	5·7
United Kingdom	1961–1962	4·2	1969	4·8

1. Abel-Smith, B. (1967) An international study of health expenditure. *WHO Public Health Paper*, No. 32. Geneva.
2. Simanis, J. G. (March, 1973) Medical care expenditure in seven countries. *Social Security Bulletin*.

Year	Cost of NHS Millions £ sterling	NHS as % of GNP
1954	564	3·53
1964	1190	4·02
1974*	3143	4·97

* Estimated at zero growth rate in GNP.

Fig. 14   Gross cost of the NHS and the cost as a proportion of Gross National Product, United Kingdom.

National groups appoint a governing committee (whatever name it is given) and a leader (whatever name he or she is given) to conduct their business. Nations vary as to the method used for appointing the top level committee and leader. Nations use various names, such as department or ministry, for the Civil Service unit that administers a particular portion of their business—food, housing, sanitation, health, social services, industry, education, recreation and transport. There are usually large central units of administration in the country's capital. Part of the work is delegated to smaller, local units of administration in a town serving an agreed surrounding area. Money raised from levies by local units is referred to as 'rates' and on the back of the rate assessment form, the local unit usually prints the headings under which it spent money in the previous year, together with the amount.

If the rankings on standardized death rates, late fetal death rate, infant mortality rate and maternal mortality rate are combined, the following rankings emerge, with Sweden having the best record.

1	SWEDEN	9	FRANCE
2	SWITZERLAND	10	WEST GERMANY
3	AUSTRALIA	11	AUSTRIA
4	NETHERLANDS	12	HUNGARY
5	CANADA	13	JAPAN
6	CZECHOSLOVAKIA	14	ITALY
7	BELGIUM	15	PORTUGAL
8 {	UNITED KINGDOM / UNITED STATES OF AMERICA		

(Source: Annual report of the Chief Medical Officer of the Ministry of Health, 1966 "On the state of the public health")

Fig. 15   Value for money.

A nation depends on its individual members for its national income and exploitation of its natural resources. The health, safety and effectiveness of these members should be a prime concern when decisions are made about spending national income. Figure 14 shows the amount of money spent by the UK Government on its National Health Service and it can be seen that the actual expenditure has more than quadrupled in 20 years; in GNP terms it has increased from 3·53 per cent to 4·97 per cent.

When spending money it is important to get the best possible value for same. Figure 15 shows the ranking for several countries using combined-standardized death rates, late fetal death rates, infant mortality rates and maternal mortality rates as a measure of effectiveness and the United Kingdom has a middle position on this particular international league table.

Each country has to ask if it is striking the right balance between

Year	Hospital Services	Pharma- ceutical Services	General Medical Services	General Dental Services	General Ophthalmic Services	Local Health Authority Services	Other	Total
1953	55·5	9·5	10·8	5·5	2·2	8·8	7·7	100·0
1963	60·1	10·1	8·3	5·7	1·6	10·0	4·2	100·0
1973	66·2	9·4	7·4	4·4	1·1	6·9	4·6	100·0

Fig. 16   Health Services as a proportion of total cost of NHS in the UK.

the money it spends on treatment and that which it spends on the prevention of disease. In the UK in 1953, 55·5 per cent of the income allocated to the National Health Service was spent in hospital services. This sum has gradually increased over the years, until in 1973 it was 66·2 per cent (Fig. 16). If we spent more on research into methods of changing attitudes, e.g. to smoking and overeating; dissemination of knowledge about healthy living, and screening the population to detect disease at an early and treatable stage, would we reduce the cost of the hospital service?

Each country has to decide its priorities and each nurse should know something about the expenditure on the health services in her own country so that she can discuss these matters and, where necessary, through her professional organization bring pressure to bear on the Government to allocate more money to deficient areas. In 1970, the Royal College of Nursing and National Council of Nurses of the United Kingdom (Rcn)* petitioned the Government to spend more money on hospitals for the mentally subnormal. The following are some questions that should be asked when deciding how money should be spent. The notes on each question pertain to the United Kingdom.

HOW MUCH MONEY SHOULD BE SPENT ON MENTAL HEALTH?
Although 45 per cent of the hospital beds in the National Health Service are for the mentally disordered, only 11 per cent of all consultants are psychiatrists and only 21 per cent of nurses work in psychiatric wards. Less than 14 per cent of National Health Service expenditure goes on hospital and community services for the mentally disordered. Less than 9 per cent of the Medical Research Council budget goes on research into mental disorder. Do we care enough for the mentally disordered?

HOW MUCH MONEY SHOULD BE SPENT ON CARE OF THE ELDERLY?
About 13 per cent of the population are now aged 65 and over. By 1981 the proportion may rise. Over 35 per cent of hospital beds are occupied by patients over 65. Do they all need to be there? Should we be spending more on better pensions, so that senior citizens could maintain themselves in better health by adequate nutrition, clothing, and heating of their homes? They continue to need money for recreation. Should we build more day centres so that senior citizens have at least one day a week spent in social activity? Could some of the elderly be maintained in dignity at home if there were more suitable housing arrangements for the elderly? Would more money spent on home helps keep more elderly people out of hospital?

HOW MUCH MONEY SHOULD BE SPENT ON HEALTH EDUCATION?
An increasing but still relatively small proportion of the National Health Service budget is allocated to the Health Education Council. Should we be spending more on health education?

Health workers are expected to be capable of playing a part in the health education of the nations as shown by the following recommendation (author's italics).

*Now the Royal College of Nursing.

25

RECOMMENDATION No. 63[225]

To the Ministries of Education

concerning

Health Education in Primary Schools

The International Conference on Public Education,
Convened in Geneva by the United Nations Educational, Scientific
and Cultural Organization and by the International Bureau of Education, meeting on the sixth of July, nineteen hundred and sixty-seven
for its thirtieth session, adopts on the fourteenth of July nineteen
hundred and sixty-seven, the following recommendation:

Considering, however, that, as teaching methods have developed,
on the one hand, and as progress has been made in preserving health,
on the other, health education can and should take the place of the
mere teaching of hygiene.

Considering that health education means the whole process which
helps to inculcate good habits, sound knowledge and an enlightened
attitude concerning the health of the individual and the community,

Considering that the scope of health education covers a person's
physical, intellectual and emotional development, and that such
education covers the health of individual, family and community,
with reference to home, school and place of work; the problem of
nutrition, mental health, sex education, accident prevention, first
aid, use of leisure, etc.,

Considering that health education given at school is an important
aspect of the general education of a child and one of the essential
ways of improving individual and public health,

Considering that such education must occupy a prominent place
throughout the child's schooling and particularly during primary
education,

Submits the following recommendation to the Ministries of
Education of the various countries:

1. Health education suited to the age, needs and interests of
pupils, first of a practical nature and subsequently of a practical and
theoretical nature, should be given by the teachers *in conjunction
with the appropriate health departments*; the form, content and
methods of this education should be determined *through consultation among the various authorities concerned.*

2. This education should not only inculcate good habits in the
pupils, likely to promote their physical and mental well-being, but
also awaken in them a sense of their individual and social responsibilities, by teaching them to respect the health and well-being of
other people, as well as their own.

3. Such education should find its natural roots in the life and
working conditions within the school. These should include the wise
planning of the school day, with a balanced proportion of work,
play and rest, together with adequate accommodation and facilities
and suitable sanitary installations.

4. The methods used in health education should involve not only the child's memory and his reasoning powers but also (and especially) his imagination; account should also be taken of his leisure activities and above all of his healthy living.

5. Health education should be related to local circumstances (urban or rural surroundings), to the climatic conditions of each country and to its economic and social development; in the light of these circumstances the teaching should emphasize the basic demands of various regions in regard to health and hygiene; teaching should also include some preparation for the probable conditions in which the child will be living and working.

6. The teachers should be prepared to give health education during their training courses by means of specialized instruction, which takes into account the aims, as well as the methods, of imparting such education.

7. Possibilities for further training should be offered to teachers in service to enable them to keep up to date with the latest developments, in the methods and means used in health education, as well as with the progress of preventive medicine.

8. Primary teachers should be provided with the materials necessary for their pupils' education (textbooks, apparatus and audiovisual aids) as well as books and periodicals and other documents from which to draw their own information; such materials, based on the country's geographical, climatic, ethnic and cultural requirements, should be produced in collaboration by those responsible for the school syllabuses *and the medical and health authorities.*

9. At the same time, steps should be taken to inform and educate parents by means of co-operation between primary school teachers *and the medical and health authorities* (e.g. through regional and local information centres).

10. Specialists in health education (*such as doctors and nurses*) who are assigned to work which is linked with health education (e.g. medical inspection) *should be prepared for these educational functions* in the course of their professional study through appropriate health and educational training.

11. Any initiative should be encouraged, whether public or private, which, in school or out of school, is taken by young people or adults and is likely to promote health education or training in first aid; this may include groups of scouts or pioneers, Red Cross or Red Crescent Societies, members of youth first aid clubs, etc. and their various activities or manifestations (lectures, competitions, periodicals, exhibitions).

12. Assistance should be given to any research of a medico-pedagogical nature which, in the matter of health education, is intended to ascertain the needs, to assess the results of the measures taken and the experiments carried out and to improve the methods and media used.

13. In the context of each country's particular planning, health education should have a place among its fundamental objectives, as an essential factor in social, economic and cultural development.

14. A permanent scheme of co-operation (as, for example, bilateral or multilateral conventions) should be established among the various countries to facilitate the exchange of specialists and of research information.

15. Use should be made, where necessary, of technical assistance provided by specialized international organizations, which may help Man, his Health and Environment the various countries to create a basis for health education, draw up syllabuses, design and produce teaching material and train staff, by means of advisory services, scholarships and material for demonstration purposes.

HOW MUCH MONEY SHOULD BE SPENT ON HELP FOR THE HANDICAPPED?
It is estimated that over three million people in private households in Britain have some physical, mental or sensory impairment. Of this total, some 760,000 women and 370,000 men can be classified as handicapped, 70 per cent of whom are elderly. Some of the services available to the handicapped are mentioned in other parts of this book: for the blind (p. 411), for the deaf (p. 424). Local authorities are encouraged to make their public buildings accessible to the physically handicapped. Should we be spending more on health and social services for the handicapped?

HOW MUCH MONEY SHOULD BE SPENT ON RESEARCH?
About £100 million per annum is being spent on medical and health care research and development, by government, industry and voluntary organizations. Is this enough? Is it being spent on the most useful projects? Is enough attention paid to implementing the results of research? Could more be done, for example to help people to stop smoking, over-eating, or over-indulging in alcohol?

HOW MUCH MONEY SHOULD BE SPENT ON BUILDINGS?
In the last 10 years five times as much hospital building was completed as in the previous 10 years. Are the new hospitals that are being planned of the right size, are they in the right places and are they for the people who need them most? Should we be spending more on better housing and more health centres and local authority residential accommodation?

HOW MUCH MONEY SHOULD BE SPENT ON HEALTH SERVICE STAFF?
Salaries account for the greatest proportion of the expenditure of each regional hospital board. Is the best use being made of this staff? Are highly trained personnel being paid to do less skilled work? Is enough attention paid to the implementation of the results of work study? Could staff perform more effectively if they had better equipment?

HOW MUCH MONEY SHOULD BE SPENT ON FAMILY PLANNING?
It is estimated by the Family Planning Association that in Britain
there are between 200,000 to 300,000 unwanted pregnancies every
year, and that the expenditure of £40 million a year on effective birth
control could save between £200 to £400 million of public spending
on maternity and child care and other services each year. At present,
expenditure on family planning services by the Family Planning   **Health Services**
Association and local authorities amounts to between £3 to £4
million per annum.

WHO IS INVOLVED?
At this point it is as well to remind ourselves of the main people
involved in defining and deciding options, priorities, policies and
objectives. They can be divided into three groups:
1. **The decision makers.** Members of Parliament
                           Members of local authorities
                           Members of health service boards, com-
                           mittees and councils
                           Members of university and other edu-
                           cation authorities
                           Members of research councils.
2. **The wielders of influence.** Civil servants
                                 Health and social service staff
                                 Staff of universities, colleges,
                                 schools
                                 Press, radio, TV
                                 Professional organizations
                                 Trade unions
                                 Voluntary organizations
                                 Consumer associations and pressure
                                 groups.
3. **The public.** The patient
                   The consumer
                   The tax-payer
                   The rate-payer
By now it must be obvious that *each person* can play a part in
deciding governmental options, priorities, policies and objectives.
Right decisions do not just happen. They are based on the following
factors:
Information—which is accurate, adequate, concise, intelligible
and up to date.
Education—schools, colleges, universities, professions, vocations.
Publicity—press, radio, television.
Discussion—at home, at work, at public and private meetings.
Does each country have appropriate machinery for decision-
making? Does each country use the machinery effectively? What
can you do to ensure that the best decisions are made?

Health is not only a personal responsibility, it is a communal responsibility, and broad policy, as we have seen, is made at a *national* level. With the speed of present-day communication and transport, nations are no longer isolated and the standards of health in one nation can affect those in another. We need to look at the *international* organizations which have some relevance to the health and well-being of the world population.

THE UNITED NATIONS ORGANIZATION

As long ago as 1851 a convention signed by several nations regulated the quarantine laws affecting international commerce and travel. This work led in 1907 to the establishment of an International Office of Public Health with headquarters in Paris. After the First World War the League of Nations was formed in the hope that it would be conducive to world peace. The League's Health Organization had its headquarters in Geneva. The members of this organization realized that if they were to reduce the amount of sickness in the world they would have to improve the conditions which precipitated disease, namely, poor diet, bad sanitation and lack of shelter. Then came the Second World War, after which the United Nations Organization came into being at San Francisco, in April, 1945. There are various councils and organizations under the auspices of the United Nations Organization. Decisions taken by these bodies are implemented and administered by the Secretariat— the permanent staff at the Headquarters at present in New York. Only those organizations which are concerned with health and living standards will be mentioned in this book.

The United Nations Educational, Scientific and Cultural Organization (UNESCO)

This body links the work of several other agencies. Its staff teach everything from simple cleanliness to higher education and works with governments that are striving to fit their peoples for life in the modern, highly complex world.

The International Labour Organization (ILO)

This organization brings together Governments, Employers and Workers from many different countries so that they can discuss mutual problems and share techniques for overcoming them. Instructors suggest ways of improving working conditions; they bring in to countries new equipment and teach people how to use it. The organization helps to train people for the kinds of jobs that a developing society needs to pursue in order to make the most of the country's natural wealth and among other things it suggests the maximum weight that a man or a woman should lift and advocates that mechanical lifting should be performed wherever possible. It also postulates the minimum standards of living, and these have relevance to health:

30

Food. Enough food every day to replace the energy used in living and working.

Clothing. Enough clothes to permit bodily cleanliness and afford protection from the weather.

Housing. Of a standard to give protection under healthy conditions.

Hygiene. Sanitation and medical care to give protection against disease and treatment in illness.

Security. Against robbery or violence, against loss of the opportunity to work, against poverty in old age.

World Health Services

Education. To enable every man, woman and child to develop to the full their talents and abilities.

Generally speaking the first three are items which a man has to provide for himself and in order to have them he must pay for them, either in money or work. The last three are generally matters for the government and other public bodies, which are paid for, usually by individual citizens, so each man must earn enough to pay his contribution to the common services, as well as to support himself and his family. The greater the amount of goods and services produced in any community, the higher its average standard of living will be.

The United Nations Children's Fund (UNICEF)
This organization attempts in a number of ways to bring health and happiness to children all over the world. It sets out the *rights of children* and claims that, without exception, they should have a heritage of freedom and grow up in peace, without fear; they should have names and a nationality of which they can be proud; enough to eat; good homes; schools to attend; games to play; doctors to give them medical care and parents or guardians to look after them.

The Food and Agriculture Organization (FAO)
The aims of this organization are to teach farmers new methods of cultivation and to wage scientific war on pests which can destroy crops. It works in close association with the Technical Assistance Board. The content of DDT and other important agricultural chemicals in food throughout the world is causing the Food and Agriculture Organization and the World Health Organization grave concern. The controversy about the use of DDT in the tropics, was brought up at the United Nations Conference on the Human Environment in Stockholm, 1972.

The World Health Organization (WHO)

This organization shares in the work of the United Nations Children's Fund (UNICEF) and the International Red Cross. The World Health Organization has its headquarters in Geneva, and has many functions. It believes that the health of all peoples is fundamental to the attainment of peace and security, and is dependent on the fullest co-operation of individuals and states.

One of its functions is to disseminate information about infectious diseases. The various countries report outbreaks of diseases such as cholera, plague, yellow fever and smallpox to the central office in Geneva and WHO alerts immigration and health officials at airports and seaports throughout the world. It advocates the immunization of individuals travelling to and from infectious areas and countries receiving individuals from infected areas are advised to keep them under surveillance until the danger of incubating the disease is past.

No project is undertaken by the World Health Organization for an individual country until the country asks for help, and can pay a proportion of the cost. War has been waged against mosquitoes so that the incidence of malaria and yellow fever decreases each year. Similarly the problems of yaws and leprosy have been tackled with effective results. Another campaign was mounted against kwashiorkor and the teams sent in to help with this problem of malnutrition did not leave until they had made a palatable protein powder from locally grown vegetables; the people had accepted the protein as an essential item in their diet, and some local people had been trained to make the powder in the newly built laboratories. Another method of helping is to give fellowships and scholarships to people from developing countries to study outside their country; and on return to the country of origin this knowledge gained is disseminated by teaching and practice.

Encouraging countries to grow more food can bring hazards to health. For example, the flooding of previously arid land can cause the formation of stagnant pools, which can encourage the breeding of mosquitoes carrying malaria or yellow fever. Or it can encourage the multiplication of snails that penetrate the skin and give rise to diarrhoea and anaemia. It can be seen that different sorts of experts need to pool their knowledge to solve the many problems that beset mankind in this twentieth century.

As yet there is not an internationally acnowledged list of food additives. In the more industrialized countries there are problems associated not with shortage of food but with food preservation. The World Health Organization and the Food and Agriculture Organization are making great efforts to produce what it is hoped will be an internationally accepted list of approved food additives.

As already mentioned the World Health Organization plays a big part in the exchange of epidemiological information concerning the occurrence of epidemics throughout the world and it also helps to disseminate information regarding modern methods of treatment, the keeping of statistics, and the control of disease—through its publications and other media of information. The national press and professional journals usually carry synopses of such publications and it is the responsibility of each health worker to keep his knowledge up to date.

Having taken a broad look at the national and international scene, let us make a general survey of the health services in our island and outline the measures taken by the local authorities to ensure that people are born healthy and have every opportunity of remaining so.

GENERAL SURVEY OF THE HEALTH SERVICE

On April 1, 1974, following the passing of the National Health Service Acts, a new integrated health service structure was established in the United Kingdom. It replaced the previous tripartite system with its Regional Hospital Boards, Executive Councils and Local Health Authorities which had remained substantially unchanged since inauguration of the 1948 Act. Although all the countries in the United Kingdom have a similar service there is some variation in nomenclature so, as an example, let us look at the current structure in Scotland.

The Secretary of State is responsible for the work of all Departments at the Scottish Office and he and his political colleagues (they change according to the party in power in the Government) are assisted by a staff of permanent Civil Servants. Of the six main Departments, health matters are dealt with mainly in one—the Scottish Home and Health Department (SHHD). The new structure created 15 Health Boards to integrate the previous three services and make a single authority responsible for the provision of a co-ordinative service in each of the Health Board areas. For administrative purposes 10 of the larger Health Boards are divided into Districts and the boundaries follow closely the regional and district boundaries drawn by the Local Government (Scotland) Act 1973 so that health and other related services can be provided within co-terminous administrative units.

Health Boards. Health Boards whose members are appointed by the Secretary of State take the major policy decisions on matters such as allocation of resources and long-term planning of the services within the area. Decisions are implemented by the Chief Officers of each Board working together as an executive group—the Chief Area Nursing Officer (CANO), the Chief Administrative Medical Officer (CAMO), the Treasurer and the Secretary. The Chief Administrative Dental Officer (CADO) and Pharmacist (CAPO) join them when matters concerning their professions are under discussion.

The executive group is responsible for the control of the hospital and community services, ascertaining the health care needs of the population in the area and assessing the effectiveness of the medical, nursing and paramedical services in meeting those needs.

Where Districts have been created a broadly similar administrative framework operates with a District Nursing Officer (DNO), District Medical Officer (DMO), Treasurer and Secretary and the DNO for example is answerable to his/her counterpart at Area level, the CANO.

Scottish Health Services Planning Council. To provide the Secretary of State with authoritative advice on the future development of the Health Service in Scotland the Scottish Health Services Planning Council was established. It consists of members appointed by the Secretary of State and also members appointed by each Health Board and each University in Scotland with a Medical School. It therefore brings together the people who help to make policy and the people who implement it so that there is the opportunity to share ideas for policy formulation.

Consultative committees. Consultative committees have been established at a national level—Medical, Dental, Therapeutic paramedical, Nursing and midwifery, Pharmaceutical, Optical, Scientific and Technical paramedical—to advise the Scottish Health Services Planning Council on the professional services with which each is most closely concerned. The nursing committee is called the National Nursing and Midwifery Consultative Committee (NNMCC). Local professional consultative committees have also been established to advise Health Boards on the provision of services within their areas.

Local Health Councils. Not only professional groups give advice on the running of the new integrated health service; public opinion is also sought. Local Health Councils have been established, usually one in each District. One third of the members are appointed by the Local Authority and the remainder by the Health Board, largely from nominations of volunteer associations and similar bodies, including for example women's organizations, trade unions and churches.

Common Services Agency. This agency has been established to service both the Scottish Home and Health Department and the Health Boards and provides a range of specialized services which are more effectively organized on a national basis such as the ambulance, blood transfusion and statistical services; legal advice; the planning and implementation of major building projects, and health education.

Health Service Commissioner. Another important feature of the new Act is the machinery for dealing with complaints about the service—there is provision for a type of 'ombudsman'. Subject to various exceptions stated in the Act, if a complaint has been investigated by the proper authority and the solution is considered by the complainant to be unsatisfactory, appeal can be made to the Health Service Commissioner. He makes a report to Parliament annually.

Other services relevant to health. Other services which are relevant to health are controlled by various committees appointed by the Local Authority to ensure that there is:

1. Provision of an adequate clean safe water supply.

2. Inspection of quality of food; control of abattoirs: cleanliness of wholesale and retail food premises, public eating places, and education of the personnel involved.

3. Provision of an adequate supply of clean, safe milk.

4. Provision of adequate housing accommodation, and education and encouragement for proper use of same by citizens.

5. Provision of adequate highways free from known hazards (adequate lighting, antiskid services).

National Health Service

6. Control of factory hygiene and safety measures.

7. Provision of safe playing fields.

8. Prevention of road accidents.

9. Street cleanliness.

10. Prevention of atmospheric pollution (there are some 'smokeless' zones in this country).

11. Prevention and destruction of pests, rats, vermin and bugs. This is especially important if the Council controls a port.[15]

12. Adequate collection and disposal of refuse.

13. Adequate collection and disposal of sewage.

14. Provision of a police force for the citizens' protection.

15. Provision of a fire brigade for the citizens' safety.

16. Inspection of cleanliness and suitability of offices, factories, shops, nursing homes and medical auxiliary services, for example chiropody.

17. Provision and maintenance of public conveniences, accessible by disabled people, especially those in wheelchairs.

18. Provision of educational facilities for children and adults.

19. Provision of suitable burial grounds and crematoria.

20. Provision of registration for the electoral roll, births, marriages and deaths.

21. Provision of public libraries and museums.

22. Public baths and laundries for those in houses that do not have these amenities.

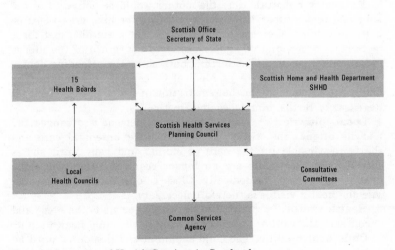

Fig. 17   Summary of Health Services in Scotland.

Each local authority co-operates with voluntary associations that provide practical help to less fortunate members of society living with some form of disability.

This is but a brief outline of the measures taken by the local authorities to ensure that people are born healthy and have every opportunity of remaining so. More may be learnt from visiting in the Community.[16]

Potential Use of Social Amenities in a Lifetime

Let us now work out how an imaginary person might use the aforementioned amenities during a lifetime.

1. **General practitioner.**[19] Pregnancy would be confirmed by a urine test and the mother's general practitioner would advise about prenatal (antenatal) care. According to availability this service could be given at a group practice surgery, a health centre or a maternity hospital.

2. **Prenatal (antenatal) care.** The majority of mothers in Britain would be advised to book a hospital delivery. A full social, medical and obstetric history would be recorded. The mother would have opportunity to learn about such things as the correct diet in the various stages of pregnancy including the procedure for obtaining cheap milk; correct clothing, footwear, posture, exercise and rest; the dangers of smoking; the availability of free dental treatment; breast and bottle feeding of an infant; parentcraft; cervical cytology; family planning and in some instances genetic counselling. The various methods available for relief of pain during childbirth would be discussed.

The mother's urine would be tested for any abnormality; her weight and blood pressure would be recorded and her ankles tested for oedema. Her blood would be examined for, among other things, its group, Rh type, anaemia and venereal disease.

3. **Maternity hospital.** Here the mother would be delivered of her baby and under supervision she would care for him, breast-feed or bottle-feed him, take care of her breasts, eat a suitable diet for a lactating mother, and do simple exercises for return of the uterus to its normal position. The birth is notified by the attending doctor or midwife to the Chief Area Medical Officer within 48 hours. The family are responsible for the registration of the birth with the local Registrar of Births, Marriages and Deaths.

4. **Local midwife.**[20,21] When a home confinement is arranged the midwife attends the mother in her home for antenatal care and she is responsible for the care of mother and baby until she is satisfied that her skills are no longer required, after which the health visitor visits periodically. Close co-operation between midwife and health visitor is therefore necessary.

5. **Health visitor.**[22,23,24,25,26] The health visitor visits the home and encourages the mother to bring the baby to the child health clinic.

6. **Child health clinic.** The child should attend this clinic until he is 5 years old and ready to go to school. The children are weighed,

and seen by a doctor at intervals. The health visitor talks with the mother about feeding, weaning and health education in general. The mothers gain much support from each other. Baby foods and drops containing vitamins A, D and C are available at less than market price.

Whilst attending in the home and at the clinic the health visitor observes the normal milestones in development (including psychological[27] and social, as well as physical). She makes simple tests of sight, hearing and intelligence. She constantly watches for any abnormality which might result from congenital defect. In unfortunate circumstances she helps the parents to adapt to any abnormality. She advises about weaning, and constantly teaches the family how to keep healthy, how to take advantage of the vaccination and immunization programmes offered in the area, how to prevent accidents in the home, etc. She observes relationships within the family and can give timely advice and support should there be any threat to good relationships, such as sibling rivalry or the father feeling neglected because the mother spends too much time on the baby; or, if an only child is not learning the normal give and take of childhood, perhaps due to an over-anxious mother who will not let him go out to play, the health visitor may advise attendance at a day nursery.

7. **Day nursery.** At the day nursery, the general health of the child, including diet, rest periods and play, are under the direct supervision of trained personnel. As well as providing stimulation for the only child, the day nursery caters for the child of a mother who is confined again, or ill or unmarried or working because of a disabled or deceased husband. The child who has been deserted by parents or who has inadequate parents may also attend and sometimes a motherless child can remain within his family if he is cared for at the nursery throughout the day time hours. Many nurseries allocate a small number of places to physically handicapped children and some accept children who have problems arising from immigration. By the time the child is five, he will be attending school and come under the care of the school health service.

8. **School health service.**[28] This service undertakes an initial medical examination followed by selective examination during school life. At more frequent intervals, observations are made to detect the presence of enlarged tonsils and adenoids. At intervals there is a cleanliness inspection, and there may be daily inspections during infectious disease epidemics. Checks are made of height, weight, sight, hearing and dental health and in some areas, a school clinic is available offering a dental service and treatment for minor ailments. At 13 years of age the child is offered a test for susceptibility to tuberculosis (Mantoux or Heaf test), and if the child shows no resistance then a protective vaccination is offered (B.C.G.). Girls are offered rubella vaccination if they have not had german measles.

The School Health Service is concerned with the assessment and subsequent care of handicapped pupils, including blind, partially sighted, deaf, partially hearing, educationally subnormal, epileptic,

37

physically handicapped, maladjusted, speech defective, and delicate children. The teachers and members of the health team work closely together, each paying attention to the physical, intellectual and emotional development of each child, thus helping him towards maturity. Where there are problems the resources of a Child Guidance Clinic are available.

Much health teaching goes on in schools nowadays,[29] all part of an introduction to the responsibilities of being a good citizen and much advice and help is available regarding the choice of a career.

9. **Occupational health service.**[30,31,32,33,34,35] Stress at work which may be conducive to mental illness, high sickness absence and low productivity, has been ignored for too long. Absenteeism, clock-watching, apathetic performance leading to poor workmanship, low productivity and a failure to meet delivery dates, either can be seen as the 'British disease', or symptomatic of a management that is inadequately trained to see the importance of creating job satisfaction and recognizing the psychological needs of the workers.

Occupational health is seen as preventive with the emphasis on safety precautions. Safety also enters into the design and siting of machinery. Health education is an important part of such a service.

Our imaginary person may well enter a factory[36] or a foundry for his livelihood. Here he will be apprenticed under the day release scheme, whereby he will spend one day a week at the nearby technical college, learning about his work. At work there will be strict regulations about cloakrooms, rest rooms, lavatories, washing accommodation, first-aid units, heating, lighting, ventilation, fatigue, protective clothing and protective measures such as salted drinks for those working in foundries.

10. **Dental service.** He should make use of this service at least annually.

11. **Ophthalmic service.** This service should be utilized if there is any impairment of vision.

All work and no play make Jack a dull boy so that in his leisure time our imaginary person may well use the following amenities:

12. **Parks and gardens**

13. **Playing grounds**   Owned by the local authority.

14. **Swimming baths**

If accident or illness befall him he may need the following services:

15. **General practitioner service.**

16. **Home nursing service** (district nurse).[37]

17. **Ambulance service.**

18. **Hospital service.**

19. **Home-help service.**[38]

20. **Chiropodist service.**

21. **Physiotherapist service.**

22. **Speech therapist** (for example following a stroke).

As he grows old he may require further facilities:

23. **Housing,** for example old-age pensioner's bungalow or sheltered housing.

24. **Health visitor** to call regularly.
25. **District nurse** attention.
26. **Meals on Wheels service.**
27. **Darby and Joan clubs.**
28. **Nursing equipment.**
29. **Laundry service** if he becomes incontinent.
30. **Night sitters.**[39,40]
31. **Eventide home.**
32. **Geriatric hospitalization.**[41]

At the last his death will have to be registered with the Registrar and arrangements made for his burial. Our man has therefore had need of at least 32 amenities in his lifetime.

THE RESPONSIBILITY OF BEING A CITIZEN

As already stated, the schools are introducing their pupils to an awareness of the responsibility of being a good citizen. This means that each individual has to learn to fit in with his environment, to respect other people and resist any desire to harm them; to respect other people's property and resist any desire to steal or harm it; to resist any desire to over-indulge in alcohol or drugs which may make him abusive, or a menace if in control of a vehicle on the roads; and to earn his own living honourably and with great personal satisfaction, thus making a contribution to the nation's economy; to pay the rates, taxes and insurances which, as we have seen, reap good dividends in the amenities they provide. Each individual should have at least an elementary knowledge of the structure and working of local government machinery and should know what is provided and how to use it wisely.

'Naderism', an American cult, preaches the gospel of responsibility of citizenship. Its belief is that we can only reform society if we first reform ourselves. It emphasises the way *we contribute* to our own exploitation, the way *we perpetuate* the injustices *we suffer* by our own apathy and compliance. It shows that each one of us plays a number of roles, that of husband, wife, bachelor, spinster; worker, tax-payer, rate-payer, consumer, and that these can be conflicting. The man at work who fails to control pollution robs himself of the requisites for enjoying leisure time, such as fresh air to breathe, clean water in which to swim and fish. A painter at a toy factory using paint of high lead content may rob his own child of health by lead poisoning.

Another example of inconsistent values is provided by a person who, as a rate- and tax-payer, demands cuts in rates and taxes so that there is less money for public expenditure, then as a parent complains about poor conditions at school, or as a patient complains about poor service, or as a commuter complains about the failure of public transport—all of them inadequately financed because of the cuts in rates and taxes and public expenditure. Naderism asks people to be *the same person all the time,* that is to think not only as the person who pays the rates and taxes—but also as the person who benefits from public expenditure; not only as the person who pollutes the air, but also as the person who breathes it. Having trained themselves to think consistently, Naderism encourages people *to act consistently,* and when necessary to report conditions which might impair the health of the workers or the health and safety of the consumers or clients.

We in this country have several consumer associations that test products and publish the results thus exposing shoddy and unsafe workmanship. But there is still room for individual challenge against conditions imposed on us, such as unsafe cars, poor repair work, unsafe working premises, shortage of staff, out-of-date working premises so that those we portend to serve get a less good service. *Complacency can be the path to disaster.* There is dynamism in Nader's creed; concern, not so much with consumer protection, as with *citizen action.*

THE IMPORTANCE OF MENTAL AND PHYSICAL HEALTH

To the individual. Mental and physical health provides the ability to choose and train for a career which will be satisfying to the individual personality. It also provides the ability to distribute the hours of each day satisfactorily between work, leisure and sleep, so that the individual is in tune with himself and with his fellow men. To summarize:

1. The individual is a member of the family group, making his contribution from his special gifts, for example gardening, and accepting other members' contributions from their special gifts, such as cooking and interior decorating.

2. He chooses a satisfactory and satisfying career in accordance with his ability and intelligence.

3. His good health prevents absenteeism and he thus contributes to the stable economy of the family unit, and indirectly to the national unit.

4. He is acceptable to any group which he may choose to join.

To the family. Much of what has been said about the importance of mental and physical health to the individual has to be reiterated when applying it to the family. It cannot be too strongly stressed that it is within the family that the first feelings of security are engendered and nurtured, so that subsequently, wherever each member may sojourn for work and play, he will do so with a confidence which is fitting to human dignity.

To the community. The importance of mental and physical health to the community is so vast a subject that I can only hope to spark off your thoughts in this direction; you should then be able to add many more from your own experience.

1. Strong, healthy members of our community will be able to till the land and produce the food without which no individual can survive.

2. Wealth in the form of iron ore, metal, stone, gypsum and coal can be mined from the land and converted into articles for the home and export market.

3. Mental and physical health cut down absenteeism, which is expensive to the national purse, for not only does it reduce productivity, thus losing orders from home and abroad, but sickness benefit has to be paid while the man is off work.

4. Health provides greater ability to resist infection and reduce the possibility of epidemics.

5. Fewer hospital beds are required for a population enjoying good mental and physical health. Those beds that are needed may well be in geriatric hospitals because of the longevity of the population.

6. In a community enjoying good health there are fewer delinquents and criminals who require expensive care.

7. A strong, healthy nation wins world-wide respect and promotes peaceful international relations.

Summary

In this section we have considered the growth of a human being, the national and international organization of financial and administrative structures whereby those services which pertain to healthy living are used to develop the citizens' sense of personal social responsibility, and those services required by the injured, disadvantaged and diseased are delivered as humanely and efficiently as possible.

The Skeleton

Throughout the years in hospital the skeleton has been familiarly known as 'Jimmy,' and as we want our story to be identified with one person and his structure, function, growth and passage through life, we could well utilize an imaginary person and call him Jimmy. Jimmy's bony system can be called skeletal system. Together with the joints formed by the bones, and the muscles covering them, the term musculoskeletal system can be used. Many people now prefer the term locomotor system which has the connotation of movement. A system consists of components; each component unit of the skeletal system is known as a bone.

LONG BONES

STRUCTURE

Many of the bones in the skeleton are classified as 'long bones' because they present a shaft and two extremities. The technical term for shaft is 'diaphysis' (diaphyses, pl.; diaphyseal, adj.) and for an extremity it is 'epiphysis' (epiphyses, pl.; epiphyseal, adj.). At least one of the extremities articulates with another bone, thus entering into the formation of a joint (a place where two or more bones meet).

In early fetal life each long bone is laid down as a rod of hyaline cartilage which is pliable. (The pliability of the bones is an important factor in allowing the safe descent of the baby through the birth canal.) Later osteoblasts in the middle of the shaft become active and deposit calcium salts. This is a complex metabolic process requiring the presence of vitamins A, D and C, various enzymes and secretion from the pituitary, thyroid and parathyroid glands at the right time. This area is known as the primary ossification centre and the conversion of cartilage into bone spreads towards each extremity. As previously stated, bone is opaque to X-ray; it therefore follows that pregnancy can be confirmed at this stage.

Shortly after birth a secondary ossification centre becomes active in each extremity, that nearest the head being referred to as the *proximal*, and that farthest from the head as the *distal* secondary ossification centre. Ossification or calcification continues to spread from these centres until it meets the same process spreading towards it from the shaft. Throughout childhood the long bones are in this changing state, but the animal matter remains in predominance to allow growth in length. This less rigid bone gives resilience against the many falls which inevitably occur as the baby learns to walk and run, and as the child learns to climb trees, ride a bicycle, etc. But for this predominance of animal matter, many would be the broken bones (fractures). When too great a blow is dealt these bones the 'sap' in them allows them to bend and crack, rather than break, as illustrated in Figure 18. The name given to the resultant injury is 'greenstick' fracture, for obvious reasons.

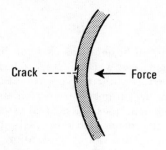

Crack ----- ◄—— Force

Fig. 18   Greenstick fracture.

As adolescence is reached there is only a layer of cartilage between each extremity and the shaft. Because it is present at each extremity (epiphysis), it is called epiphyseal cartilage. Can you remember the term used for the one nearest to the head and the one farthest from the head? Osteoblasts continue to lay down calcium at the shaft (diaphyseal) side of the epiphyseal cartilages, so that growth in length ensues, and when maximum height is reached (partly dictated by the genes), the epiphyseal cartilages become completely ossified and further growth in length (or height) is impossible. Until this takes place a fortunately rare injury can occur; an epiphysis (extremity) can become displaced from the diaphysis (shaft), and this condition is known as a slipped epiphysis, about which you will learn more in your orthopaedic classes.

The long bones reach stability in composition in the adult, there being a predominance of animal matter, that is two-thirds to one-third mineral matter. Because of the wear and tear to which the articulating ends of long bones are subjected throughout life, they are covered with hyaline cartilage, a durable, resilient tissue. Surrounding the cylindrical shaft the strong membrane called periosteum remains throughout life. Can you remember its functions? (p. 17). Inside the periosteum of the shaft and the hyaline cartilage at the ends is a layer of compact bone tissue which is actually thickest in the shaft. (Details of the wonderful architecture of this tissue can be found in *Applied Anatomy for Nurses*.[42]) Inside the compact bone the ends are filled in with cancellous bone tissue containing reticulo-endothelial cells that have a phagocytic action (Fig. 4). The spaces are filled with red bone marrow, in which red and white blood cells and platelets are made. There is also a layer of cancellous bone tissue filled with yellow bone marrow throughout the shaft, which has a central canal known as the medullary cavity to lighten the bone and allow transmission of blood and lymphatic vessels, and nerves. These vessels enter and leave via the nutrient foramina which can be seen to pierce the long bone.

In advancing age yet more calcium is deposited within the bones, so that finally it is in predominance over the animal matter. This makes the bones brittle and easily broken by a force which they would have been able to withstand in earlier adult years.

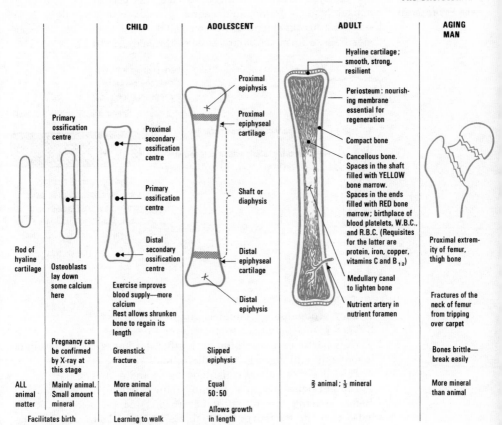

Fig. 19   Pictorial summary of Jimmy's long bones.

FUNCTIONS[43]

1. It participates in joint formation.
2. It gives attachment to muscles and ligaments.
3. It acts as a lever in weight movement.
4. It makes red blood cells, leucocytes and platelets.
5. A long bone acts as a storehouse for calcium.

FLAT BONES

STRUCTURE

In early fetal life the flat bones start off as sheets of strong white fibrous tissue in the shape of the bone to be formed. In late fetal life osteoblasts become active in the central ossification centre. Spicules,

45

arranged rather like a spider's web, spread out to the periphery, and along these the osteoblasts deposit calcium, gradually converting the white fibres into bone. Examples are the breast and cranial bones, and at birth they contain more animal than mineral matter, to allow 'moulding' through the birth canal.

In adult life the outer nourishing membrane remains, to supply the underlying bone with blood. Within this periosteum a layer of compact bone is found, and centrally a layer of cancellous bone tissue, the arrangement being somewhat like a sandwich. The flat bones help to form the boundaries of the body cavities.

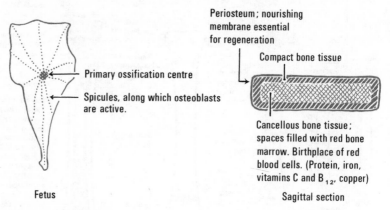

Periosteum; nourishing membrane essential for regeneration

Compact bone tissue

Primary ossification centre

Spicules, along which osteoblasts are active.

Cancellous bone tissue; spaces filled with red bone marrow. Birthplace of red blood cells. (Protein, iron, vitamins C and B_{12}, copper)

Fetus

Sagittal section

Fig. 20   A flat bone.

FUNCTIONS

1. They protect delicate underlying organs.
2. They give attachment to muscles and ligaments.
3. They participate, in union with other flat bones, in formation of immovable and slightly movable joints.
4. They make red blood cells, leucocytes and platelets.
5. The flat bones act as a storehouse for calcium.

Before leaving the actual structure of bone, we need to be reminded that the requisites for its formation must be included in the diet, about which you will learn more later. The most important factor to learn at the moment is that mineral salts are essential, especially those of phosphorus and calcium, vitamin D being necessary for the absorption and utilization of the latter. All three items are present in meat, eggs, cheese, butter, cream and milk. Vitamin A, D and C drops for babies and one-third of a pint of free milk at school play a large part in building up healthy skeletons in the British population, and in cutting down the incidence of rickets.

After these generalized considerations of long and flat bones, we are now ready to build up Jimmy's skeleton, commencing with the skull.

THE SKULL

There is only one royal road to the successful study of the skull, and that is to spend the time handling the actual bones, identifying the landmarks, and understanding the application of this knowledge to the patients.

The skull comprises 22 bones and is divided into two: the cranium, made up of eight bones, and the face, made up of 14 bones.

Cranium

Most of the cranial bones are flat ones, arranged as a box; in fact it can be called the brain box for the brain is the sole content of the cranial cavity.

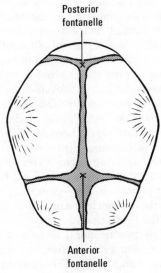

Fig. 21 The fontanelles of the cranium.

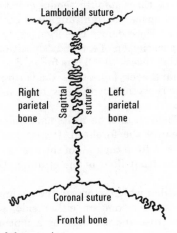

Fig. 22 The sutures of the cranium.

47

Parietal bones. These are two square bones which form the roof, meeting in the midline and stretching down to form part of the sides. All four edges of each bone are serrated to fit into another bone, thus forming immovable joints called sutures. Superiorly the parietal bones join in the sagittal suture. Anteriorly they join the frontal bone in the coronal suture. Posteriorly they join the occipital bone in the lambdoidal suture, so called because of its likeness to the Greek letter λ.

At birth there is incomplete ossification where the three bones meet back and front, and the spaces are called the posterior and the anterior fontanelle. The incompletely joined sutures and these two spaces allow considerable moulding of the baby's head during its descent down the birth canal. The sutures unite shortly after birth; the posterior fontanelle closes at the end of 3 months and the anterior fontanelle at the end of 18 months. Delay in any of these closures calls for investigation. Whilst the anterior fontanelle remains open it can be seen to pulsate, and indeed the pulse can be taken here since the venous blood immediately beneath this fontanelle transmits the pulsation from the arteries immediately underlying it. In illness fluid can be introduced into this venous blood, so you have already learned one site for intravenous (intra = into; venous = vein) infusion in a young baby. Looking at the under surface of the parietal bones several well-defined grooves will be seen. In life these contain arteries and it is not difficult to appreciate that a blow on the head can rupture one of these arteries, which will then bleed and the blood will compress the brain tissue.

Temporal bones. The two temporal bones not only complete the sides of the box, but a triangular portion juts inwards to form part of the floor. The areas you must be able to identify are as follows:

1. The external ear canal (auditory meatus).

2. The space at the end of the canal (the middle ear).

3. A minute set of canals shaped like a snail's shell (cochlea) in the triangular portion.

4. Another set of three minute semicircular canals set in three different planes of space, at right angles to each other, also in the triangular portion. These house the organ responsible for the sense of balance and appreciation of the body's position in space.

5. The zygomatic process which juts forward to meet the zygomatic process of the malar (cheek) bone, thus completing the zygomatic arch.

6. A socket for reception of the jaw bone, forming the temporomandibular joint.

Frontal bone. The frontal bone, as its name implies, forms the front of the box, in reality the forehead. It also has a horizontal plate jutting backwards to form the roof of the eye cavity (orbit), which is simultaneously the front floor of the cranium. Other features to be identified are:

1. The orbital ridge which supports the eyebrows.

2. The frontal sinuses, one on either side, which communicate with the nasal cavities, and are lined by a continuation of the nasal mucosa, which can become inflamed with the common cold.

Occipital bone. The occipital bone forms the back of the box and curves inwards to form part of the posterior floor of the cranium. Four fossae should be noted on the inner surface; the two upper ones house the posterior, inferior portion of the large brain (cerebrum); the two lower ones house the two hemispheres of the small brain (cerebellum). The large hole is called the foramen magnum, and it provides a passage for the spinal cord coming from the brain. Inferiorly and on either side of this hole are two condyles which articulate with the first bone of the spine (atlas) forming the joint which gives nodding movement to the head.

Sphenoid bone. The sphenoid bone, said to be the shape of a bat with outstretched wings, completes the main portion of the floor of the box. Anteriorly it joins the horizontal plates of the frontal bone; laterally it joins the temporal bones; posteriorly it joins the occipital bone. The body contains a large air sinus called the sphenoidal sinus, which communicates with the nose. On its upper surface is a depression—the pituitary fossa—which houses the pituitary gland. The wings contain many foramina for the transmission of cranial nerves. The optic foramina and their proximity to the pituitary fossa should be noted, so that it can be appreciated that any abnormality in this region can manifest with eye complaints (Fig. 24).

There is still a small portion of the floor of the box unaccounted for. It lies between the horizontal plates of the frontal bone, and it is filled in by a spongy plate belonging to the ethmoid bone.

Ethmoid bone. The ethmoid bone presents three portions:

1. A perforated, spongy, horizontal plate, oblong in shape, which lies between the horizontal plates of the frontal bone and completes the anterior fossa of the cranial floor. It allows passage for the nerves of smell (olfactory), the end plates of which are distributed in the mucous membrane of the upper portion of the nasal cavities.

2. A perpendicular plate which forms the posterior portion of the bony nasal septum (Figs 23 and 26).

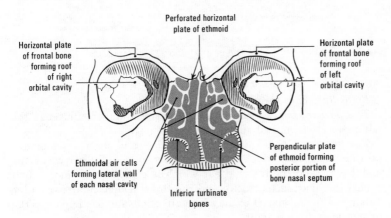

Fig. 23   Diagrammatic representation of ethmoid bone.

3. Two spongy portions which form the lateral walls of each nasal cavity. These ethmoidal air cells are covered with nasal mucosa, which easily becomes inflamed with the common cold.

'Fractured base of skull' is a term you will hear in the wards, and it means that there is a crack in the cranial floor which, if you look down upon it, is divided into three fossae (Fig. 24).

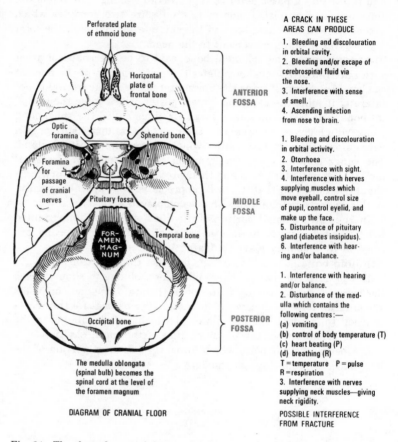

Perforated plate
of ethmoid bone

Horizontal
plate of
frontal bone

Optic
foramina

Sphenoid bone

Foramina
for
passage
of cranial
nerves

Pituitary fossa

FOR-
AMEN
MAG-
NUM

Temporal bone

Occipital bone

The medulla oblongata
(spinal bulb) becomes the
spinal cord at the level of
the foramen magnum

DIAGRAM OF CRANIAL FLOOR

ANTERIOR
FOSSA

MIDDLE
FOSSA

POSTERIOR
FOSSA

A CRACK IN THESE
AREAS CAN PRODUCE

1. Bleeding and discolouration
in orbital cavity.
2. Bleeding and/or escape of
cerebrospinal fluid via
the nose.
3. Interference with sense
of smell.
4. Ascending infection
from nose to brain.

1. Bleeding and discolouration
in orbital activity.
2. Otorrhoea
3. Interference with sight.
4. Interference with herves
supplying muscles which
move eyeball, control size
of pupil, control eyelid, and
make up the face.
5. Disturbance of pituitary
gland (diabetes insipidus).
6. Interference with hearing and/or balance.

1. Interference with hearing
and/or balance.
2. Disturbance of the medulla which contains the
following centres:—
(a) vomiting
(b) control of body temperature (T)
(c) heart beating (P)
(d) breathing (R)
T = temperature   P = pulse
R = respiration
3. Interference with nerves
supplying neck muscles—giving
neck rigidity.

POSSIBLE INTERFERENCE
FROM FRACTURE

Fig. 24   The three fossae of the cranial floor.

SUMMARY OF THE CRANIUM

Eight flat bones form a box.

Two parietal bones form the roof and sides, and they meet in the midline at the sagittal suture.

Two temporal bones complete the sides, and form a triangular portion of the floor which contains the external auditory meatus, the middle ear, and the internal ear which is composed of snail's shell canals for hearing (cochlea), and the three semicircular canals for balance. They form part of the zygomatic arch, and participate in the temporomandibular joint.

One frontal bone forms the forehead, the horizontal plate of which

forms the roof of the orbits which is the floor of the cranium. It
thickens to form the orbital ridges, and it contains the frontal sinuses
which drain into the nasal cavity.

One occipital bone forms the back of the box. The two upper
indentations are known as the cerebral fossae. It contains the foramen magnum, on either side of which are the condyles forming the
nodding joint with atlas.

One sphenoid bone forms the floor. Its body contains the sphenoidal
sinus which drains into the nose and on its upper surface is the
pituitary fossa, into which fits the pituitary gland. The foramina for
the cranial nerves are to be found in the wings, the most important
being the optic foramina.

One ethmoid bone completes the anterior fossa of the cranial floor,
and allows a passage for the nerves of smell.

So much for the cranium, to which we now need to append a face.

The Face

A collection of 14 bones make up the face. Try to learn them in an
order which leads methodically from one to the next, starting from
the top and working downwards.

Nasal bones. Two nasal bones form the bridge of the nose. They are
small, oblong and meet in the midline. They join the frontal bone
above and are separated from the lacrimal bone by part of the
maxilla on either side. They manifest not only racial but family
characteristics. They can be destroyed by the germ causing syphilis,
giving rise to the classical 'saddle nose,' and their deformation
contributes to the 'snuffles' of the congenitally syphilitic baby, now
fortunately a rarity. Can you explain this? (p. 36).

Lacrimal bones. Two lacrimal bones lie on either side separated
from the nasal bone by the frontal process of maxilla. They form part
of the medial orbital wall. They are thin and grooved, so that the
orbital cavity has communication with the nasal cavity on either
side. In life, each groove contains a membranous duct—the tear
duct—via which tears that have washed the front of the eye are
conveyed to the nose.

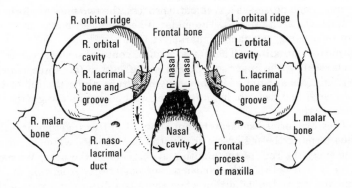

Fig. 25   The nasal and lacrimal bones.

The vomer. Keeping to the midline in this area, you will find a thin plate of bone arising from the middle line of union of the bony palate and passing posteriorly to the inferior surface of the sphenoid. Superiorly it joins the perpendicular plate of ethmoid and thus completes the bony nasal septum. Many textbooks describe it as 'plough-shaped,' but since many of your generation have never seen a plough, you will have to describe its shape in your own words.

Inferior turbinate bones. Lying inferiorly to the ethmoidal air cells (Fig. 23) these two delicate, scroll-like bones jut out from the lateral nasal walls, increasing their surface area. Being spongy in texture they can become infected, which sometimes necessitates their removal by surgical operation (turbinectomy). (Take every opportunity of building up these words which are new to you, by learning and using the prefixes and suffixes.[44])

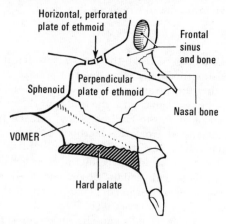

Fig. 26   The vomer and nasal septum.

Malar bones. Lying laterally to the two nasal and the two lacrimal bones are two malar (cheek) bones (Fig. 25). They form not only the prominence of the cheeks, but part of the floor of the orbit. Examine carefully and you will find the lines of union. The bar of bone jutting posteriorly is called the zygomatic process. Can you remember which cranial bone possesses a zygomatic process, the two together forming a complete arch? (p. 48). The malar, like the nasal bones, manifest not only racial but family characteristics.

Superior maxillae. These two bones contain sockets (alveolar processes) for the upper set of teeth, and a 'gutter' (air sinus, maxillary sinus or antrum of Highmore) which runs along the top of the side teeth and has a hole on its inner wall communicating with the nasal cavity.

They are formed separately, but should be fused together at birth, thus forming the major portion of the bony palate which is simultaneously the roof of the mouth and the floor of the nose. We have already learned that the vomer passes posteriorly from this union. Can you name the condition when there is a failure of union of the two maxillae (maxilla, sing.)?

52

Palatal bones. These two L-shaped bones complete the posterior portion of the bony palate, the posterior lateral nasal walls, and the posterior orbital cavity.

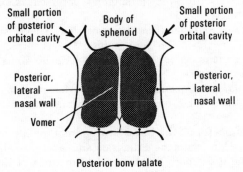

Small portion of posterior orbital cavity

Body of sphenoid

Small portion of posterior orbital cavity

Posterior, lateral nasal wall

Posterior, lateral nasal wall

Vomer

Posterior bony palate

Fig. 27   Diagram of a palatal bone viewed from behind.

Mandible. This is the only movable bone in the skull and it lies below all those already mentioned. It forms the lower jaw, and the main points are noted in Figure 28.

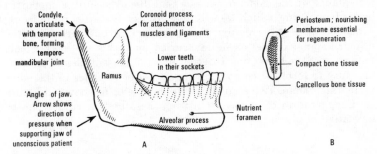

Condyle, to articulate with temporal bone, forming temporomandibular joint

Coronoid process, for attachment of muscles and ligaments

Periosteum; nourishing membrane essential for regeneration

Lower teeth in their sockets

Compact bone tissue

Ramus

Cancellous bone tissue

'Angle' of jaw. Arrow shows direction of pressure when supporting jaw of unconscious patient

Nutrient foramen

Alveolar process

A

B

Fig. 28   A, Diagram of right lateral view of mandible. B, Sagittal section of mandible.

Functions. The mandible has four main functions.

1. It articulates with the temporal bone forming the temporomandibular joint, a hinge (condyloid) joint.
2. It gives attachment to muscles and ligaments.
3. It gives attachment to the tongue.
4. It contains the teeth.

SUMMARY OF THE FACE

The face is composed of 14 bones.

Two nasal bones form the bridge of the nose.

Two lacrimal bones connect the orbital and nasal cavities on either side.

The vomer is a thin plate of bone forming the posterior nasal septum. It arises from the superior surface of the middle line of union of the bony palate and terminates at the inferior surface of the sphenoid.

53

Two inferior turbinate bones are scroll-like and spongy in texture. They jut out from the lateral nasal walls to increase their surface area.

Two malar bones form the prominence of the cheek and part of the floor of the orbit.

Two superior maxillae form the upper jaw. They fuse together in the midline and form the bony palate. They contain the alveolar processes for the upper set of teeth and a large air sinus which opens into the nasal cavity.

Two palatal bones complete the posterior portion of the bony palate, the posterior lateral nasal walls and the posterior orbital cavity.

The mandible forms the lower jaw and is the only movable bone in the skull. Can you make a diagram and label it to show the main points of the mandible? What are the functions of the mandible?

Jimmy's skull is now complete, but it has to be supported on his spine.

BACKBONE, SPINE, SPINAL COLUMN, VERTEBRAL COLUMN

One glance at the skeleton tells us that Jimmy's skull is supported by his bony spine, which we will now discuss. It started off by being a collection of 33 irregular bones, curved in one direction (the primary spinal curve) as he lay curled up in his mother's womb. Later five of the bones fuse together to form the sacrum and another four fuse to form the coccyx, giving a total of 26 bones in the adult vertebral column. Two secondary curves develop as shown in Figure 29. Exaggeration of or deviation from these normal curves can be indicative of disease.

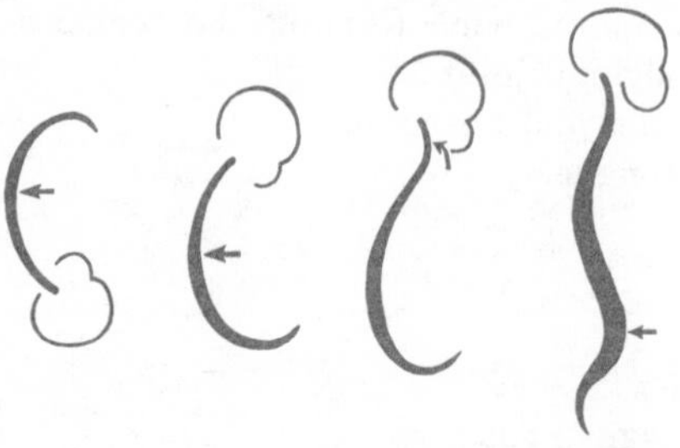

IN UTERO	AFTER BIRTH	ON SITTING UP	ON WALKING	LAY TERM FOR AREAS	TECHNICAL TERM FOR AREAS	NO. OF BONES IN CHILD	NO. OF BONES IN ADULT
				Neck	Cervical	7	7
				Chest	Thorax Thoracic Dorsal	12	12
				Loin	Lumbar	5	5
				Pelvis	Sacrum	5	1
					Coccyx	4	1
Primary curve, concave anteriorly	Primary curve, concave anteriorly	First secondary curve, convex anteriorly	Second secondary curve, convex anteriorly				
					TOTAL	33	26

Fig. 29   Diagram of normal spinal curves.

The lay terms for this area are the backbone and the spine. The technical term is the spinal or the vertebral column. The term 'vertebral column' arises because each bone in the column is called a vertebra (vertebrae, pl.). The first part of the plan is to master the names of the parts making up each vertebra, and they are set out in Figure 30. The 26 bones are 'strapped' together by the anterior and posterior spinal ligaments.

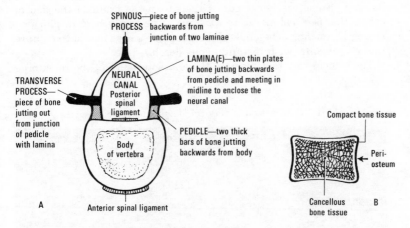

SPINOUS—piece of bone jutting
PROCESS backwards from
junction of two laminae

LAMINA(E)—two thin plates
of bone jutting backwards
from pedicle and meeting in
midline to enclose the
neural canal

TRANSVERSE
PROCESS—
piece of bone
jutting out
from junction
of pedicle
with lamina

NEURAL
CANAL
Posterior
spinal
ligament

Body
of vertebra

PEDICLE—two thick
bars of bone jutting
backwards from body

Compact bone tissue

Peri-
osteum

A

Anterior spinal ligament

Cancellous
bone tissue

B

Fig. 30   A, Diagram of a typical vertebra. B, Cross section.

It is impossible in this diagram to show the two articulating surfaces, one for the body of the vertebra above and one for the body of the vertebra below. It is also impossible to show the four articulating facets which enter into the formation of gliding joints. They lie at the junction of the pedicle with the lamina on either side, two above and two below. They can, however, be imagined from Figure 31 showing a side view of several vertebrae.

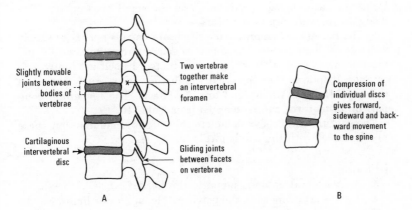

Slightly movable
joints between
bodies of
vertebrae

Two vertebrae
together make
an intervertebral
foramen

Compression of
individual discs
gives forward,
sideward and back-
ward movement
to the spine

Cartilaginous
intervertebral
disc

Gliding joints
between facets
on vertebrae

A

B

Fig. 31   A, Diagram of vertebral cartilaginous (slightly movable) and gliding joints. B, Diagram of compression of intervertebral discs.

Of the seven cervical vertebrae, the first two are not 'typical' (as illustrated in Fig. 30). The first one is merely a ring of bone with two fossae to receive the occipital condyles, forming a condyloid (hinge) joint that allows nodding movement of the head. It is called atlas, as from Greek mythology he was supposed to carry the world on his shoulders.

The second one, the axis, has a peg of bone protruding from its tiny body, which fits into the superior ring of bone (atlas) forming its body. Since the nerve cord passes from the brain in the neural canal, the 'peg' belonging to axis has to be firmly held in place by strong ligaments. (The click of the hangman's rope is said to rupture these ligaments, with instantaneous death.)

The easiest way to remember the characteristics of the remaining

Table II. Comparison of cervical, dorsal and lumbar vertebrae.

	CERVICAL	DORSAL	LUMBAR
BODY	Smallest	Larger than cervical Smaller than lumbar 4 facets for ribs	Largest
NEURAL CANAL	Largest	Smaller than cervical Larger than lumbar	Smallest
SPINOUS PROCESS	Bifid	Pointing downwards	Broad and square
TRANSVERSE PROCESS	Foramina for passage of blood vessels and nerves	Facets for ribs	Thin with no foramina or facets

5 cervical, the 12 thoracic and the 5 lumbar vertebrae is to grasp two principles.

1. When building a column with play bricks of varying size the large ones are placed at the base and the small ones on top. The 'body' of the vertebra is the brick.

2. The nerve cord coming from the brain carries nerves to supply every bit of body tissue. These nerves emerge from the neural canal via the intervertebral foramina, thus the cord itself gradually gets thinner. It follows therefore that the further down injury occurs to the cord, the less widespread are the effects. The remaining comparison of features can be learned from Table II.

The sacrum and coccyx complete the vertebral column, but these are discussed with the pelvic girdle (p. 59).

FUNCTIONS

1. The vertebral column supports the head.

2. It gives attachment to the pelvic girdle; via lower limbs, weight is transmitted to the ground.

3. It gives attachment to the rib cage.

4. It protects the spinal cord, contained in the neural canal, and gives safe passage to the spinal nerves via the intervertebral foramina.

5. It protects delicate structures in the thoracic, abdominal and pelvic cavities.

6. It stores calcium.

7. It provides two-thirds of the body's red blood cells. It also makes leucocytes and platelets.

8. It gives attachment to muscles and ligaments.

9. Via its slightly movable and gliding joints, it provides movement. There are also the nodding and side-to-side movements of the head.

10. It forms a rigid central support and is the main structure in the maintenance of good posture.

Posture

In the *Concise Oxford Dictionary* this word is defined as carriage; attitude of body or mind; condition; state. Given a normal skeleton clothed with healthy muscle tissue, and controlled by a sound brain, we know that posture is under the direct control of the will, for we can stand to attention, or relax in an armchair, as occasion demands. We are not constantly 'willing' ourselves to maintain good posture, so this means that it must be taken over at an 'automatic' level. It is in fact the *habitual* way we hold ourselves, and the habit of mind that maintains good posture can be taught and reinforced. In this process appeal to pride and self-regard gains the best result. Posture is closely related to our *mental outlook* on life. Do we not say: 'He is down and out'? Do we not brace our shoulders and hold up our heads when we feel ready to face the world and whatever may come?

Posture cannot be considered as an entity, but in close relation to adequate nutrition and metabolism of the tissues, and maintenance of muscle tone, and this surely involves exercise. The maxim for good standing posture is head up, shoulders straight, 'tail' tucked in, heels together. When standing for any length of time move the toes at frequent intervals to avoid stasis (stagnation) of blood in leg

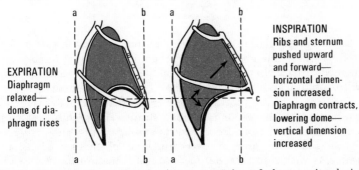

EXPIRATION
Diaphragm relaxed— dome of diaphragm rises

INSPIRATION
Ribs and sternum pushed upward and forward— horizontal dimension increased. Diaphragm contracts, lowering dome— vertical dimension increased

Fig. 32 Diagram illustrating maximum expansion of chest cavity during inspiration and maximum retraction during expiration.

vessels. Bad things about posture are round shoulders, sagging (bulging) abdominal wall, slouching back, drooping head and fallen arches. Frequent practice at controlling the muscles so as to avoid these bad points is necessary, and muscle 'tone' improves with learning to use muscles actively.

Correct posture when sitting or standing allows maximum expansion of the chest cavity during inspiration, ensuring maximum oxygen for metabolism in the tissues.

Similarly correct posture allows maximum retraction of the chest cavity during expiration, ridding the body of waste products— excess carbon dioxide and water—and preventing stagnation of secretions in the base of the lungs, which encourages chest infections. A few deep breaths taken daily in the open air is a health habit well worth acquiring.

Blood vessels supplying the soft organs (liver, stomach, bowels) of the abdominal cavity are transmitted in folds of peritoneum attached to the posterior wall (Fig. 33). Posture which allows the organs to fall forwards and/or downwards will make the folds of peritoneum taut so that they compress the blood vessels to the liver, stomach and bowels, which will then function less efficiently.

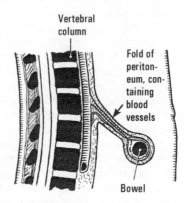

Fig. 33   Blood supply to the abdominal organs.

Good muscle tone in the anterior abdominal wall is essential to good posture and good health. School games and exercises are planned with this in mind. 'Cycling on the back' for five minutes daily will maintain an efficient abdominal 'corset' throughout life.

The arches of the foot are so arranged that the body weight is transmitted to the ground without strain. The arches are maintained by slings formed from tendons and ligaments, and only good muscle tone can prevent over-stretching of these tendons and ligaments with resultant 'flat feet'. Muscle tone in this area can be maintained by describing a circle, clockwise and anticlockwise, with the big toe several times daily.

Since it is customary for British people to wear shoes on their feet, any consideration of posture needs to be combined with sensible footwear (p. 71).

It cannot be too strongly stressed that our posture is closely related to our mental attitude to life, and that environment can be used to assist this attitude, e.g. an armchair suggests relaxation; the classroom and the library suggest study.

Whilst you are nursing you will require to develop correct posture in lifting and moving patients,[45] etc. A good, well-illustrated account of this is to be found in *Protective Body Mechanics in Daily Life and Nursing*.[46]

Having paused to consider posture in relation to the maintenance of health, we must now complete Jimmy's vertebral column with a description of the sacrum and the coccyx. It is important for you to realize that a description of the vertebral column or of the pelvic girdle is not complete without the sacrum and coccyx.

PELVIC GIRDLE

One glance at the skeleton shows that this is rather like a basin with the sacrum and coccyx at the back and an innominate bone on either side (Fig. 34); the female pelvis being wider and more shallow than the male pelvis.

The word 'anatomy' involves the position, relationships and structure of the part being described. It is useful to establish good habits of description, which will now be applied to the sacrum.

Sacrum
POSITION AND RELATIONSHIPS

This bone forms the posterior boundary of the pelvic cavity. The fifth lumbar vertebra lies above it, forming the lumbosacral joint

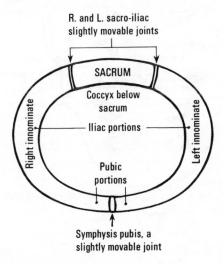

Fig. 34   Diagram of pelvic girdle.

which is a slightly movable joint. The coccyx lies below, forming the sacrococcygeal joint which is also slightly movable and of importance during pregnancy. On either side lies an innominate bone forming the right and left sacro-iliac joints. Again these are slightly movable joints and are of importance during pregnancy. The bowel as it descends towards the rectum lies in front of the sacrum; behind there is muscle fascia, subcutaneous tissue and skin. You will notice that we have put something above, below, at either side, in front and behind the sacrum. This is a useful list for checking that you thoroughly understand the position and relationships of any part.

STRUCTURE

The sacrum starts off as five separate vertebrae which eventually fuse into one bone. The marks of union can be seen on the anterior surface, and the rudimentary spinous processes remain on the posterior surface. The neural canal continues as the sacral canal, centrally within the bone. The pairs of intervertebral foramina can also be observed, these giving safe passage to the sacral (spinal) nerves.

BLOOD SUPPLY

Branches from the internal iliac artery.

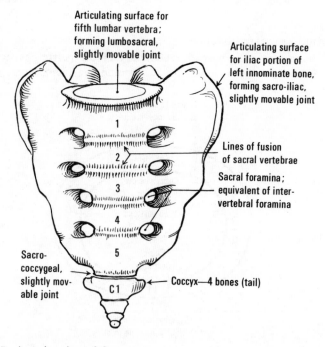

Articulating surface for fifth lumbar vertebra; forming lumbosacral, slightly movable joint

Articulating surface for iliac portion of left innominate bone, forming sacro-iliac, slightly movable joint

Lines of fusion of sacral vertebrae

Sacral foramina; equivalent of inter-vertebral foramina

Sacro-coccygeal, slightly movable joint

Coccyx—4 bones (tail)

Fig. 35 Anterior view of the sacrum.

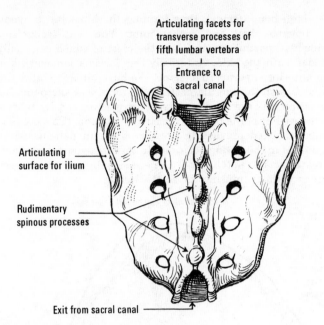

Articulating facets for
transverse processes of
fifth lumbar vertebra

Entrance to
sacral canal

Articulating
surface for ilium

Rudimentary
spinous processes

Exit from sacral canal

Fig. 36   Posterior view of the sacrum.

VENOUS DRAINAGE

Branches to the internal iliac vein.

FUNCTIONS

1. It gives protection to the lower abdominal and pelvic organs.

2. It allows safe passage of sacral nerves via sacral canal and foramina.

3. It gives attachment to muscles and ligaments.

4. It forms the following slightly movable joints: lumbosacral, right and left sacro-iliac, sacrococcygeal.

5. It takes the weight of the body when lying on the back. (In nursing lessons you will learn how to prevent pressure sores arising in this area.)

Coccyx

This is found below the sacrum and is often called the tail bone. It is made from four rudimentary vertebrae fused together. As already stated, it superiorly forms the sacrococcygeal, slightly movable joint.

Innominate Bone

Handling this bone you need to realize that it was made in three parts which fused together in the socket (acetabulum) to receive the head

61

of the thigh bone (femur), thus forming the hip joint, a synovial, freely movable, ball and socket joint. You will recognize the superior 'flat' portion as *the ilium*, the crest of which can easily be felt underneath the skin. Anteriorly this forms a landmark known as the anterior superior iliac spine. We have already stated that it articulates with the sacrum posteriorly, forming a sacro-iliac joint.

The ischium is the thickest, strongest portion and its tuberosity takes the weight of the body in the sitting position. (This needs to be remembered when preventing pressure sores in patients who are being nursed in the sitting position.) In this area you need to identify the greater sciatic notch, for it is necessary to give intramuscular injections without injuring the sciatic nerve.

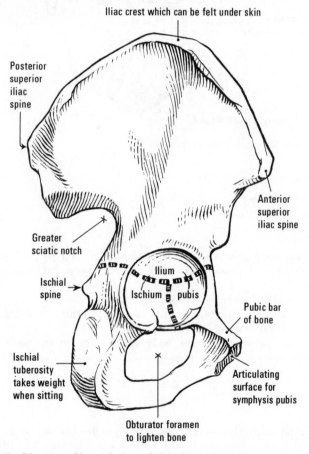

Iliac crest which can be felt under skin

Posterior superior iliac spine

Anterior superior iliac spine

Greater sciatic notch

Ischial spine

Ilium

Ischium pubis

Pubic bar of bone

Ischial tuberosity takes weight when sitting

Articulating surface for symphysis pubis

Obturator foramen to lighten bone

Fig. 37　Diagram of lateral view of right innominate bone.

The pubis lies anteriorly and is composed of two bars of bone enclosing a large hole (obturator foramen) which serves to lighten the bone. It meets its fellow centrally, forming the slightly movable joint called symphysis pubis, which allows slight increase in size of the pelvic cavity during pregnancy.

Branches from the internal iliac artery.

Branches to the internal iliac vein.

The Skeleton

FUNCTIONS
1. It gives protection to the lower abdominal and pelvic organs.
2. It protects the sciatic nerve.
3. It forms the following joints: right and left sacro-iliac (slightly movable), symphysis pubis (slightly movable), the hip which is a synovial, freely movable, ball and socket joint.
4. It gives attachment to muscles (iliopsoas, gluteals, quadriceps extensor femoris, p. 102) and ligaments.

When writing about the functions of the *pelvic girdle,* it should be added that in the female it is capable of slight expansion under hormonal influence during pregnancy.

SUMMARY OF VERTEBRAL COLUMN AND PELVIC GIRDLE

A *typical vertebra* consists of a body, two pedicles and two laminae, enclosing a neural canal which contains the spinal cord. A spinous process projects posteriorly from the union of the laminae, and a transverse process projects laterally from the union of a pedicle with a lamina on either side. The body has an articulating surface above and below for formation of slightly movable joints. A vertebra presents four articulating facets for formation of synovial gliding joints. Two vertebrae together complete a pair of intervertebral foramina.

The *vertebral column* is composed of 33 bones in a child and 26 in an adult.

The *atlas* is a complete ring of bone which receives the occipital condyles to give nodding of head. The atlas is the first cervical vertebra.

The *axis* is the second cervical vertebra. It has a peg projecting from its body up into the lumen of the atlas; this peg acts as the 'body' of the atlas and is held in position by strong ligaments.

The *remaining five cervical vertebrae* can be identified by their small bodies, large neural canals, bifid spinous processes and foramina in the transverse processes.

The *12 dorsal or thoracic vertebrae* can be identified by facets on the body and transverse processes for articulation with the ribs. The spinous processes are long and look downwards. The bodies are larger than those of the cervical vertebrae, but smaller than the lumbar. The neural canals are smaller than those of the cervical vertebrae, but larger than the lumbar.

The *five lumbar vertebrae* are easily identified as they have the largest bodies and the smallest neural canals. The spinous processes are square and jut posteriorly in a horizontal plane. The fifth lumbar vertebra articulates with the sacrum below, forming the lumbosacral joint.

The sacrum is triangular in shape and results from the fusion of five sacral vertebrae. The lines of fusion, the rudimentary spinous processes, the sacral foramina and canal can all be identified. Laterally the sacrum articulates with the right and left ilium to form the right and left sacro-iliac, slightly movable joints; below lies the sacrococcygeal, slightly movable joint. The sacrum takes the body weight when lying on the back.

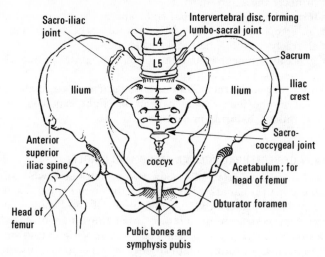

Fig. 38   Anterior view of pelvic girdle.

The coccyx is the tail bone and results from the fusion of four coccygeal vertebrae.

The pelvic girdle is composed of the sacrum, coccyx and two innominate bones.

The innominate bone is formed in three parts which meet in the acetabulum, a socket that receives the femoral head forming the hip joint.

The ilium is the broad, flat portion, the crest of which can be felt underneath the skin terminating at the anterior superior iliac spine—a landmark.

The ischium is the thickest, strongest portion; the tuberosity takes the weight in the sitting position. The greater sciatic notch transmits the sciatic nerve.

The pubis is two bars of bone completing the obturator foramen. It meets its fellow centrally forming the symphysis pubis, a slightly movable joint.

Branches from the internal iliac arteries supply the sacrum, coccyx and innominate bones with blood which is drained into the internal iliac veins.

Can you write out the functions of the sacrum and the pelvic girdle?

LOWER EXTREMITY

The pelvic girdle and lower extremity are built to give stability, so that the upright position can be assumed and maintained by a human being.

Bones cannot be taught or learned from a book, so again I advise you to get out the bones, handle them, use a well-illustrated anatomy book from the reference library, and try to identify and learn the following, continuing to teach yourself to be methodical in description.

Femur (thigh bone)

POSITION AND RELATIONSHIPS

Superiorly it articulates with the acetabulum of the innominate bone to form the hip, a synovial, freely movable, ball and socket joint.

Inferiorly it articulates with the tibia and patella (knee cap) to form the knee, a synovial hinge joint.

The surrounding muscles are the gluteals and the quadriceps extensor femoris (p. 102). The nearby blood vessels are the femoral artery and the deep femoral vein. Lymphatic vessels also lie near the bone.

STRUCTURE

Refresh your memory about the basic structure of a long bone and the changes which occur in it throughout life by turning to page 43 then add the following facts to your existing knowledge.

Proximal extremity or epiphysis. The head is more than a hemisphere and is covered with hyaline cartilage. What is the function of hyaline cartilage? The fovea is a central depression on the head which gives attachment to the ligamentum teres which limits movement. A blood vessel lies near the ligamentum teres. Torsion of the ligament can interfere with the blood supply to the head of the femur and cause necrosis. The head is connected to the shaft at an angle of 120 degrees by the surgical neck, at the lower extremity of which are the greater (lateral) and lesser (medial) trochanters. Joining these two are the anterior and posterior inter-trochanteric lines, which give attachment to the capsule of the hip joint to keep the bones in apposition.

Shaft or diaphysis
1. Linea aspera, a ridge down the back for attachment of muscles.
2. Nutrient foramina transmit blood vessels.

Distal extremity or epiphysis.
1. Condyles for articulation with the tibia.
2. Smooth surface on the front, covered with hyaline cartilage for the patella.
3. A deep notch at the back for attachment of muscles and ligaments, and protection of blood vessels.

65

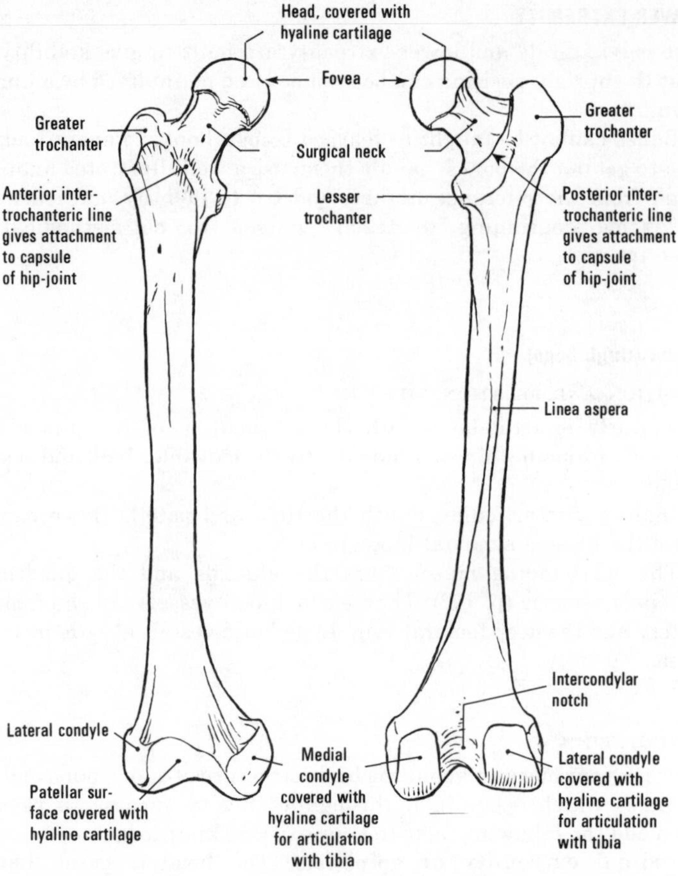

Head, covered with
hyaline cartilage

Fovea

Greater
trochanter

Greater
trochanter

Surgical neck

Anterior inter-
trochanteric line
gives attachment
to capsule
of hip-joint

Lesser
trochanter

Posterior inter-
trochanteric line
gives attachment
to capsule
of hip-joint

Linea aspera

Intercondylar
notch

Lateral condyle

Medial
condyle
covered with
hyaline cartilage
for articulation
with tibia

Lateral condyle
covered with
hyaline cartilage
for articulation
with tibia

Patellar sur-
face covered with
hyaline cartilage

A   B

Fig. 39   The right femur. A, Anterior view. B, Posterior view.

BLOOD SUPPLY

Femoral artery.

VENOUS DRAINAGE

Deep femoral vein.

FUNCTIONS

1. The head articulates with the acetabulum of the innominate bone to form the hip joint, a freely movable, synovial, ball and socket joint.

2. The distal extremity articulates with the tibia and patella to form the knee joint, a freely movable, synovial hinge joint.

3. It gives attachment to muscles (iliopsoas, gluteals, quadriceps extensor femoris) and ligaments.

4. It acts as a lever in weight moving.

5. It helps to transmit weight to the ground.

6. It supports the pelvic girdle.

Tibia

This is the shin bone, which lies medially in the foreleg.

POSITION AND RELATIONSHIPS

Superiorly lies the patella, and the tibia articulates with the femur.
Inferiorly it articulates with talus (ankle bone). Laterally lies the
fibula. The surrounding muscles are the tibialis anticus, gastrocnemius and soleus. The nearby blood vessels are the anterior and
posterior tibial arteries and the deep anterior and posterior tibial
veins. Lymphatic vessels lie near the bone.

STRUCTURE

Proximal extremity or epiphysis.

1. Two half-moon shaped, concave articular surfaces covered with
hyaline cartilage, for articulation with the femur, forming the knee,
a synovial hinge joint.

2. Ridge between them which gives attachment to cruciate
ligaments.

3. Tubercle for insertion of patellar ligament.

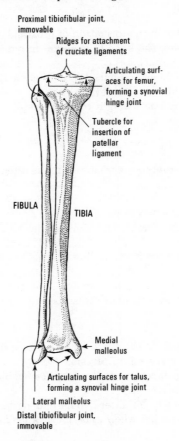

Proximal tibiofibular joint,
immovable

Ridges for attachment
of cruciate ligaments

Articulating surf-
aces for femur,
forming a synovial
hinge joint

Tubercle for
insertion of
patellar
ligament

FIBULA

TIBIA

Medial
malleolus

Articulating surfaces for talus,
forming a synovial hinge joint

Lateral malleolus

Distal tibiofibular joint,
immovable

Fig. 40   Anterior view of right tibia and fibula.

67

4. Laterally, facet for fibula, forming an immovable joint (proximal tibiofibial).

Shaft or diaphysis.

1. Triangular in shape.

2. Front of △ called crest or 'shin'. This lies so near the skin that fractures of the tibia easily become compound.

3. Nutrient foramina transmit blood vessels.

Distal extremity or epiphysis.

1. Projection of medial malleolus.

2. Two-sided articulation surface for talus.

3. Articulating facet laterally for fibula forming an immovable joint (distal tibiofibial).

BLOOD SUPPLY

Branches from anterior and posterior tibial arteries.

VENOUS DRAINAGE

Branches into deep anterior and posterior tibial veins.

FUNCTIONS

1. It participates in the formation of proximal and distal tibiofibial joints (immovable), the knee (a synovial hinge joint) and the ankle (a synovial hinge joint).

2. It gives attachment to muscles and ligaments.

3. It acts as a lever in weight moving.

4. It helps to transmit weight to the ground.

Fibula

This is a thin, medial foreleg bone which is often called 'brooch pin' bone.

POSITION AND RELATIONSHIPS

Superiorly lies the femur, and the patella with which it does not articulate. Inferiorly lies the talus and remaining tarsal bones. Medially lies the tibia. Surrounding muscles are the tibialus anticus, the gastrocnemius and the soleus. The nearby blood vessels are the anterior and posterior tibial artery, and the deep anterior and posterior tibial vein. Lymphatic vessels lie near the bone.

STRUCTURE

Are you remembering the basic structure of a long bone? (p. 43).

Proximal extremity or epiphysis.

1. Articulating facet medially for immovable joint with tibia (proximal tibiofibial).

2. Groove laterally for lateral popliteal nerve. (Pressure on this nerve must be avoided when applying a knee bandage or a below-the-knee plaster.)

Shaft or diaphysis. Nutrient foramina transmit blood vessels.

Distal extremity or epiphysis.

1. Lateral malleolus.
2. Articulating facet for immovable joint with tibia (distal tibiofibial).
3. Larger articulating facet for talus.

BLOOD SUPPLY
Branches from the anterior and posterior tibial arteries.

VENOUS DRAINAGE
Branches into the deep anterior and posterior tibial veins.

FUNCTIONS
1. It participates in the formation of proximal and distal tibio-fibial (immovable joints) and the ankle (a synovial hinge joint).
2. It gives attachment to muscles and ligaments.
3. It helps to transmit weight to the ground.

The Foot

This is also called the tarsus. It is made up of 7 tarsal bones, 5 metatarsal bones and 14 phalanges, 2 in the big toe and 3 in each of the other toes. They are arranged as in Figure 41.

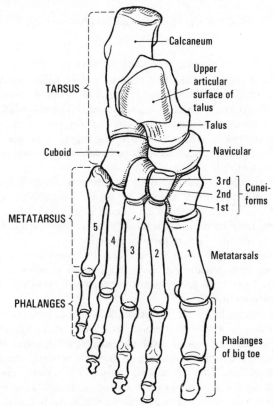

Fig. 41 Right foot from above.

69

The architecture of the foot is such that the weight of a 15-stone man can be satisfactorily transmitted to the ground. This is no mean feat.

The position of the arches is maintained by the ligaments and tendons, and they in turn depend upon good tone in the muscles of which they are a part. These muscles can only be kept in good tone by constant exercise (p. 108).

Realising the importance of our feet, let us now consider their care in the maintenance of health.

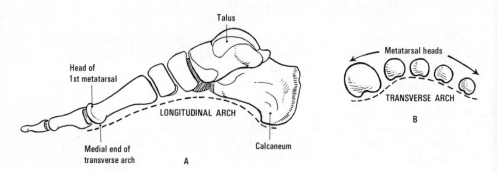

Fig. 42   A, Medial side of right foot showing longitudinal arch. B, Transverse arch.

CARE OF THE FEET IN THE MAINTENANCE OF HEALTH [47,48]

The bones and tissues of the feet need to be well nourished through-out life, and accordingly we need to take an adequate diet, containing sufficient phosphorus, calcium and vitamin D. Feet need regular exercise to strengthen and maintain the support of the natural arches; describing a circle with the big toe is excellent for this purpose. The toe nails need to be kept clean and cut straight across to avoid ingrowing toe nails.

The feet should be washed daily to prevent odour from decom-posing perspiration; they should be dried thoroughly, especially between the toes, and dusted with talcum powder. If perspiration is troublesome they should be swabbed with surgical spirit for extra cooling; a deodorant talcum powder is useful.

It is a good plan to walk in bare feet for some part of the day, provided the area is safe, e.g. no stray pins or needles, no broken glass on the beach: no hookworms, for they can penetrate the skin and cause anaemia.

The feet should be suitably clothed, and medical advice should be sought if there is any interference with the health of the feet.

Choice and care of socks and stockings. Socks and stockings should fit the feet comfortably and should not restrict movement or impede circulation. They should be made of absorbent 'non-shrink' material, which will withstand frequent washing, as clean ones should be worn daily. It is a good plan to establish the habit in older children of washing their own stockings each night. Stockings should

70

be suspended from the waist; garters should be avoided as they impede the return of blood from the feet, in the superficial veins, to the heart, and this may contribute to the formation of varicose veins. Pantihose should have sufficient stretch to prevent foot compression when the wearer sits or bends.

From an early age encouragement should be given to choice and buying of suitable socks and stockings, either to match, or tone with, the outfit with which they are to be worn.

Choice and care of footwear. Footwear should be suitable for the occasion on which it is going to be worn. Shoes should be supple and should fit the foot's natural contour, give sufficient support to the arches and should not throw the body's weight forward on to the toes. Modern fashion does not always consider these basic rules, therefore it is important that the 'fashion follower' does possess one pair of shoes which do, and that she wears them for some part of each day. It is a very good plan to change shoes during the day and to 'air' them between each wearing. If there is any trace of odour in the shoes, sprinkle the inside with deodorant talcum powder.

Some leathers and plastics are hard and incapable of bending with the natural foot movement, thus friction is encouraged and blisters result. It is therefore false economy to buy shoes which are not supple. It is also often false economy to pass on an older child's shoes to a younger child, because the creases already made in the uppers may not be in the natural folds of the younger child's foot and may produce friction with blister formation. Whilst speaking of children's shoes and blisters, it is wise to remember, when buying children's shoes, that they must fit at the heels, but the sole of the shoe should be half to three-quarters of an inch longer than the sole of the foot. Some shops provide a pedascope to ascertain this, but medical opinion states that such machines should be used as little as possible, for there is some relationship between exposure to X-ray and leukaemia. Some shops now supply a visual fitting gauge, so that parents can tell when children's shoes are too small.

Suitable footwear for mentally handicapped people is difficult to procure. Problems come from incontinence, from abnormal foot shapes and from scuffing and kicking. The Disabled Living Foundation are investigating these problems.

Plimsoles, with canvas tops and crepe soles, are widely used for physical exercise in many forms, as they are light and do not restrict movement. It is unwise to wear them without socks, as there is increased perspiration with the increased muscular activities. After use they should hang in the open air for thorough drying, and should not be stuffed back into a shoe bag.

Bedroom slippers are mostly comfortable and relaxing. They should fit, however, and be kept in good repair, otherwise they can be the cause of accidents in the home. It is a bad habit for housewives to wear them throughout the day.

Lighter shoes which allow ventilation are best for summer wear; these are now made in many shapes and shades, and the development of 'taste' can be encouraged by matching, and toning, colour schemes.

Opinion is divided about foot exercise sandals. Walking is one of the best methods of stimulating the proper function of the leg and foot muscles.[49,50,51,52,53]

Outdoor shoes should be waterproof; for the country walkers, extra wear and protection can be afforded by having them 'nailed'.

Boots have a place in our national footwear. There are the more decorative forms for travelling to and fro in our inclement weather. Boots are the standard footwear for several sports. There are also the strong protective boots for miners and outdoor and factory workers. In heavy industry the men are required to wear steel toe-plates as a safeguard against crushed toes. 'Wellington' boots are useful if muddy pathways have to be traversed. An extra pair of woollen socks will help to absorb the perspiration, and the inside of the wellingtons should be thoroughly dried before the boots are worn again.

Where heavy outdoor footwear has to be worn, it is wise to have lighter footwear for indoors. If this changing is established as a habit from childhood onwards, there is much less dirt brought indoors with consequently less wear and tear on floors and carpets.

All footwear should be kept in good repair. 'Down at heel' does not describe the smart, alert, successful person. Shoes should be cleaned after each wearing, for not only does good polish preserve leather, but a shining surface discourages the adherence of germs, and well-polished shoes manifest our personal standard of cleanliness and attention to detail to all with whom we come into contact.

All people should be encouraged to think about the suitability of footwear for the occasion for which they are getting ready, e.g. if the hostess is proud of her parquet floor, or lawn, then it is rank unkindness to wear high stiletto heels; when visiting in hospital or nursing home, wear rubber heels to manifest your acquaintance with the problem of hospital noise [from the daily press] and your willingness to co-operate in avoidance of 'preventable' noise.

CONDITIONS WHICH CAN INTERFERE WITH THE HEALTH OF THE FEET

Blisters. A collection of fluid between the superficial and true skin. As a rule this fluid will absorb, but if the overskin is punctured the fluid will escape, and germs may enter this moist area to flourish and multiply, with resultant sepsis.

Protect an unpunctured blister with a clean, preferably sterile dressing; dab a punctured blister with a reliable disinfectant such as Dettol (2 per cent), or iodine, and keep it covered until it is healed.

Callosity. A local hardening of the skin in response to pressure and/or friction. Rubbing with a pumice stone or saturated solution of salt and olive oil helps to remove same.

Corn. A cone-shaped thickening of the epidermis, with the point of the cone in the deepest layers. It arises as the result of friction and/or pressure. One arising over a toe joint is a 'hard' corn and one arising between the toes is a 'soft' corn because of the apposition of the two skin surfaces enclosing a film of perspiration. If a corn is not removable by the application of a proprietary corn plaster, a chiropodist's service is required.

Wart[54] (verruca, sing.; verrucae, pl.). A flat wart occurring on the sole of the foot, and called verruca plantaris. It is highly contagious, and it is thought that many infections are acquired from the swimming baths. At some public baths the bathers must wear rubber 'slip-ons' on their feet.

The only reason for insisting on 'gym' shoes for gymnastics in schools, colleges, etc., is prevention of spread of foot infections.

Bathroom floors and also bath mats are always suspect. A cork bath mat should be swabbed with one per cent formalin solution, which will vaporize and penetrate the cork between bathers; otherwise a clean towelling mat is safest.

Expert treatment is required and the afflicted person should visit his own doctor.

Athletes' foot (ringworm of the foot—tinea pedis). This is also called dermatophytosis, since dermatophytes are a group of fungi that invade the superficial skin; in the case of the feet they invade between the toes. Dermatophytes are classified as external vegetable parasites. The skin becomes soggy and white and peels off, leaving a raw area. Frequent, meticulous washing and drying of the area and the socks and stockings, with application of an antifungal (antimycotic) powder will quickly clear the condition. Should it fail to do so there is an antibiotic, griseofulvin, which is active against fungal organisms. The dose is one to two grams orally, daily, until there is local improvement, when the dose is reduced but continued for four weeks.

The fungi in the shoes must be killed at the same time, and placing them in a biscuit tin with a small vessel containing 10 per cent formalin solution for 24 hours will accomplish this.

As the condition is contagious, more than one member of a family is usually affected, and it can be troublesome where many people live together, such as in boarding schools, hotels, hospitals, etc. Education regarding bathroom floors and bath mats is necessary to avoid the condition.

Scabies. A condition caused by the *Acarus (Sarcoptes) scabiei* (itch mite) which is parasitic to the human skin. The condition is contagious, and can become a social problem in overcrowding, as in refugee camps and air-raid shelters. Acarus scabiei is classified as an external, animal parasite. The mite favours those areas where the skin is thin, such as between the fingers and toes. The female burrows underneath the skin, and at the end of the 'tunnel' (visible to the naked eye as a tiny black line) she lays her eggs, then dies. The eggs hatch in a few days, crawl out on to the skin causing intense irritation, mate with the male mites which remain on the skin, and the process starts all over again. Itching induces scratching, and trauma to the skin leaves a route for secondary infection to enter, often in the form of impetigo.

The treatment is a warm bath, after which an emulsion of benzyl benzoate or Quella is applied from the neck to cover the whole skin.

73

This is left to dry before the afflicted person dresses. He is given a bottle of emulsion which he is told to apply after necessary washing, but the object is to leave the first application for 48 hours, after which a cleansing bath is taken, clean clothes put on, and the infection should be cured. The clothes taken off should be washed immediately. Monosulfiram (Tetmosol) is the most recent topical application. It is a 25 per cent alcoholic solution which is diluted with three parts of water immediately before use. Systemic toxic effects occur if alcohol is taken during treatment.

Overweight (Obesity). This can interfere with the health of the feet inasmuch as it is accompanied by a general flabbiness, so that the arches are not sufficiently strong to deal with the extra weight, and flat feet can result.

Flat feet. Fallen arches which give a characteristic lack of spring in the gait can be congenital and painless. Any pain experienced in feet exhibiting fallen arches must be reported to a doctor without delay, as much can be done by simple measures, such as wedging of the shoe's inner border and intensive exercises to strengthen the muscles.

Chilblain (Erythema pernio). Congestion in the tissues resulting in discoloration and swelling attended with severe itching and burning sensation, in reaction to cold. It is thought that there are several general factors contributing to the local condition, and these need to be assessed and treated by a doctor. In prevention in susceptible subjects, they should wear shoes one size larger than their normal, and two pairs of stockings, preferably woollen ones next to the skin, with the outer ones of pure silk. A good diet, plenty of foot exercise, and avoidance of draughts on the feet and legs all help to prevent chilblains.

Ingrowing toe nail. In this condition the lateral nail border has pierced the skin groove, which it normally traverses. Separation of the nail from the skin groove with a wisp of cottonwool soaked in surgical spirit will allow healing to take place. The foot must be kept specially clean to avoid secondary infection with pus-producing germs. The latter condition requires medical attention. Some people benefit from cutting a V in the middle of the nail; shoe pressure on the V raises the lateral border from its skin groove and further facilitates healing.

Hammer toes. These can interfere with the health of the feet as, apart from the pain, a corn often develops on the upper clawed prominence. Throwing the weight forward on to the toes, by wearing too high a heel, predisposes to this condition.

Bunion. Displacement of the big toe towards the foot's midline. Friction and pressure over the produced prominence set up an accompanying condition of bursitis. Narrow-toed shoes contribute to this deformity. Footprinting 1,500 school children in Aberdeen[55] showed that more boys than girls have bunions.

There are many more troubles that can afflict the feet. It is important to keep your feet in good condition and to teach other people to do so.

Jimmy is now complete to his feet, but he has no protection for his thoracic and upper abdominal organs, so these we must enclose with a *breast bone* and *a rib cage.*

THORACIC CAGE
Sternum (breast bone)

This is a flat bone. The formation of such bones has already been described (p. 45).

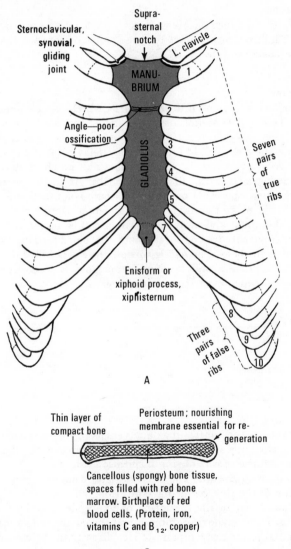

A

Thin layer of compact bone

Periosteum; nourishing membrane essential for regeneration

Cancellous (spongy) bone tissue, spaces filled with red bone marrow. Birthplace of red blood cells. (Protein, iron, vitamins C and B_{12}, copper)

B

Fig. 43   Diagram showing A, Anterior view of sternum showing the attachment of the ribs. B, Cross section of sternum.

The sternum forms part of the anterior wall of the thoracic cavity. *Anteriorly* lies subcutaneous tissue and skin.

Posteriorly lies the heart and the large blood vessels entering and leaving this organ. External pressure on the sternum of a collapsed patient lying flat on a hard surface will depress the sternum 4 to 5 cm (1½ to 2 inches) which squeezes blood out of the heart into the general circulation.

Superiorly lie the clavicles (collar bones) and anterior neck.

Inferiorly lies the abdominal wall, and there is attachment to the diaphragm.

Laterally the sternum gives direct attachment to seven pairs of ribs and indirect attachment to three pairs. In between the ribs lie the intercostal muscles.

Man, his Health and Environment

STRUCTURE

The sternum is divided into *three parts,* which can be likened to a sword. The *upper portion* is the handle, the *manubrium sterni.* Laterally on its superior surface are two hyaline-covered depressions to receive the clavicles, to form the sternoclavicular joints. Between these two depressions is the suprasternal notch. Laterally the manubrium receives the first pair of ribs. At its union with the *middle portion*, the blade or *gladiolus,* there is a prominence anteriorly referred to as the angle of the sternum. Ossification remains poor in this area, and here the sternum can be punctured and red bone marrow withdrawn, to estimate the blood-forming processes going on in the body. The *lowest portion,* the tip, is called the *xiphoid* or *ensiform process,* and it gives attachment to a portion of the diaphragm. It also never completely ossifies.

BLOOD SUPPLY

The sternal artery.

VENOUS DRAINAGE

The sternal vein.

FUNCTIONS

1. It protects the heart, lungs and vessels in the thoracic cavity.

2. It moves upwards and forwards to increase the horizontal dimension of the thorax during inspiration (Fig. 32).

3. It moves downwards and backwards to decrease the horizontal dimension of the thorax during expiration.

4. It takes part in the formation of joints, i.e. the right and left sternoclavicular (synovial gliding), and the costosternals (cartilaginous joints) joining seven pairs of ribs directly to the sternum.

5. It gives attachment to muscles (pectoralis major and minor, sternocleidomastoid) and ligaments.

6. It provides red bone marrow, easily available for examination.

A Rib

This is also a flat bone. It is flatter anteriorly, where it becomes cartilaginous for its attachment to the sternum. Posteriorly it has a facet, covered with hyaline cartilage, for articulation with the body of a dorsal or thoracic vertebra. This end is sometimes called the head, and below it lies a slightly constricted portion called the neck. Below this lies yet another projection covered with hyaline cartilage, for articulation with a transverse process of a dorsal or thoracic vertebra.

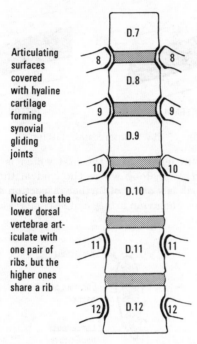

Fig. 44   Articulation of ribs with dorsal vertebrae.

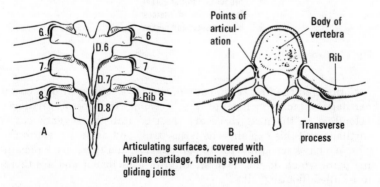

Fig. 45   A, Articulation of ribs with transverse processes of the dorsal vertebrae (posterior view). B, Articulation of ribs with transverse processes and body of a dorsal vertebra (superior view).

77

The narrow inferior surface is oblique and grooved for protection of the intercostal (inter = between, costa = rib) blood vessels and nerves. Along this inferior border the rib gives origin to the external intercostal muscle and insertion to the internal intercostal muscle. On the rib's superior surface the opposite happens—it gives insertion to the external and origin to the internal intercostal muscles.

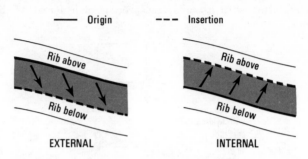

Fig. 46   Relationship of ribs to intercostal muscles.

The ribs vary in length from the first which is roughly 7·5 cm (3 in), to 30 cm (12 in), long where the conical thoracic cavity is widest. The first rib is worthy of further inspection as it is the bone against which the subclavian artery can be compressed to stop bleeding in first-aid work.

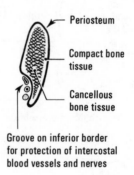

Fig. 47   Cross section of a rib.

Notice that the first seven pairs of ribs have each their own costal cartilage attaching them to the sternum; the costal cartilages belonging to the next three are inserted into the seventh costal cartilage and this explains the terms 'true' and 'false' in relation to ribs. We have accounted for 10 pairs of ribs, and there are 2 remaining pairs which do not have an anterior attachment and are therefore called 'floating' ribs.

BLOOD SUPPLY
Intercostal arteries.

Intercostal veins.

FUNCTIONS

1. They protect the organs and vessels in the thorax and upper abdomen (Fig. 48).

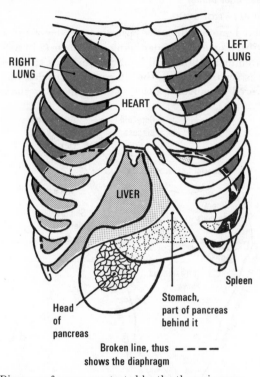

Fig. 48   Diagram of organs protected by the thoracic cage.

2. They take part in the following joint formation: synovial (gliding) with the bodies of the dorsal vertebrae; synovial (gliding) with the transverse processes of the dorsal vertebrae; non-synovial (cartilaginous) with the sternum.

3. They give origin and insertion to the intercostal muscles.

4. They move upwards and forwards to increase the horizontal dimension of the thorax during inspiration (Fig. 32).

5. They move downwards and backwards to decrease the horizontal dimension of the thorax during expiration.

It is important to remember that if you are describing the boundaries of the thoracic cavity, in addition to the ribs and sternum, the thoracic vertebrae, the intercostal muscles and the diaphragm need description.

So much for Jimmy complete with his rib cage, but you will agree that he is somewhat useless without a shoulder girdle and upper extremity, so let us hasten to describe these.

79

SHOULDER GIRDLE

This girdle, attached by strong muscles to the upper rib cage, is composed of *two shoulder blades* (scapula, sing.; scapulae, pl.) and *two collar bones* (clavicles).

The Scapula (shoulder blade)

POSITION AND RELATIONSHIPS

This triangular bone overlies the posterior surface of the second to the seventh ribs, its medial border being 5 to 7·5 cm (2 in to 3 in) from the vertebral column. Its lateral boundary extends to the shoulder. Anteriorly it is separated from the rib cage by muscles, and posteriorly it is covered by muscles, fascia, subcutaneous tissue and skin.

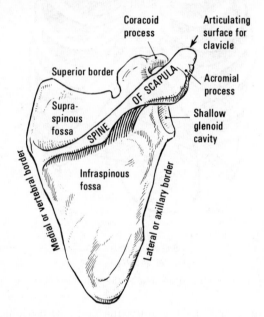

Fig. 49   The right scapula. Posterior view.

STRUCTURE

Again you are presented with the opportunity for revision of the structure of a flat bone (p. 45).

The anterior surface of the scapula is smooth, and shaped to fit the posterior contour of the rib cage. It gives attachment to strong muscles which bring about movement in the arm, neck and back.

The posterior surface is divided into a small upper area and a large lower area by the *spine of scapula.* It traverses from the medial border, getting bigger until it ends laterally in a large prominence, the *acromial process,* which lies above the shoulder joint and prevents its upward dislocation. Medially this acromial process has an articular facet for the lateral end of clavicle, with which it forms the *acromioclavicular,* a synovial, gliding *joint.*

80

Anterior to the acromial process is another projection lying in front of the shoulder joint, which is called the *coracoid process*. It prevents forward dislocation of the shoulder joint and gives attachment to strong muscles of the arm and chest.

Lying inferiorly to these two processes is a shallow socket, the *glenoid cavity*, covered with hyaline cartilage to receive the head of the humerus to form the shoulder, a freely movable, synovial, ball and socket joint.

BLOOD SUPPLY

Branches from the scapular artery.

VENOUS DRAINAGE

Branches into the scapular vein.

FUNCTIONS

1. Each scapula gives attachment to muscles and ligaments.
2. The right scapula protects the right lung and the left scapula protects the left lung.
3. Each scapula takes part in the formation of an acromio-clavicular, synovial, gliding joint; and a shoulder joint which is a freely movable, synovial, ball and socket joint.

The Clavicle (collar bone)

This is the 'brace' to the shoulder.

POSITION AND RELATIONSHIPS

Perched superiorly on the anterior rib cage, it is easily felt underneath the skin from its attachment to the sternum along to its articulation with the acromial process of scapula. The large blood vessels arising out of, and entering the thorax (notably the subclavian—sub = under; clavian = clavicle) are to be found near the medial end of the clavicle. It is surrounded by muscle tissue.

STRUCTURE

This bone has two slight curves, a downward one at its medial extremity and an upward one at its lateral extremity. Medially it has an articular facet for the sternum, with which it forms the sterno-clavicular, a synovial, gliding joint. Laterally there is another facet for the acromial process of scapula, forming the acromioclavicular, a synovial, gliding joint. The anterior surface is smooth, the posterior surface roughened for the attachment of muscles.

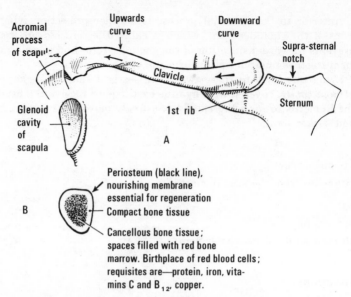

Fig. 50   The right clavicle. A, Anterior view. B, Cross section.

BLOOD SUPPLY
The clavicular artery.

VENOUS DRAINAGE
The clavicular vein.

FUNCTIONS
   1. Each clavicle gives attachment to muscles (sternocleidomastoid, pectoralis major and minor) and ligaments.
   2. Each clavicle acts as a brace to the shoulder on its side. (When a clavicle is broken the shoulder on that side slumps forward in a characteristic fashion.)
   3. Each clavicle takes part in the formation of a sternoclavicular, synovial, gliding joint, and an acromioclavicular, synovial, gliding joint.

UPPER EXTREMITY
The shoulder girdle and the upper extremity are built to give free mobility, in contrast to the pelvic girdle and the lower extremity which give stability in maintenance of the upright position.

Humerus
This is a long bone and it forms the upper arm.

Superiorly it articulates with the glenoid cavity of the scapula. Inferiorly it articulates with the radius laterally and the ulna medially. The surrounding muscles with which a nurse requires to become acquainted are the deltoid, biceps and triceps. The nearby blood vessels are the brachial artery and the deep brachial vein. Lymphatic vessels lie near the bone.

STRUCTURE

Proximal extremity or epiphysis. Not forgetting the basic structure of a long bone, and the changes that occur in it through life, and armed with the skeleton, the humerus and a well-illustrated anatomy book, try to learn and identify the following, continuing to be methodical in description:

1. Head, less than a hemisphere, covered with hyaline cartilage.
2. Immediately below the head is a thick anatomical neck.

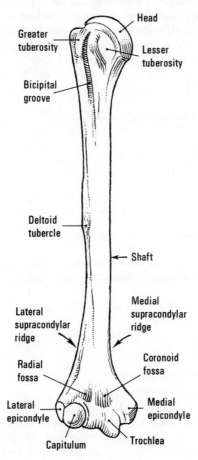

Fig. 51   The right humerus. Anterior view.

83

3. Laterally lies the greater tuberosity, and medially the lesser tuberosity.

4. Between them lies the bicipital groove for transmission of the nerve of the same name.

5. Surgical neck, a slightly constricted portion below the tuberosities.

Shaft or diaphysis.

1. Deltoid tubercle laterally at midshaft; gives insertion to the deltoid muscle.

2. Radial or musculospiral groove, for transmission of the nerve of that name.

3. Nutrient foramina for blood vessels.

Distal extremity or epiphysis.

1. Laterally lies the capitulum, an articulating surface covered with hyaline cartilage for the radius.

2. Medially lies the trochlea, an articulating surface covered with hyaline cartilage for the ulna.

3. Above the trochlea, anteriorly, is the coronoid fossa, to receive the coronoid process of the ulna during flexion of the forearm.

4. Above the trochlea, posteriorly, is the olecranon fossa, to receive the olecranon process of the ulna during extension of the forearm.

5. Medial and lateral epicondyles.

BLOOD SUPPLY

Branches from the brachial artery.

VENOUS DRAINAGE

Branches into the deep brachial vein.

FUNCTIONS

1. It gives attachment to muscles (deltoid, triceps, biceps) and ligaments.

2. Superiorly it articulates with the glenoid cavity of scapula to form the shoulder joint, a freely movable, synovial, ball and socket joint.

3. Inferiorly it articulates with the radius and ulna to form the elbow, a synovial, hinge joint.

Radius (rotating forearm bone)

POSITION AND RELATIONSHIPS

Superiorly it articulates with the capitulum of humerus. Medially it lies on the ulna; inferiorly lie the scaphoid and lunate bones of the wrist. Amongst the surrounding muscles is the biceps. The nearby blood vessels are the radial artery and the deep radial vein. Lymphatic vessels lie near the bone.

84

Proximal extremity or epiphysis.
1. Disc-shaped head covered with hyaline cartilage to articulate with capitulum of humerus superiorly (elbow joint of the synovial, hinge variety), and with the radial notch of ulna medially (proximal radio-ulnar, a synovial, pivot joint).
2. Slightly constricted neck.
3. Radial tuberosity for insertion of the biceps muscle.

Shaft or diaphysis. Nutrient foramina for blood vessels.
Distal extremity or epiphysis.
1. Broadens out and inferior surface articulates with scaphoid and lunate bones of wrist to form a synovial, double hinge joint.

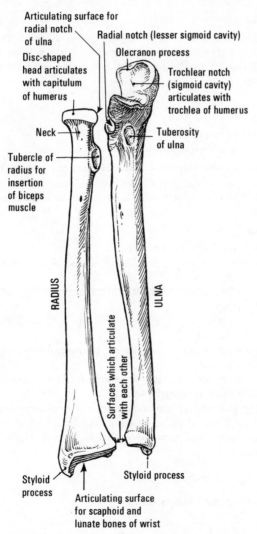

Fig. 52 Right radius and ulna. Anterior view.

85

2. Articulating surface medially for ulna, forming the distal radio-ulnar, a synovial, pivot joint.
3. The radial styloid process inferiorly and laterally.

BLOOD SUPPLY

Branches from the radial artery.

VENOUS DRAINAGE

Branches to the deep radial vein.

FUNCTIONS

1. It gives attachment to biceps and other muscles, and ligaments.
2. It takes part in the elbow, a synovial, hinge joint.
3. It takes part in the proximal and distal radio-ulnar, synovial, pivot joints.

Ulna (stationary forearm bone)

POSITION AND RELATIONSHIPS

Superiorly lies the humerus. Medially lies the radius. Inferiorly lie the carpal bones of the wrist. The nearby blood vessels are the ulnar artery and the deep ulnar vein. Lymphatic vessels lie near the bone.

STRUCTURE

Proximal extremity or epiphysis.
1. Olecranon process which fits into olecranon fossa of humerus in extension of forearm.
2. Coronoid process which fits into coronoid fossa of humerus in flexion of forearm.
3. Sigmoid cavity lying between these two; it articulates with the trochlea of humerus.
4. Lesser sigmoid cavity (radial notch) lying laterally for articulation with the radius.
Shaft or diaphysis. Nutrient foramina for blood vessels.
Distal extremity or epiphysis.
1. Button-shaped end which does not enter into movement with the carpal bones from which it is separated by a pad of cartilage.
2. Articular surface covered with hyaline cartilage laterally for the radius.

BLOOD SUPPLY

Branches from the ulnar artery.

VENOUS DRAINAGE

Branches to the deep ulnar vein.

86

FUNCTIONS

1. It gives attachment to muscles and ligaments.
2. Superiorly it enters into formation of the elbow, a synovial, hinge joint.
3. Superiorly it enters into the formation of the proximal radio-ulnar, a synovial, pivot joint.
4. Inferiorly it enters into the formation of the distal radio-ulnar, a synovial, pivot joint.

The Hand

This organ, capable of fine delicate movement and yet at the same time possessed of great strength, is made up of eight carpal bones arranged in two rows of four; five metacarpals for the palm and back of the hand; 14 phalanges, 2 for the thumb and 3 for each remaining finger. The Medical Defence Union advise the labelling of the fingers as shown in Figure 53.

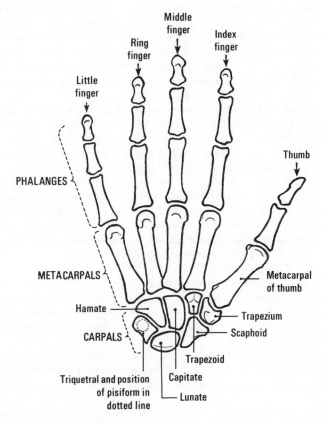

Fig. 53   The right hand. Anterior view.

87

Skin, when intact, acts as a barrier to germs. Germs (streptococci, staphylococci and *Bacillus coli*) are so frequently found on the skin (from which they can be cultured) that they can be considered its natural flora. Germs split into two by mitotic division in approximately 20 minutes. The body's defence mechanism can deal with germs in small numbers, but cannot withstand the onslaught of thousands or millions. It need hardly be stressed that frequent washing of the skin is the only way to keep its germ population within reasonable limits. Hands must be washed frequently, especially *after* visiting the lavatory and *before* preparing or eating food.[56] To withstand frequent washing the skin should be thoroughly dried to prevent chafing, and replacement of natural grease should be achieved by frequent application of a good emollient lotion or cream. The wearing of cotton gloves after the last application before sleep will ensure maximum benefit to the skin.

Phosphorus, calcium and vitamin D which is necessary for the absorption of calcium, are essential for the formation of good strong nails. Finger nails are best rounded to the contour of the finger ends, should preferably be short and always clean. There are fashion phases during which it is considered smart to have long pointed nails. Some occupations, e.g. baking, hairdressing, nursing, etc., cannot be performed safely for the client by a person with such nails, and therefore the employers are impelled to state that their employees shall have short, clean nails, shaped to the finger ends. If an adolescent finds it impossible to obey such a rule she must be careful not to choose such work.

Cuticles should be kept in good repair, and if necessary a nourishing cuticle cream should be used. Excessively thickened cuticle should not be cut, but treated with a blunt instrument and cuticle remover. If the cuticle does break there is a 'hang nail' which, until healing is complete, should be covered by a washable dressing.

Decorative nail varnishes of every hue have become increasingly popular over the years. Bacteriologically they do not present a hazard when intact, but immediately they are chipped germs can be cultured from the margins. Since it cannot be guaranteed that varnished nails will not chip it is safer for the patients if nail varnish is not worn during duty hours by hospital staff.

Hands can be protected from:

1. **Cold.** By wearing mitts and gloves in inclement weather. These should be sufficiently large to allow free movement and good circulation. Fur and fur-lined mitts and gloves will be the warmest, but many people prefer wool as they can be purchased in gay colours to brighten a winter's day.

2. **Dirt.** (*a*) By using a 'dry' barrier cream before performing household chores, and washing it off immediately afterwards. (*b*) By wearing cotton household gloves and remembering to wash these after use.

3. **Water.** (*a*) By using a 'wet' barrier cream. (*b*) By using rubber gloves. The lined gloves are more comfortable as the lining absorbs

the perspiration. They should be 'aired' between each wearing to prevent smell.

4. **Chemicals.** Again we have the choice of rubber gloves and barrier creams. Of the latter, Cenitect gives protection against wet and dry compounds; is non-greasy; does not interfere with the handling of tools, equipment, etc. Men in industry need a great deal of encouragement to use the protective measures provided for them.

5. **Heat.** This mainly applies to industry and protection is accomplished by wearing asbestos gloves; domestically there is the oven-cloth and the kettle-holder.

CONDITIONS WHICH CAN INTERFERE WITH THE HEALTH OF THE HANDS

Thumb sucking. Though this will not harm the thumb itself, it can interfere with the general health should the thumb be covered with millions of germs. When it persists into childhood it is thought to be a form of regressive behaviour. The way to tackle eradication of the habit is to give the child something to do with his hands, without mentioning the thumb. The more he is scolded the worse the condition will become.

Nail biting. Since this leaves a route for secondary infection, it is important to break the bad habit. The most successful way is to employ the hands in constructive work and give praise for unbitten nails.

Blisters and callosities. These can occur on the hands as on the feet (p. 72).

Warts. Those found on children's hands are called verruca plana juvenilis. Verruca vulgaris is the common brown wart with a rough, pitted surface. The treatment is to use caustics, and accordingly it is much safer to have them treated by a doctor.

Ringworm of the nail (tinea). This is caused by an external, vegetable parasite, a fungus, which can be of human or animal origin. Dogs and cats that are to be household pets should be carefully examined. Local treatment with antifungal (antimycotic) powder or ointment has been variable in result, but now oral treatment is preferred and continued for at least four weeks. Griseofulvin, 1 to 2 grams daily, is given, gradually reducing as clinical response occurs. It is an antibiotic active against fungal organisms.

In the early stages great care must be taken to see that toilet articles, towels, gloves, etc., are kept separately to avoid spread of the condition.

Scabies and chilblains. These can occur on the hands as on the feet (pp. 73 and 74).

Cuts, pinpricks and splinters. The last named should, if possible, be removed by using tweezers. The area should then be thoroughly washed and dabbed with a reliable skin disinfectant, such as Dettol, diluted to 2 per cent, or iodine.

Where sharp instruments are used, e.g. in butchery, the personnel should be taught to cut in a direction away from themselves. Each unit must have a well-equipped first-aid box, and at least one person needs to be conversant with practical first aid.

Sharp machinery such as circular saws should be guarded, and again the men need much encouragement to put the guard in place before starting.

Whitlow (paronychia). This most commonly results from a hang-nail and invading germs that produce pus in the tissue. Frequent soaking in hot water encourages extra blood to the area to combat the infection. Prevention is better than cure.

Eczema.[57] This is a reaction to an irritant on an already susceptible skin.

Dermatitis. Dermatitis is an inflammation of the skin. The terms 'eczema' and 'dermatitis' are sometimes used synonymously, but the only advice to give anyone with any skin reaction is that they must see a doctor. While not accepting that enzyme detergents are more likely than other detergents to cause dermatitis, manufacturers agreed in 1971 to print a health warning on packets.

SUMMARY OF THE SKELETON

The skeleton is composed of 206 bones. Most of these are long or flat bones; a few are classified as irregular bones.

The skull is composed of the cranium and the face, made up of 22 bones.

The cranium is composed of eight bones—two parietal, two temporal, one frontal, one occipital, one sphenoid and one ethmoid bone.

The face is composed of 14 bones—2 nasals, 2 lacrimals, 1 vomer, two inferior turbinates, two malars, two superior maxillae, two palatals and one mandible.

The vertebral column is composed of 33 vertebrae in a child, 26 in an adult. These are divided into 7 cervical, 12 thoracic or dorsal and 5 lumbar vertebrae. The five sacral vertebrae of the child fuse into the sacrum and the four coccygeal vertebrae into the coccyx in an adult.

The pelvic girdle is composed of the sacrum, the coccyx and two innominate bones.

The innominate bone is formed from the ilium, the ischium and the pubis.

The lower extremity is built for stability and is composed of the femur, the patella, the tibia and fibula, seven tarsal bones, five metatarsals and 14 phalanges.

The breast bone and rib cage give protection to the thoracic and upper abdominal organs.

The ribs, of which there are 12 pairs, are divided into true, false and floating ribs.

The shoulder girdle is composed of two scapulae and two clavicles, and is built to give mobility to the upper limbs.

The upper extremity is composed of the humerus, the radius and ulna, 8 carpal bones, 5 metacarpals and 14 phalanges.

When looking at 'Jimmy' in the classroom you might be tempted to think of bone as 'dead'. In life it is a complex substance in which there is constant metabolic activity, i.e. complex substances are constantly being broken into simple substances with the *release of energy* (p. 7), and simple substances are constantly being built into complex substances *using up energy*. After absorption of calcium from the bowel in the presence of vitamin D, calcium is deposited in the proteinous matrix (foundation substance) of bone—again in the presence of vitamin D. It is released to the blood as needed, and is released in excessive amounts should the blood become excessively alkaline. Other elements are absorbed from food and laid down in bone. These activities are controlled by various *enzymes*, particularly alkaline phosphatase. Enzymes are substances, usually of a proteinous nature, that facilitate change in other substances without themselves being changed in the process. Various *hormones*— growth hormone from the anterior pituitary gland, thyroxine secreted by the thyroid gland and calcitonin, probably secreted by both the thyroid and parathyroid glands—all play their part in the growth and structure of bone at any one moment in time, throughout a person's life. Hormones are 'chemical messengers' secreted in a gland, and absorbed into the blood passing through that gland to go to those tissues in the body where they have a specific effect—in this instance bone. All structures—bones, organs and systems, are *interdependent*. Learning about the body and its functions is rather like doing a jigsaw puzzle. One can only put one piece in place at a time, and appreciation of the full significance of each part is only obvious when the last piece is put in its place.

Jimmy is now complete as far as his skeleton is concerned, but he is so inert and inactive that he needs to be clothed in muscle tissue, the primary function of which is to bring about movement.

Muscular System

MUSCLE TISSUE

This is a suitable place for the revision of a cell, for the name 'tissue' is given to a collection of cells. From Figure 13 and page 15 epithelial and connective tissues can be revised, before proceeding to an elementary consideration of muscle tissue, which brings about movement because of its contractility.

Skeletal, Striped, Voluntary Muscle Tissue

The muscle tissue which lies around the human body (and indeed clothes the skeleton) is called skeletal, striped, voluntary muscle tissue. Each muscle is made up of millions of thread-like fibres up to 5 cm in length and about the thickness of a hair. Under the microscope their cytoplasm is 'striped' transversely, contains nuclei, and is bounded by a special connective tissue called sarcolemma. Several fibres are bound by more connective tissue into a small bundle, and several small bundles are similarly bound into a large bundle, until

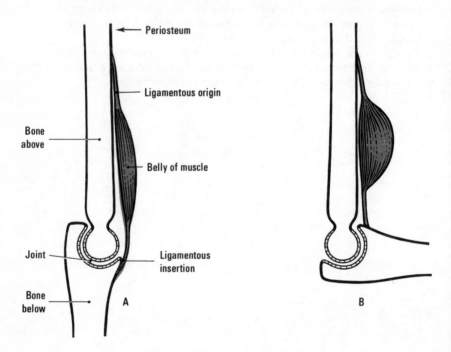

Fig. 54   Skeletal, striped, voluntary muscle. A, Relaxed. B, Contracted.

the 'muscle' is formed, and enclosed within its own muscle fascia. Each fibre has its own nerve ending, via which it receives a stimulus, in reaction to which it contracts, so that the total muscle in contraction is much shorter and thereby the two bony attachments of its extremities are brought closer together. At each bony attachment the muscle becomes a ligament, blending with the periosteum to give strength and stability. The attachment which does not move is called the origin; the one which does move is called the insertion. The 'fleshy' portion (belly) of most skeletal muscles passes over a joint, which is reasonable when we remember that it is by our muscles that we 'move.' A skeletal muscle can be represented diagrammatically as in Figure 54.

One-third of the fibres in a muscle are always 'on duty,' i.e. contracted, to give muscle its 'tone' or 'at the ready'. They work on 'shifts' so that no one set of fibres gets exhausted, for we have to maintain 'muscle tone' even during sleep. However, when we 'will' (in our brains) an action to take place at least two-thirds of the fibres receive the stimulus and respond to it by contracting, thus bringing the ligamentous insertion nearer to the origin.

You will notice from Figure 54, A and B that the 'relaxed' muscle is long and thin and the 'working' (contracted) muscle is short and fat. The 'action' of the muscle is therefore to produce movement by drawing the insertion nearer to the origin; provided you know the origin and insertion of a muscle you can work out its action.

A muscle can only act in response to a stimulus brought to it by a nerve fibre, and accordingly some knowledge of the nervous system is required. Familiarity with the nervous system will only be acquired when all the other systems have been mastered.

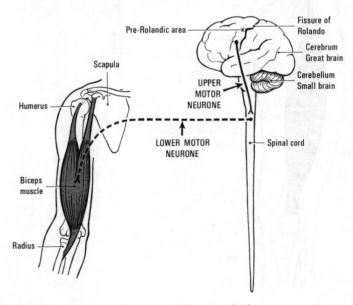

Fig. 55   Motor nerve supply to muscle in upper limb.

There is a portion of the brain (the pre-Rolandic area, Fig. 55) that deals with 'willing' muscular action. The stimulus produced by 'willing' passes along a nerve fibre (an upper motor neurone) out of the brain and into the spinal cord in the neural canal of the vertebral column. It leaves the spinal cord in another nerve fibre (lower motor neurone), a filamentous ending of which is found in each muscle fibre making up the total muscle.

When studying diseases of the nervous system it is necessary to remember this nerve supply to skeletal muscle, in order to understand the difference between spastic and flaccid paralysis.

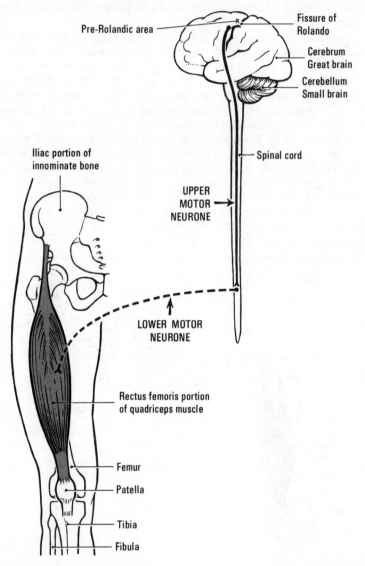

Fig. 56   Motor nerve supply to muscle in lower limb.

To perform all the work briefly mentioned a muscle requires an excellent blood supply; in other words it is highly vascular tissue. For this reason drugs can be introduced into it by injection (intramuscular injection), knowing that they will be rapidly absorbed into the blood stream to produce their specific effect. The blood brings glucose, fat and oxygen to the muscle, which already has insoluble glucose in the form of glycogen stored within its fibres. When the nerve stimulus arrives at the muscle, a complicated enzymic and chemical reaction occurs, resulting in the production of heat, energy and waste products—carbon dioxide, water and mildly acid bodies.

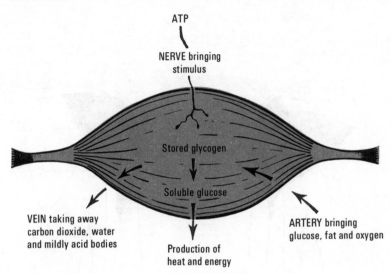

Fig. 57   Elementary physiology of muscle tissue.

A muscle rarely acts independently, but as many as a dozen muscles may be required to produce the movement involved in picking up a pencil. When the muscles contract synchronously to produce a smooth graceful movement, we speak of the presence of 'muscle co-ordination.' Absence of this function results in awkward, jerky movements which you will learn to recognize when working in the medical or neurological ward.

There are hundreds of muscles in our bodies, and it is better to learn a few properly and look up the others as your experience develops and you require more knowledge of muscles, e.g. when working in an orthopaedic ward. I have selected only those that have a distinct bearing on general work, and I hope that you will develop a methodical description of any other muscles that you may need to learn.

Deltoid. This forms an epaulette over the shoulder.

Origin. Clavicle and spine of scapula.

Joint. Shoulder.

Insertion. Deltoid tubercle on lateral aspect, midshaft of humerus.

(The vascular portion of this muscle into which an injection can be given is in the upper third of the upper arm.)

Main Action. To raise arm at a right angle to the body.

Nerve Supply. White upper motor neurones from pre-Rolandic area in brain, to appropriate level in spinal cord. White lower motor neurones from appropriate level of spinal cord to muscle. White sensory fibres from muscle to brain, for interpretation of sensation (kinaesthetic sense). All these fibres belong to the central nervous system.

Blood Supply. Deltoid branch of acromiothoracic artery.

Venous Drainage. Cephalic and deep brachial veins.

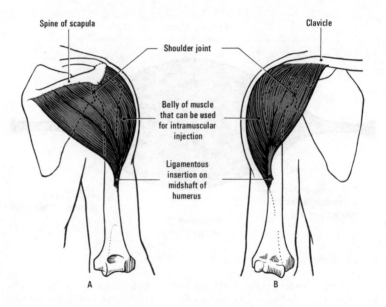

Fig. 58   The right deltoid muscle. A, Posterior view. B, Anterior view.

Sternocleidomastoid. This forms a strong strap on either side of the neck. A knowledge of this muscle is needed to treat the condition of wry neck (torticollis).

Origin. Sternum and clavicle.

Insertion. Mastoid process of the temporal bone.

Main Action. When contracted simultaneously they draw down the face. When contracted separately, each turns the head to the opposite side.

Nerve Supply. White upper motor neurones from pre-Rolandic area in brain to appropriate level in the spinal cord. White lower motor neurones from appropriate level in the spinal cord to muscle. White sensory fibres from muscle to brain, for interpretation of sensation (kinaesthetic sense). All these fibres belong to the central nervous system.

Blood Supply. Sternomastoid branches from occipital and superior thyroid arteries.

Venous Drainage. Branches into the jugular veins.

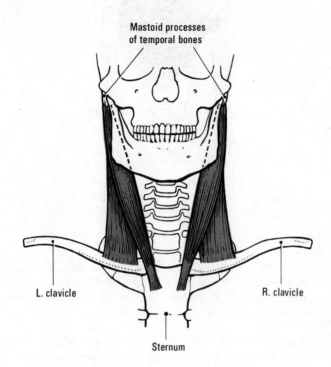

Mastoid processes of temporal bones

L. clavicle

R. clavicle

Sternum

Fig. 59   The right and left sternocleidomastoid muscles.

Pectoralis major. On this muscle lies the mammary gland (breast). Patients who have had a breast removed (mastectomy) also lose a considerable portion of this muscle.

Origin. Sternum, ribs and clavicle.

Joint. Shoulder.

Insertion. Outer edge of bicipital groove on humerus.

Main Action. Drawing arm downwards and across front of chest.

Nerve Supply. White upper motor neurones from pre-Rolandic area in brain, to appropriate level in spinal cord. White lower motor neurones from appropriate level in the spinal cord to muscle. White sensory fibres from muscle to brain, for interpretation of sensation (kinaesthetic sense). All these fibres belong to the central nervous system.

Blood Supply. Pectoral branches from acromiothoracic artery.

Venous Drainage. Branches to acromiothoracic vein.

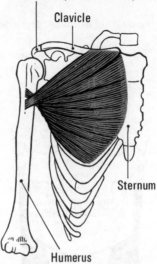

Acromion process of scapula

Clavicle

Sternum

Humerus

Fig. 60   The right pectoralis major muscle.

ABDOMINAL WALL

Many times in the operating theatre you will witness the incision of this muscular sheet, which is made up of four muscles.

Rectus abdominis. A long strap-like muscle originating from the fifth, sixth and seventh costal cartilages. It is inserted into the pubic bone.

External oblique. This is the outermost muscle of the abdominal wall. It originates on the outer surface of the lower ribs, the fibres passing downwards and forwards to become a strong sheet of white fibrous tissue known as an aponeurosis (aponeuroses, pl.; aponeurotic, adj.), which fuses with its fellow of the opposite side down the midline from sternum to pubis. This line of fusion is the linea alba or white line. During pregnancy it becomes pigmented and is called the linea nigra. The lower aponeurotic border has a free, rolled edge from pubis to iliac crest, known as the inguinal ligament (of Poupart), at the medial border of which the outlet is known as the external inguinal ring. This 'tunnel' gives passage to the spermatic cord in the male and the much smaller round ligament of the uterus in the female, and as these structures emerge from the tunnel through the external inguinal ring there is an anatomical weakness which is greater in the male. Consequently, inguinal hernia, about which you will be learning in surgery, occurs more often in men.

Just below the inguinal ligament, lateral to the external inguinal ring and medial to the femoral vein, is the femoral ring, another anatomically weakened area, because of the passage of vessels to and from the trunk. This area is greater in the female, and therefore femoral hernia occurs more often in women.

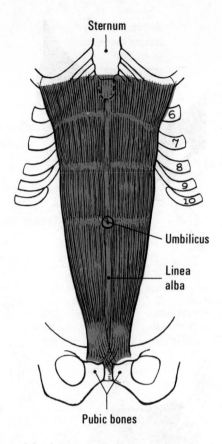

Sternum

6
7
8
9
10

Umbilicus

Linea
alba

Pubic bones

Fig. 61   The rectus abdominis.

Internal oblique. This muscle lies underneath the external oblique, the fibres of which it crosses at right angles. It originates below from the inguinal ligament and iliac crest, and is inserted above on to the costal cartilages and along the midline aponeurosis, which splits into two to enclose the rectus muscle, except in the lower third where the splitting does not take place, so that the rectus is deprived of its strong posterior sheath, and this again constitutes an anatomical weakness.

Transversus abdominis. As its name implies, the transversus abdominis fibres run transversely, this being the innermost muscle of the abdominal wall. It also enters into the commencement of the inguinal canal at the internal inguinal ring. It originates along the iliac crest, lumbar vertebrae and lower costal margin, the fibres passing round to the front to be inserted into the aponeurosis with those of the internal oblique, and it behaves in the same fashion in the lower third, thus adding its quota to the existing anatomical weakness and contributing to the possibility of an umbilical hernia.

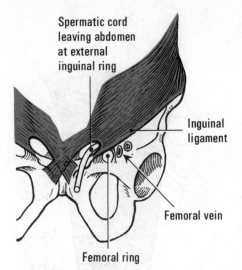

Spermatic cord
leaving abdomen
at external
inguinal ring

Inguinal
ligament

Femoral vein

Femoral ring

Fig. 62   The left inguinal ligament (male) and femoral ring.

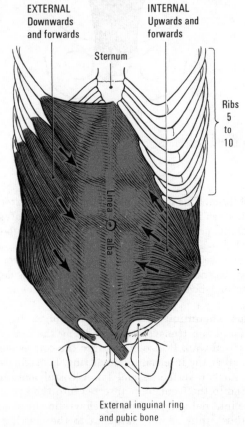

EXTERNAL
Downwards
and forwards

INTERNAL
Upwards and
forwards

Sternum

Ribs
5
to
10

Linea alba

External inguinal ring
and pubic bone

Fig. 63   The anterior abdominal wall.

100

Concerted action of the abdominal wall.

1. It acts as a 'corset' to hold in the abdominal contents and improve the blood supply to same (Fig. 33).

2. It can compress the abdominal contents and so assist in the expulsive movements of (a) defaecation, (b) micturition, (c) vomiting and (d) parturition.

3. It can contract vigorously to protect against blows.

4. By stabilizing the pelvis (especially the rectus muscle), it contributes to good posture.

Blood supply to the abdominal wall. The rectus muscle is supplied by the superior and inferior epigastric arteries, and the other three muscles by the deep circumflex iliac and lumbar arteries.

Venous drainage. By corresponding veins.

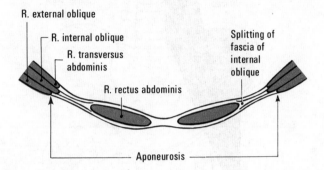

Fig. 64   Section of abdominal wall.

Nerve supply. White upper motor neurones from the pre-Rolandic area in brain, to appropriate level in spinal cord. White lower motor neurones from appropriate level of spinal cord to muscle. White sensory fibres from muscle to brain for interpretation of sensation (kinaesthetic sense). All these fibres belong to the central nervous system.

MUSCLES OF THE HIP

Iliopsoas muscle. This is composed of the posas major and minor and the iliacus.

Origin. Lumbar vertebrae and iliac crest.

Joint. It crosses the front of the hip joint.

Insertion. Lesser trochanter of femur.

Action. It raises the thigh.

Blood Supply. Lumbar and deep circumflex iliac arteries.

Venous Drainage. By corresponding veins.

Nerve Supply. White upper motor neurones from pre-Rolandic area in brain, to appropriate level in spinal cord. White lower motor neurones from appropriate level in cord to muscle. White sensory fibres from muscle to brain for interpretation of sensation (kinaesthetic sense). All these fibres belong to the central nervous system.

Later in your course you may wonder why a patient with tuberculosis of the lower spine should present an abscess in the groin. The answer lies in remembering the iliopsoas muscle and realizing that the pus tracks down its fibres.

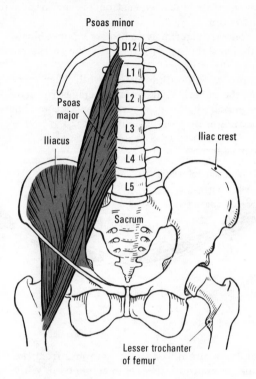

Fig. 65   The right iliopsoas muscle.

Gluteal muscles. These cross the back of the hip in a fan shape, forming the buttock (breech, rump).

Origin. Posterior surface of ilium and sacrum.

Joint. They cross the hip joint posteriorly.

Insertion. Greater trochanter of femur.

Action. Extends the hip, as when standing. Gluteals are relaxed and at their longest when you are sitting down.

Blood Supply. Superior gluteal artery.

Venous Drainage. By corresponding vein.

Nerve Supply. White upper motor neurones from pre-Rolandic area in brain to the appropriate level in the spinal cord. White lower motor neurones from appropriate level in cord to muscle. White sensory fibres from muscle to brain for interpretation of sensation (kinaesthetic sense). All these fibres belong to the central nervous system.

The main reason for learning the gluteal muscles is that they are a site for intramuscular injection.[58] Remembering that the sciatic

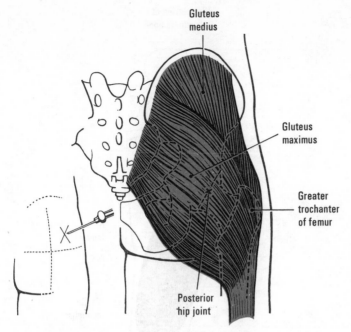

Fig. 66　The right gluteal muscles. Insert on left shows the site for
intramuscular injection.

notch on the innominate bone transmits the sciatic nerve, that the
best blood supply is in the 'belly' of the muscle, and realizing that it is
dangerous to inject drugs into aponeurotic, ligamentous, or fatty
tissue, large areas of the gluteal muscle are ruled out, and only the
upper outer quarter can be used with safety.

Quadriceps muscle. The quadriceps is made up of four parts and
covers the front and side of the thigh.

Origin. Ilium and femur.

Joint. Hip and knee.

Insertion. Via the patellar tendon into the tibia.

Action. Extension of the knee, flexion of the hip.

Blood Supply. Lateral circumflex femoral artery.

Venous Drainage. Via corresponding veins into the deep femoral
vein.

Nerve Supply. White upper motor neurones from pre-Rolandic
area in brain to the appropriate level in the spinal cord. White
lower motor neurones from appropriate level in cord to the muscle.
White sensory fibres from the muscle to the brain for interpretation
of sensation (kinaesthetic sense). All these fibres belong to the
central nervous system.

Again the quadriceps is a site for intramuscular injection. From
Figure 67 you will see that the 'belly' of the rectus femoris is entered
from the anterolateral middle third of the thigh, and the 'belly' of
the vastus lateralis or externus is entered from the lateral-medial
middle third of the thigh.

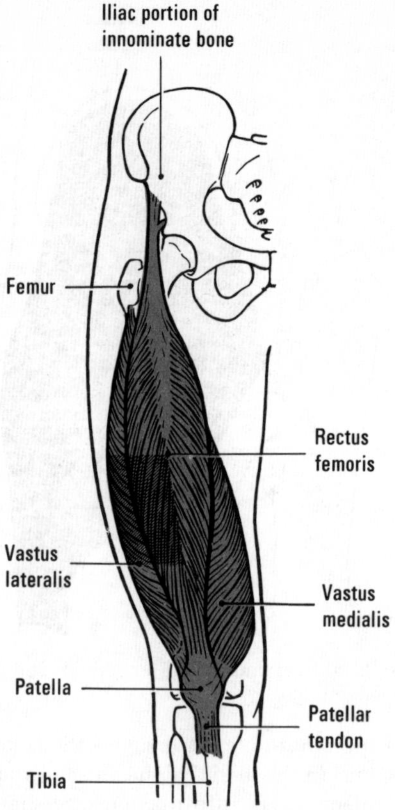

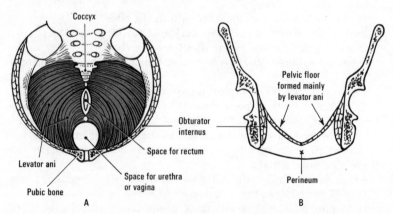

Iliac portion of
innominate bone

Femur

Rectus
femoris

Vastus
lateralis

Vastus
medialis

Patella

Patellar
tendon

Tibia

Fig. 67   The right quadriceps muscle.

PELVIC FLOOR

The pelvic floor is often called the pelvic diaphragm. It supports the
pelvic organs and maintains their normal position.

Coccyx

Pelvic floor
formed mainly
by levator ani

Obturator
internus

Space for rectum

Levator ani

Space for urethra
or vagina

Pubic bone

Perineum

A

B

Fig. 68   The pelvic floor.   A, Superior view.   B, Cross-section.
The apex of the △ is shown between the vagina and rectum as a small circle.

Levator ani. This is the main component muscle; it originates from the bony boundary of the base of the pelvis and is inserted into the fibres from the opposite side. This line of insertion is pierced in the female by the urethra, vagina and rectum, and in the male by the urethra and rectum. Other component muscles of the pelvic floor run from the pubis in front to the ischium, sacrum and coccyx behind.

Perineum. Below the levator ani muscles is a diamond-shaped area, the perineum, bounded by the rectum posteriorly and the genitalia anteriorly. Deficiency of, or trauma to, this area results in prolapse of the organs passing through the pelvic diaphragm.

The diaphragm, internal and external intercostal muscles will be discussed with the respiratory system (p. 359).

Having spent considerable time discussing the skeletal, striped, voluntary muscle that clothes the human skeleton, we must pass on to the second type of muscle tissue found inside the body.

Internal, Plain, Involuntary Muscle

This consists of spindle-shaped fibres, each possessing a nucleus, and their cytoplasm is of one density; it therefore does not exhibit bands of darker colour as does skeletal muscle. It is from this that the words plain, smooth, unstriped and non-striated have been used in the description of involuntary muscle.

The tissue, unlike that of skeletal muscle, is capable of a weak rhythmical contraction without innervation from a nerve. However, with innervation from a nerve the contractions are more forceful and efficient. The statement that it is 'involuntary' assumes some sort of automatic control, and this is provided by grey neurones belonging to the autonomic nervous system, derived from the central nervous system and divided into two parts, the sympathetic portion and the parasympathetic portion, each having opposite action in the same organ.

In those organs that are tubular, for the purpose of conveying something along the tube, the internal, plain, involuntary muscle is arranged in two layers, an inner circular one and an outer longitudinal one. When the inner circular fibres contract, the lumen of

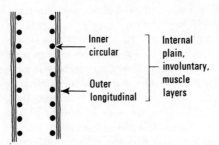

Fig. 69   Arrangement of involuntary muscle fibres in a tube.

the tube is narrowed; when they are completely relaxed the lumen of the tube is normal. Contraction of the longitudinal fibres shortens, and thus widens the tube. A bolus of food, therefore, can be pushed along (by contraction of circular fibres) into an area immediately below, which has just widened to receive it by contraction of its longitudinal fibres.

Peristalsis (peristaltic, adj.) is the name given to this special movement, and you can see why many books explain it as a wave of contraction preceded by a wave of relaxation.

We now come to the third type of muscle tissue.

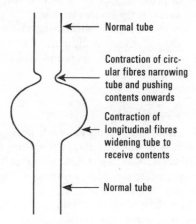

Fig. 70   Diagram of peristalsis.

Cardiac, Striped, Involuntary Muscle (Myocardium)

A moment spent comparing these names gives the clue as to why the heart muscle has to have a category of its own. It is dispersed around the heart chambers as illustrated in Figure 71, which shows that it is thickest where there is most work to do. (Again that jigsaw puzzle. You will learn more about the heart when we come to the circulatory system.)

However, you will have to accept for the moment that cardiac muscle is thickest around the left ventricle, thinner around the right ventricle, and thinnest around the two atria (atrium, sing.).

Cardiac muscle has a highly specialized property of being able to contract automatically and rhythmically at a slow rate, independent of its nerve supply, but this rate is too slow to maintain a sufficient circulation to the tissues, therefore acceleration is provided by grey neurones belonging to the sympathetic portion of the automatic nervous system, and deceleration is accomplished by the parasympathetic portion (vagus nerve). The heart's activity is finely adjusted to meet the body's changing needs.

From all that has been said about the muscle tissue making up the human body we ought to collect together the functions of muscle

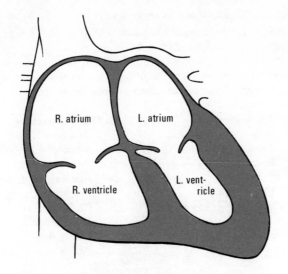

Fig. 71   Arrangement of cardiac muscle tissue.

before passing on to a more generalized consideration of exercise and the part it plays in the maintenance of health.

Functions of Muscle Tissue

1. Muscle tissue clothes the skeleton and gives 'shape' to the human body.

2. It is the end means by which voluntary movement can be effected.

3. It is the end means by which the automatically controlled functions of the body can take place, e.g. heart beating, transport of blood, urine, faeces, etc.

4. It produces heat to keep us 'warm blooded'.

5. It produces energy via which we do our work.

6. It maintains posture.

7. It massages blood back to the heart.

SUMMARY OF MUSCULAR SYSTEM

A tissue is a collection of like cells.

An organ is a collection of tissues.

A system is a collection of organs.

You should be familiar with the following words: ligament, tendon, fascia, aponeurosis, origin, insertion, muscle tone, muscle co-ordination, highly vascular.

Account in a simple manner for the physiology occurring in muscle.

107

Found around the human skeleton, and it is the means by which voluntary movement is produced.

Most muscles have their origin from one bone, pass over at least one joint to be inserted into another bone, so that on contraction of the muscle the bone of insertion is brought nearer to the bone of origin.

Man, his Health and Environment

Apply this formula:

Origin
Joint
Insertion
Action
Blood supply
Venous drainage
Nerve supply

To these muscles:

Deltoid
Sternocleidomastoid
Pectoralis major
Abdominal wall
Iliopsoas
Gluteals
Quadriceps
Pelvic floor

INTERNAL, PLAIN, INVOLUNTARY MUSCLE

This plain muscle, not controlled by the will, is found in all hollow viscera. It is supplied by the grey neurones from the autonomic nervous system, being stimulated by one group of fibres and inhibited by the other group; these two groups are the sympathetic and the parasympathetic fibres belonging to the autonomic system. In tubes where something has to be moved along, the muscle fibres are arranged in an inner circular and an outer longitudinal layer to allow peristalsis.

CARDIAC, STRIPED, INVOLUNTARY MUSCLE

Found in wall of heart forming its middle layer, which is thickest in the left ventricle, thinner in the right ventricle, and thinnest in the two atria. It can contract automatically and rhythmically at a slow rate independent of its nerve supply, but this rate is too slow to maintain a sufficient circulation to the tissues, so the rate is accelerated by grey neurones from the sympathetic portion of the autonomic nervous system and decelerated by parasympathetic neurones. Cardiac muscle is referred to as the myocardium.

EXERCISE

Having considered the physiology of active muscle in the previous section, it seems appropriate to turn our thoughts to the effect produced throughout the body by exercise. It may seem arbitrary to discuss it here, since it has an effect on so many other bodily functions, not yet discussed. This is yet another example of the interdependence of the body systems.

The forms of exercise of which a nation partakes are in part culturally and socially determined, and in part climatically determined. This may account for some of the forms of energetic dancing that are practiced in hot climates, dancing being an evening activity. For those who live near the sea, not only does swimming provide attractive exercise, but it is also cooling to the hot body, especially if the water is not wiped off, but allowed to evaporate from the skin. School programmes in all countries include 'physical education' and many of the exercises are performed to music, encouraging an individual to enjoy the feelings engendered by his moving body. Movement to music has been found beneficial for many mentally and physically handicapped people.

Starting with the active muscle using up the glucose, fat and oxygen brought to it by the blood, such active muscle requires still more blood to bring it more of these substances and take away the waste products of metabolism to the appropriate excretory organs, consequently the heart and blood vessels have to work more efficiently under the increased demand. To get more oxygen into, and the extra carbon dioxide out of the blood, the lungs and the muscles of respiration need to work more efficiently, and do we not take deeper breaths, and indeed sometimes become 'breathless' in the attempt? We seldom use our lungs to capacity when at rest, and aeration of the whole lung, especially the bases, during increased activity is an excellent prevention of stagnation in those bases.

The nervous system becomes more active, as alert and accurate sight and sound has to be interpreted, decisions made and voluntary muscles put into action to achieve the desired decisions.

Exercise is thought to play a part in the prevention of coronary thrombosis.

The 'massage' action of the active leg muscles against the nearby veins aids the return of venous blood to the heart, thus preventing stasis with possible formation of varicose veins. A similar action of the abdominal muscles assists in the prevention of bowel stasis, thus preventing flatulent distension, constipation and the subsequent formation of piles (varicosity of the rectal veins).

With the increased blood supply, the sweat glands of the skin become more active to help dissipate the extra heat. Some of the extra waste matter is removed from the body by the skin.

Having used up some of the body's food reserve, the balancing mechanism is such that we feel in need of more food—the appetite is improved.

Being healthily tired from exercise, the body is more ready for a good sound sleep.

With all this increased activity throughout the body in response to exercise, there is a feeling of exhilaration and well-being, which has to be fostered for the maintenance of health. The feelings engendered by jealousy, envy, hatred, etc., have to be coped with to maintain 'well-being'. The energy released by such emotions can be used up in physical exercise to good effect, e.g. digging the garden instead of kicking the furniture.

Exercise must therefore be part of each individual's daily programme. Indeed it was Jimmy's first response, in the form of a cry, when he was ejected into this world to live his own separate existence. He gradually partook of more exercise as he learned to kick in his pram, splash in his bath, crawl, then walk, run and ride a tricycle. Natural curiosity kept him busy and active throughout his toddler's waking hours, so that more blood with its bone-forming requisites allowed growth in height.

At school he will learn to take part in organized games which, as well as providing exercise and the development of skilled movements, will help to develop his corporate spirit, whereby he can be a good loser as well as take a proper pride in winning. Many of these competitive games are played in the open air, which is an added advantage.

By adolescence he should have attained the discretion whereby those forms of exercise that especially appeal to his individual personality and temperament are retained as hobbies. There are many opportunities for older school children and adolescents for exercise with a goal in view: working for the Duke of Edinburgh Award, and taking part in courses at Outward Bound Schools are two examples. The Physical Education Society also runs many types of holiday centres offering pony treking, canoeing, boating etc. The challenge offered by these activities can be an important component of mental and physical health (p. 195).

It is important during adult years to *plan towards old age,* so that when physical ability is naturally declining, a form of more gentle exercise that gives true pleasure and satisfaction can be carried on.

FATIGUE

The general appreciation of fatigue is that of exhaustion, and there *is* a point of physiological *exhaustion of muscle* when the tissue is slightly acid. Normal physical fatigue induced by a day's work is dispelled by a night's rest. *Accumulative fatigue* increases through the working week and is not relieved by one day's rest from work, which is traditionally for the Christian world on the Sabbath day.

There are many theories about the production of fatigue, but we are more sure of its effect in the body.

SIGNS AND SYMPTOMS

Reduced vitality

Lack of enthusiasm

Lowered resistance to infection

Nervousness, irritability and insomnia

Reduced efficiency

Difficulty in concentration

Distorted viewpoints.

Granted that there is some physiological base for fatigue, it is closely related to boredom and nervous tension. The following may prove the point:

A nurse comes off duty, believing herself to be 'dead tired' and saying to herself, 'My feet are killing me!' Her friend calls along the corridor, 'Will you make a foursome up for tennis?' and in no time she is in her shorts and out on the tennis court, to all appearances full of vitality, and playing a very good game.

There is still a responsibility for those in charge of the lives of others to see that the conditions provided do not directly contribute to early or excessive fatigue. There are six main contributory factors:

1. *Housing or Living Conditions which preclude Rest and Sleep.* We have already mentioned consideration of others in the Nurses' Home.

2. *Transportation to and from Work.* This can be very tiring, and makes compulsory residence for student nurses reasonable in some areas.

3. *Habits of Living.* These may preclude suitable recreation and facilities for maintenance of physical fitness. Most hospitals try to provide some form of social club for their staff, and it is up to the staff to support it and make it a flourishing concern.

4. *Hours of Work too long.* The Royal College of Nursing and National Council of Nurses of the United Kingdom (Rcn), and other representatives on the staff side of the Whitley Council have worked hard to reduce the hours. From January, 1972, they are reduced to 40 per week.

5. *Conditions of employment.* Bad conditions such as poor ventilation, too high a room temperature, poor lighting, badly sited rooms are still present in some of the older hospitals. However, more money is being spent on the building of new hospitals and renovation of parts of the old ones, and we hope to see many improvements in the not too distant future.

6. *Methods of Work.* Much has come to light regarding methods of doing time-honoured tasks more easily, and during your training many methods of today may well be considered 'old fashioned,' so that from the beginning you must keep an open and critical mind in this area. 'Always let your head save your feet' is still a wise motto.

Joints, or Articular System

With the skeleton successfully built up and suitably clothed with muscle tissue, and consideration given to exercise and fatigue, we are now ready to learn about the types of movement of which the human body is capable through the mechanism of its joints.

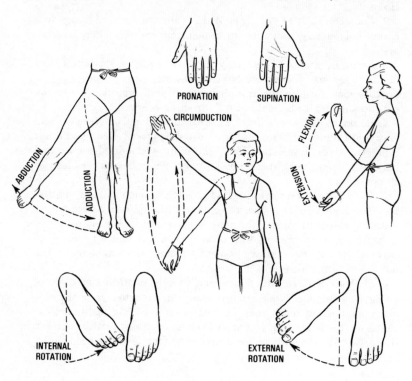

Fig. 72   Illustration of movements.

MOVEMENTS

Abduction, away from the midline.
Adduction, towards the midline.
Flexion, bending.
Extension, straightening.
Circumduction, a combination of the above four movements.
Internal rotation, turning on its axis, towards the midline.
External rotation, turning on its axis, away from the midline.
Pronation, turning the palm downwards.
Supination, turning the palm upwards.

112

A *joint* has been defined as a place where two or more bones meet. It is therefore not illogical (as it appears at first sight) to classify them as immovable, slightly movable and freely movable.

Immovable joints. This term denotes the strong union of one bone with another in such a manner as to preclude movement. We have met all the examples in our study of the skeleton, so this is an exercise in revision.

Sutures of skull. Can you name them?

Teeth in their sockets.

Innominate bone. Can you remember the names of the component parts?

Proximal and distal tibiofibial joints.

The joining of the coccygeal bones.

The joining of the sacral bones.

The union of the mandible.

The union of the superior maxillae. Failure of union results in cleft palate.

Slightly movable joints. We have already discussed these at considerable length, and only need to underline the fact that the articulating surfaces are covered with hyaline cartilage and are separated from each other by a pad of fibrocartilage. Can you name the slightly movable joints we have already discussed?

Symphysis pubis.

Right and left sacro-iliac joints.

Those between the bodies of the vertebrae. If you use the word intervertebral, then you must qualify it with slightly movable as there are the intervertebral, gliding joints also.

Lumbosacral joint.

Sacrococcygeal joint.

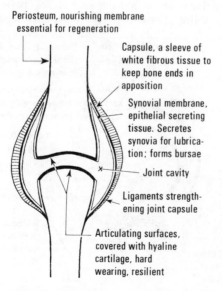

Periosteum, nourishing membrane essential for regeneration

Capsule, a sleeve of white fibrous tissue to keep bone ends in apposition

Synovial membrane, epithelial secreting tissue. Secretes synovia for lubrication; forms bursae

Joint cavity

Ligaments strengthening joint capsule

Articulating surfaces, covered with hyaline cartilage, hard wearing, resilient

Fig. 73 Characteristics of an imaginary, synovial, freely movable joint.

113

Freely movable, synovial joints. These are further classified into:
Ball and socket
Pivot
Gliding
Hinge
Double hinge.

Joints are more easily learned if you master the characteristics of an imaginary one and then apply them to those of a human skeleton.

CHARACTERISTICS OF AN IMAGINARY SYNOVIAL, FREELY MOVABLE JOINT

I am not much in favour of mnemonics, but the following one is offered for what it is worth; it is quickly written on your rough paper in an examination, and ensures mention of the essential parts— JACC SLIMM.

J. Here mention that it is synovial, freely movable, and then give the subdivision.

A. Name the articular surfaces and cover them with hyaline cartilage, stating the functions of hyaline cartilage.

C. Describe the capsule which is a 'sleeve' adherent to the bone above and the bone below. It is for the purpose of keeping the bone ends in apposition, and is made of strong white fibrous tissue.

C. Cavity—the preceding structure encloses the joint cavity.

S. Synovial membrane, an epithelial tissue covering all structures within the joint cavity, except hyaline cartilage. It secretes synovia for lubrication. In the large joints it forms bursae (bursa, sing. = water cushion) for protection.

L. Ligaments of strong white fibrous tissue, blending with the capsule to give added strength.

I. Intracapsular structures, such as a muscle passing through the joint cavity, or lips of cartilage to deepen the socket, etc.

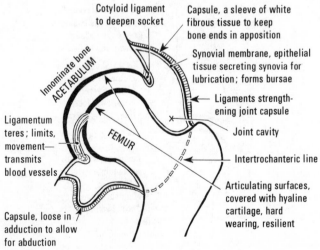

Fig. 74   Diagram of left hip joint.

M. Movements. We have already discussed these and you need to learn those specific to each joint.

M. Muscles. A joint is useless without the surrounding or extra-capsular muscles acting upon it, as witnessed in the paralysed person. Mention should therefore be made of these surrounding muscles, even if you cannot remember their individual names.

Blood supply. Tissue cannot survive without this.

Venous drainage. The waste products have to be taken away.

We are now ready to apply the knowledge from the previous section to each joint in turn.

THE HIP JOINT

This is a synovial, freely movable *joint* of the ball and socket variety.

The *articulating surfaces* are the acetabulum of the innominate bone and the hemispherical head of the femur. These are both covered with hyaline cartilage, which is strong and durable and gives resilience.

The *capsule* is made of strong white fibrous tissue, placed like a sleeve over the two bone ends and adherent to the circumference of the acetabulum above and the anterior and posterior intertrochanteric lines below.

The capsule encloses the joint *cavity,* which is lined with *synovial membrane,* an epithelial tissue covering all structures within the cavity, except the hyaline cartilage. It secretes synovia for lubrication and forms several bursae for protection.

Ligaments of strong white fibrous tissue blend with the capsule to give added strength.

The *intracapsular* structures are:

1. Ligamentum teres, a very short ligament stretching from the base of the acetabulum to the femoral head. It allows passage for blood vessels and restricts movement.

2. The cotyloid ligament, strap-like, attached to the circumference of the acetabulum to deepen same.

The *movements* occurring at the hip are flexion, extension, abduction, adduction, circumduction, internal and external rotation.

There are many extracapsular *muscles,* but keeping to those we have learned in this course, there are the gluteals, iliopsoas and quadriceps.

Blood supply. Femoral artery.

Venous drainage. Deep femoral vein.

THE KNEE JOINT

This is a synovial, freely movable *joint* of the hinge variety.

The *articulating surfaces*—femoral condyles, tibial tuberosities, posterior patella—are all covered with hyaline cartilage, which is strong and durable and gives resilience.

The *capsule* is made of strong white fibrous tissue, forming a sleeve

115

over the joint, being adherent to the femur above and to the tibia and fibula below.

The capsule encloses the joint *cavity*, which is lined with *synovial membrane*, an epithelial tissue covering all structures within the cavity, except the hyaline cartilage. It secretes synovia for lubrication, and forms bursae for protection, the most important being the prepatellar bursae, inflammation of which is called housemaid's knee (bursitis).

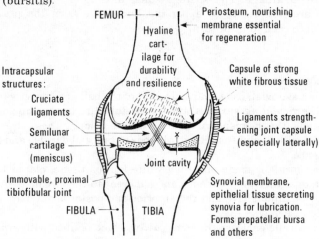

FEMUR

Hyaline cartilage for durability and resilience

Periosteum, nourishing membrane essential for regeneration

Intracapsular structures:

Cruciate ligaments

Semilunar cartilage (meniscus)

Immovable, proximal tibiofibular joint

FIBULA

TIBIA

Joint cavity

Capsule of strong white fibrous tissue

Ligaments strengthening joint capsule (especially laterally)

Synovial membrane, epithelial tissue secreting synovia for lubrication. Forms prepatellar bursa and others

Fig. 75   Diagram of right knee joint.

Ligaments of strong white fibrous tissue blend with the capsule, especially laterally, to give added strength.

The *intracapsular* structures are:

1. Two semilunar cartilages (menisci, pl; meniscus, sing.) placed like a slice of orange on the superior surface of each tibial tuberosity to deepen same for the reception of the femoral condyles. You have already read in the newspapers from time to time of famous athletes having cartilage trouble, and in the wards you will learn to care for these men after they have had a meniscectomy.

2. The patella (knee-cap), a sesamoid bone, i.e. one formed in a tendon, and this one is formed in the patellar tendon of the quadriceps muscle.

3. Two cruciate ligaments which are short and make a St Andrew's cross.

The main *movements* occurring at the knee are flexion and extension. There are many extracapsular *muscles*—the quadriceps, hamstrings and gastrocnemius are but three of them.

Blood supply. Popliteal artery.

Venous drainage. Popliteal vein.

THE ANKLE JOINT

A synovial, freely movable *joint* of the hinge variety.

The three-sided, superior *articulating surface* is formed by the malleoli laterally, and the tibia superiorly. Into this is wedged the

116

three-sided articulating surface of the talus or ankle bone. Each three-sided surface is covered with hyaline cartilage for durability and resilience.

The *capsule* is made of strong white fibrous tissue, forming a sleeve over the joint, being adherent to the tibia and fibula above and the talus (ankle bone) and calcaneum (heel bone) below.

The capsule encloses the joint *cavity*, which is lined with *synovial membrane*, an epithelial tissue covering all structures within the

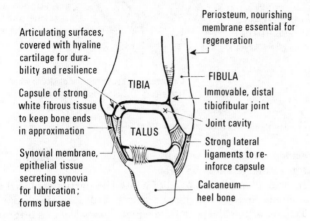

Articulating surfaces, covered with hyaline cartilage for durability and resilience

Capsule of strong white fibrous tissue to keep bone ends in approximation

Synovial membrane, epithelial tissue secreting synovia for lubrication; forms bursae

Periosteum, nourishing membrane essential for regeneration

TIBIA

FIBULA
Immovable, distal tibiofibular joint

Joint cavity

TALUS

Strong lateral ligaments to reinforce capsule

Calcaneum— heel bone

Fig. 76   Diagram of left ankle joint.

cavity except the hyaline cartilage. It secretes synovia for lubrication and forms bursae for protection.

Ligaments blend with the capsule, especially laterally, to give added strength.

The main *movements* occurring at the ankle are dorsiflexion raising the toes; plantar flexion, raising the heel or pointing the toe.

Extracapsular *muscles* present around the capsule produce movement at the joint.

Blood supply. Anterior and posterior tibial arteries.

Venous drainage. Corresponding veins.

THE SHOULDER JOINT

A synovial, freely movable *joint* of the ball and socket variety.

The *articulating surfaces* are the glenoid cavity of the scapula and the head of the humerus, both covered with hyaline cartilage to give durability and resilience.

The loose *capsule* is made of strong white fibrous tissue and is placed like a sleeve over the joint, being adherent to the circumference of the glenoid cavity above and the humerus below.

The capsule encloses the joint *cavity*, which is lined with *synovial membrane*, an epithelial tissue covering all structures within the cavity except the hyaline cartilage. It secretes synovia for lubrication and forms bursae for protection.

117

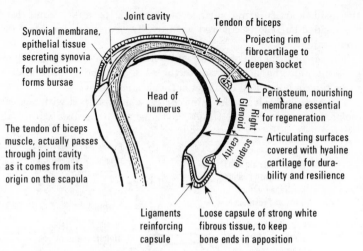

Joint cavity

Synovial membrane,
epithelial tissue
secreting synovia
for lubrication;
forms bursae

Tendon of biceps

Projecting rim of
fibrocartilage to
deepen socket

Head of
humerus

Periosteum, nourishing
membrane essential
for regeneration

Right
Glenoid
scapula
cavity

The tendon of biceps
muscle, actually passes
through joint cavity
as it comes from its
origin on the scapula

Articulating surfaces
covered with hyaline
cartilage for dura-
bility and resilience

Ligaments
reinforcing
capsule

Loose capsule of strong white
fibrous tissue, to keep
bone ends in apposition

Fig. 77   Diagram of right shoulder joint.

Ligaments blend with the capsule to give added strength.
Intracapsular structures are:
1. Labrum glenoidae, a band of white fibrocartilage placed around
the circumference of the glenoid cavity to deepen the socket.
2. Tendon of the biceps muscle which passes from its origin on the
scapula through the joint cavity.
The *movements* are flexion, extension, abduction, adduction, a
combination of these known as circumduction, internal and external rotation.
The extracapsular *muscles,* mentioning only the biceps, triceps,
deltoid and pectoralis major, produce the movement.
Blood supply. Axillary artery.
Venous drainage. Axillary vein.

THE ELBOW JOINT

This is a synovial, freely movable *joint* of the hinge variety.
The *articulating surfaces* are the lateral capitulum of humerus
with the head of radius; the medial trochlear surface of humerus
with the sigmoid cavity of the ulna. These are all covered with
hyaline cartilage for durability and resilience.
The *capsule* is made of strong white fibrous tissue forming a sleeve
over the joint, being adherent to the humerus above and the radius
and ulna below.
The capsule encloses the joint *cavity,* which is lined with *synovial
membrane,* an epithelial tissue covering all structures within the
joint cavity except hyaline cartilage. It secretes synovia for lubrication and forms bursae for protection. Tennis or student's elbow is
inflammation of one of the superficial bursae (bursitis).
Ligaments blend with the capsule, especially laterally to give
added strength.
There is an annular *intracapsular* ligament which holds the

118

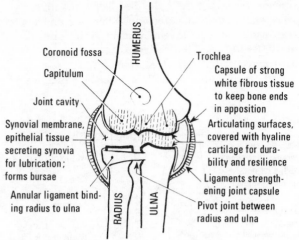

Coronoid fossa

Capitulum

Joint cavity

Synovial membrane, epithelial tissue secreting synovia for lubrication; forms bursae

Annular ligament binding radius to ulna

HUMERUS

RADIUS

ULNA

Trochlea

Capsule of strong white fibrous tissue to keep bone ends in apposition

Articulating surfaces, covered with hyaline cartilage for durability and resilience

Ligaments strengthening joint capsule

Pivot joint between radius and ulna

Fig. 78   Diagram of right elbow joint.

button-shaped head of radius against the radial notch of ulna, as you will see from Figure 78. This means that enclosed in the joint cavity of which we are speaking there is also a pivot joint between the radius and ulna.

As well as the *movements* of flexion and extension permitted by the hinge portion of this joint, there is also supination (turning the palm upwards) and pronation (turning the palm downwards) permitted by the pivot portion of the elbow joint.

The extracapsular *muscles* help in the production of these movements.

Blood supply. Brachial artery.

Venous drainage. Brachial vein.

THE WRIST JOINT

A synovial, freely movable *joint* of the double-hinge variety.

The superior *articulating surfaces* are the base of the radius laterally and a disc of cartilage below the ulna. The scaphoid and lunate carpal bones articulate with the radius, the triquetrum articulates with the inferior surface of the disc, the ulna articulates with the superior surface of the disc, and enclosed within this same cavity is the distal radio-ulnar, pivot joint. Figure 79 shows this clearly.

The *capsule,* made of strong white fibrous tissue, is adherent to the radius and ulna above and to the carpal bones below.

The capsule encloses the joint *cavity,* which is lined with *synovial membrane,* an epithelial tissue covering all structures within the joint cavity except those covered with hyaline cartilage. It secretes synovia for lubrication and forms bursae for protection.

Ligaments blend with the capsule anteriorly, posteriorly and laterally to strengthen same.

119

The *intracapsular* structure is the disc below the ulna.
The *movements* are flexion, extension, adduction and abduction.
The extracapsular *muscles* give movement to the joint.
Blood supply. Radial and ulnar arteries.
Venous drainage. Radial and ulnar veins.
These are the main joints of the human body, which we will now summarise.

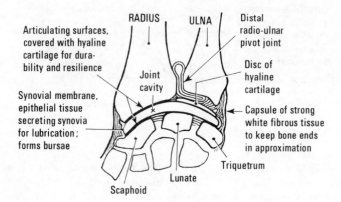

Fig. 79   Diagram of right wrist joint.

SUMMARY OF JOINTS

Define joint.
Define immovable joint. Give examples.
Define slightly movable joint. Give examples.
Define freely movable, synovial joint. Give a further classification of these.

Apply this formula:	To these joints:
Joint—type	
Articulating surfaces	
Capsule	
Cavity	Hip
Synovial membrane	Knee
Ligaments	Ankle
Intracapsular structures	Shoulder
Movements	Elbow
Muscles	Wrist
Blood supply	
Venous drainage	

The Skin

Having built up the skeleton, clothed it with muscle tissue and produced some movement in it by discussing joints, we must now envelop it with skin.

This remarkable organ provides a considerable portion of our exterior, which we present to the world at large. In your training you will become acutely aware of the observations which are constantly necessary as you survey the skin of your patient. It will

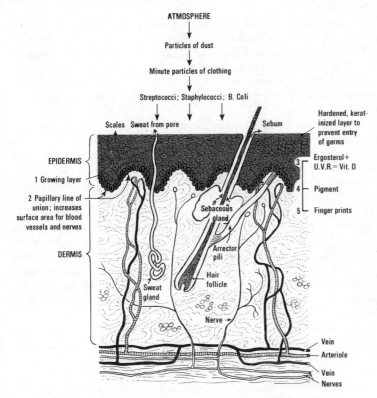

Fig. 80   Diagram of the skin.

portray his racial characteristics; it will portray his genetic inheritance—it may be pink indicating a thin epidermis (especially over the cheeks) allowing blood near the surface, or it may be naturally pale indicating a thick epidermis; it may have a greyish pallor from shock, or a waxen-like pallor from loss of blood; it may be flushed with fever, tinted yellow from jaundice, or blue, indicating a shortage of oxygen; it may be moist, or dry and wrinkled from dehydration; it may be wrinkled with *anno Domini*. It acts like a thermostat maintaining the body at a constant temperature in all seasons and climates. Many of the complex metabolic processes essential to life only function within a narrow temperature range. The skin can pale from fear or anger; it can blush from shyness, embarrassment or guilt; the hair (which is an appendage of the skin) can 'stand on end' with fright, and so one could go on, but this is sufficient for you to realize that the skin is an interesting and important organ in health and in sickness.

It is divided into two main layers, the *epidermis* or outer skin, and the *dermis* or true skin.

The Epidermis

You learned a little about this as you studied epithelium. Is it a simple or compound epithelium? What does this mean? In some parts of the body there are as many as 60 layers of cells, e.g. the palms of the hands and the soles of the feet, as they are exposed to more friction; in some areas it is thin, notably in the skin folds—the axillae, under the breasts, the groins, and between the fingers and toes.

The deepest layer of cells, i.e. those adjoining the dermis, are arranged in a wavy (papillary) line; they have definite shape and possess a nucleus. For what purpose is a nucleus essential? The cells nearer the surface have lost their shape but retained their nucleus, whereas those nearest the surface have also lost their nucleus and become hardened, so that in fact they form a layer of 'scales' and millions of them are shed from the surface each minute. This hardened, keratinized layer protects the living cells underneath from friction, too much drying and heat loss, and prevents the entry of germs. The epidermis is an active tissue, for the cells 'grow' at the base and are eventually shed as scales. It is not, however, richly supplied with blood as one might expect, blood vessels being absent, but the deeper layers are bathed in *lymph*, the fluid part of blood.

The papillary line of union between the epidermis and the dermis increases the surface area for distribution of blood vessels in the dermal tissue, from which the lymph for the epidermal tissue is derived. Similarly, there is a greater area for the distribution of nerve endings of touch, temperature, pain and pressure. This line gives to the skin of the hands its finger-prints, of which no two persons are identical—an amazing fact when we consider the millions of world inhabitants. The latest finger-printing method, called dermatoglyphics, is being used in the medical world for detection of disease. Along this line ergosterol is deposited, and when activated by the sun's ultra-violet rays vitamin D is produced. Pigment is also deposited so that here lies the true reason for being fair or dark-skinned. Pigment protects from the sun's rays, and when more is laid down in response to sunshine we say we become 'tanned.' The papillary line gives origin to the arrectores pilorum muscles.

The nails. The nails are an appendage of the skin and arise from special infolding of the epidermis. The nail continues to grow from underneath the proximal fold—the nail-bed; the two lateral folds transmit the nail to the finger end. Nails are present on all the digits of the hands and feet and serve for protection of same. The care of them has been discussed on pages 70 and 88.

The Dermis

This is a layer of loose connective tissue supporting the sebaceous glands, the hair follicles and the small muscles attached to them, the sweat glands, blood vessels, grey secretory and motor neurones, and white sensory neurones. Below the dermis is a layer of subcutaneous tissue, in the base of which fat is deposited as a store and to act as an insulator. The dermis blends with the connective tissue enclosing the muscles.

The hair follicle. This is a tubular structure, placed obliquely, with a nipple-like projection into its base for the entry of a blood-vessel. The follicle rises through the dermis and epidermis and terminates at the surface of the latter. Growing from each nipple-like projection and passing through each hair follicle to project beyond the epidermal surface is a hair. You already know that these are stronger and more numerous on some parts of the body. The colour of the hair is dictated by the amount and type of pigment in the papillary base. Absence of pigment and the presence of air bubbles makes a hair appear grey or white.

Sebaceous gland. Surrounding each hair follicle and pouring an oily secretion (sebum) into it, is a sebaceous gland. This is well supplied with blood vessels to enable the secretion of sebum, a nourishing, protective, bactericidal substance that coats individual hairs, preventing them from becoming dry, brittle and lustreless. Inserted into the lower portion of the hair follicle, and originating from the papillary line of union of dermis with epidermis, is a strap of internal, plain, involuntary muscle tissue, the arrector pili or 'raiser of hair'. Remembering the meaning of the words *origin* and *insertion* with regard to muscle tissue, and looking at Figure 80, this phenomenon becomes clear. The nerve supply to internal, plain, involuntary muscle is always grey motor neurones from the autonomic nervous system. When we have 'goose flesh' the arrector muscles are active in an attempt to produce heat for us. They 'burn' glucose, some of the released energy is used for work; most of it appears as heat. The raised hairs trap a layer of air which, being a bad conductor of heat, limits heat loss from the skin.

The sweat gland.[59] This is a fine tube appearing on the epidermal surface as a 'pore'. The tube descends through the epidermis and the dermis, and eventually is coiled up like a ball around blood vessels, from which it makes its secretion—sweat. It is supplied with grey neurones from the autonomic nervous system which respond to any slight change in blood temperature in the regulating centre in the

123

brain (hypothalamus). With an increased temperature more sweat is poured on to the skin surface. Heat is taken from the skin to change the liquid state of the sweat into gas, with resultant cooling of the skin. This heat, used to bring about a change of state, not of temperature, is latent heat.

Sometimes the air is still and humid and cannot absorb sweat as it arrives at the surface. The sweat is visible as beads of perspiration—called sensible perspiration. It is for this reason that operating theatres and dressing rooms have to be adequately ventilated to avoid sweat dropping into wounds. Insensible perspiration is a steady, slow loss of water through the skin—not from sweating, but related to diffusion and osmosis, the sweat coming up through the epidermis. This was discovered by Sanctorius Sanctorius who lived 1561–1636. After living in a chair balance in a glass enclosure he defined a steady water loss by evaporation from skin as 'insensible perspiration!' Thermal sweating occurs when the skin is in contact with warm air and occurs on the brow and upper lip, neck and chest. Emotional sweating produced mainly by anxiety and fear, occurs in the palms, soles and axillae.

Sweat. This is composed of 98 per cent water with 2 per cent solids (mainly sodium chloride) dissolved in it. We mentioned earlier that in sweating occupations salted drinks had to be provided—now you can see the reason why.

The blood vessels. These are numerous and branch into each dermal papilla. They play an important part in the heat regulation of the body. One of the functions of blood is to distribute heat around the body. When heat production is great and the body needs to lose some the skin blood vessels dilate, so that a greater amount of blood flows near the surface from whence heat is lost by conduction, convection and radiation. Simultaneously the sweat glands are making more sweat which is poured as a film on to the surface, from whence it is evaporated with further cooling of the skin. If this process is prolonged you may well wonder if the body will become short of fluid, and the answer is, not in health, because the balancing mechanism is such that less urine is secreted and the feeling of thirst is a safety valve.

When the body is not making sufficient heat to maintain its normal temperature of 36·6°C. (98·4°F.) the skin vessels contract and less blood flows near the surface, so that the minimum loss from conduction, radiation and evaporation takes place, and the kidneys deal with the water balance and excrete more urine. When the body is really cold, then all those tiny hair-raising muscles come to our aid and we resort to shivering, or do some physical work to produce heat. (See p. 388, hormone secretion in a cold body.)

Now we are at the stage where we can collect together the functions of the skin.

Functions of the Skin

1. By the secretion of sweat it assists in the regulation of body temperature.

2. It is a sensory organ via which we can become aware of our environment.

3. When intact it acts as a barrier to germs.

4. It excretes sebum, a bactericidal substance that nourishes the hair and prevents drying of the skin.

5. It helps in the maintenance of the body's fluid balance.

6. Via its pigment it gives protection from the sun's rays.

7. Via its ergosterol it produces vitamin D.

8. It has some power of absorption of medicaments applied to it.

9. It acts as a storehouse for fat.

10. It grows appendages, e.g. nails to protect the digits, eyebrows to prevent forehead sweat running down into the eyes.

I hope you will now agree that the skin is indeed a wonderful organ. However, on its surface there is always a collection of particles of dust and soot from the atmosphere, minute particles from clothing, sebum and sweat which are constantly being secreted, scales which are constantly being rubbed off, and bacteria (streptococci, staphylococci and *Bacillus coli*) which can be cultured, even from freshly washed skin. We therefore have a great responsibility with regard to the cleanliness of the skin.

Care of the Skin in the Maintenance of Health

The special areas of the hands and feet and their appendages, the nails, have already been dealt with, and the further appendage, the hair, is dealt with at the end of this section, along with the face, since it is cared for separately and needs special consideration in the maintenance of 'health'. The ensuing remarks therefore apply to the remaining skin.

The substances that accumulate on the skin's surface require to be removed before they decompose and produce an unpleasant odour; daily removal will suffice in temperate climates, but more frequent removal is advantageous in tropical climates. A bland soap[60] is required as a grease solvent, and soft water is preferable to hard. Removal leaves the skin free to function adequately, and is attended by a feeling of well-being.

When Jimmy was born he was covered with vernix caseosa, which was swabbed away shortly after birth. His umbilical cord was securely tied 10 cm (4 in) from his abdominal wall, and cut. Some authorities advocate a keyhole dressing for the cord, which is powdered with an antiseptic until its separation in about three to six days. Some authorities do not establish daily bathing[61,62] of the infant until the demonstration bath by the midwife, which is followed by the mother bathing her infant with supervision from the midwife, on one or more occasions, before the midwife's services are withdrawn after the birth. It is via the skin that the baby receives the first contact with his environment. He arrives naked, and spends more time naked during babyhood and childhood than at any other time during his life. Sensations experienced during this time are

125

built into the 'unconscious' part of his mind and will determine his adult attitude to naked flesh and how much handling of his flesh he finds comfortable. A tremendous amount of learning is experienced through the skin—tactile learning—things are hard, soft, round, square, etc. The health visitor can continue to give the mother any help or support that she requires in the care of her baby's skin. It is important for the mother to have confidence in, and derive pleasure from, bathing her baby, so that there can be an unfettered development of the feeling of security within the growing baby.

The skin areas that need special attention at this time are behind the ears, especially the upper portion which can become 'cracked', under the axillae, the groins, the skin fold between the buttocks and the remaining napkin area. These should be thoroughly dried with a soft, absorbent towel, and sprinkled with a reliable talcum powder specially prepared for a baby's skin.

A special bath should be reserved for the baby and the water prepared at body temperature for such sensitive skin.

The toddler can progress to the family bath, where there is ample space for him to enjoy sailing his duck and boat. Care still needs to be taken with the temperature of the water, and the cold should always be run first to avoid overheating of the bath itself and scalding of an always curious child.[63] Encouragement should be given, so that the child masters the technique of bathing himself, keeping his face and body cloth separately, returning all articles used to their appointed places, and leaving the bathroom clean for the other family members. When a daily bath is impossible the child should be taught to sponge himself thoroughly from head to foot.

Many people's conception of 'having a bath' is reclining in the water, but there are some advantages in the 'shower'. In this activity the body is moistened, well soaped and the resultant 'emulsion' removed by the spraying water, so that it immediately goes down the drain and does not float around the bather, then have to be forcibly removed as it adheres to the bath. Also, two showers can be installed in the space taken by one bath, a consideration in dormitories, etc. With reference to world shortage of water, showers are advocated as they take less water than the majority of baths—though modern baths are smaller than their predecessors. In industry involving coal dust, tar, soot and other possible cancer-producing agents, showers are more efficacious and are encouraged before the workmen go home.

Returning to Jimmy, as adolescence is reached, under-arm perspiration may become a problem, and he will need to add a deodorant and/or antiperspirant to his toilet articles for daily application to these areas. There are shirts made from material that neutralizes the odour of perspiration. No matter how often the material is washed, the anti-odour property remains.

As adult years are reached the reasons for and time of having a bath may vary. The vigorous athlete may take a tepid bath in the early morning to tone up the muscular system. After a round of intensive training he may bathe to remove the excessive waste products and help to prevent stiffness. The more timid and anxious may take a hot bath immediately before retiring to bed, to withdraw blood from the brain and leave that organ less active, thus encouraging sleep.

In Jimmy's declining years he may have increasing difficulty getting in and out of the bath. Many gadgets are now available to give older people confidence and enable them to maintain their high standards of personal cleanliness throughout life.

In caring for the skin, we must not forget prevention of burning from excessive sunshine. More pigment is laid down in response to sunshine, but the body must be allowed time to accomplish this, and exposure for five minutes, increased by five minutes each time, is a foolproof sunbathing routine. Some people do not lay down more pigment in response to sunshine and they must take their fresh air in the shade. (Albinos have no pigment, and often exposure to ordinary light is hazardous to their skin.)

Though we chose a man around which to build our story, we cannot forget the special hygiene pertaining to women. Approaching puberty, not only should the girl be mentally prepared for the onset of menstruation, but she should understand that all her glands are more active during menstruation, and the daily bath even more necessary. If this is taken at night she will need local washing and use of a deodorant powder each morning to freshen her for the day. America is having trouble at present with allergy to 'feminine' sprays. Such sprays are now on sale in this country. Are the public being conned? The doctors say these sprays do not do anything that soap and water cannot do. She should understand that sanitary towels need to be changed two-hourly to prevent odour. She should have a healthily modest, as opposed to blasé attitude to buying, storing and disposing of same, since many girls are living in a household which includes father and brothers. She should budget for this monthly expense so that she is not caught unawares when she has run short of pocket money. There are still people who cannot afford to buy sanitary pads. They make them from old clothes, bed sheets etc., and they wash and re-use them. Used pads should be soaked in cold water to remove blood as heat coagulates protein. They should then be washed, boiled and hung in the fresh air to dry.

Internal sanitary protection in the form of tampons has been in more frequent use of recent years. It has obvious advantages, comfort, convenience, absence of odour, particularly important for social occasions. The instructions for insertion and removal should be carefully followed, but their use by the very young is not advocated. Tampons should not be used where there is inflammation of the vagina and/or cervix, or during the early weeks of the puerperium.

CARE OF THE HAIR

Hair reflects to some extent the general health and nutrition of the body. Nourishment for the hair (as for all tissues, skin, nails, etc.)

comes from the blood, so that a well-balanced diet is a prerequisite for the health of any part of the body.

A person who looks after his hair and is aware of its attractiveness is more confident and feels much better than the person with matted, greasy hair; or the person with dry, lank hair and a permanent shower of dandruff on the shoulders. Hair acts as a filter and needs to be washed according to environmental conditions. It is advisable to cover the hair when dusting and to keep hats well brushed, sponged and aired.

Never before have such a variety of lotions been available for shampooing the hair. Cream shampoos or those with oily additives do not completely overcome the problem of 'dry' hair, as the contained detergent removes the incorporated grease as well as the small amount produced naturally. Dryness and brittleness of hair is only reduced by *hydration*. Application of animal, vegetable or mineral oil after hair washing, protects the hair, reduces evaporation and imparts a pleasant sheen. Weekly washing is sufficient for most people to keep the hair healthy and well groomed. However, *daily* washing of brushes and combs is essential.

As well as the scalp being kept clean it should be free from dandruff—the result of a virus infection that can trigger off acne. Many efficient lotions are available for ridding the scalp of dandruff. The manufacturer's instructions should be followed implicitly.

The danger of head lice in the 60s and 70s is less than it was in the 30s and 40s, but the current long hair styles adopted by young men and women are blamed for an increasing incidence of head infestation. Long hair is a danger in laboratories and near machinery. Boys and girls are asked to wear hair nets to avoid such danger.

'Invisible hair nets' in the form of sprays are now popular. Inhalation of particles is thought to produce lung pathology. It is recommended that hair spray is used sparingly, not more than once a day, and the spray directed away from the face. The dried 'spray' must be brushed out of the hair daily using a clean brush, because flakes of dried spray look like nits in the hair.

Food handlers should have all their hair covered so that dandruff cannot contaminate the food, nor can the worker touch his hair and then deposit the germs in the food.

Nurses used to wear their caps as a protection against lice from their patients; as this is no longer a danger, why do we continue to wear them? There is still one area of our work in which the hair needs to be completely covered, and that is when we do dressings or attend in the operating theatre.

CARE OF THE FACE

You will spend much of your time observing people's faces, and you have already learned how to care for the skin on your own face, but do you realize that the expression portrayed on your face is under your voluntary control? When a mistake is pointed out to you, do you accept it graciously and determine never to make the same mistake

128

again, or do you express annoyance or belligerence? When you enter an examination room do you look at the examiner as if you wished she were not there, or do you smile graciously, so helping to establish a good relationship, thus making her difficult task more easy?

Conditions which can Interfere with the Health of the Skin, Hair and Face

Blackheads (comedo, sing.; comedones, pl.) These are composed of solidified sebum occupying the outlet from a sebaceous gland on to the skin surface. This material becomes 'black' from oxidation. Comedones are a part of the condition known as acne vulgaris.

Emulsifying agents should be used to soften the sebum, which renders it removable by a suitable grease solvent such as 1 per cent Cetavlon. Much injury and permanent marking of the skin can result from too vigorous attempts at manual removal of blackheads. Where their removal is desirable a comedo extractor should be used.

Acne vulgaris. This is a skin condition common in adolescence. The sebaceous glands and hair follicles become pustular and blackheads and septic spots are visible on the skin, mainly of the face, neck and upper part of the chest and back. All persons suffering from this condition must be advised to visit their family doctor, as it requires expert attention. Adolescence is a time of self-consciousness and self-preoccupation, so that when it is combined with an outbreak of acne, the family need to play their part in reinforcing the background security and developing a sympathetic yet optimistic outlook so necessary in the successful treatment of this condition. In November 1971, money was given for a three-year study, in Glasgow, of androgen metabolism in human skin and its relation to acne vulgaris.

EXTERNAL ANIMAL PARASITES

The pediculi that can infest the human skin are of three types, named according to the site which they choose to infest.

Pediculus capitis (head louse). Small greyish parasite, the female being slightly larger than the male. Its six legs have claws with which it hangs on to hair. Its head has a sucker to puncture skin and suck blood from its host. The female lays tiny white eggs called 'nits,' and cements them to the hairs near the scalp especially behind the ears. These hatch out in one week and are fully grown in three. During a life span of five weeks they live on human blood. More than one million British children and their families are still affected by head lice.[64]

They cause intense irritation with consequent scratching; if this is performed with dirty finger nails, impetigo may be caused, and the resulting dried discharge and matted hair makes detection and disinfestation difficult. There is loss of sleep with lowering of the vitality; the sores may give rise to enlarged lymphatic nodes, with possible abscess formation (p. 175).

129

Treatment. Bend the head forward over a wash-basin, cloth or sheet of brown paper, and allow the hair to fall forward. Make several partings and drop a little Suleo (DDT emulsion) into each until one tablespoonful has been used. Throughout 1970 there were reports of resistance to a 4 per cent solution of DDT among head lice. Carbaryl, a new insecticide, has been used with success in South America and South Africa and is now being used in Britain. Another insecticide, Prioderm (malathion lotion) is on the British market. Massage the scalp gently all over, ensuring even distribution of the emulsion through the hair nearest the scalp where lice lodge and nits are deposited, giving special attention to the back of the neck and behind and above the ears. Allow a few minutes for the Suleo to dry out, then tidy the hair normally with comb and brush, not vigorously, as this may drive the lice away from the scalp. If desired the hair may be washed 24 hours after, but it will remain protected for at least 14 days against any nits which may hatch out later. In cases of severe infestation, or where there is a danger of reinfestation, Suleo may be applied at weekly intervals. Lorexane No. 3 shampoo and Esoderm medicated shampoo can also be used for the removal of head lice. Instructions are issued with each shampoo. The World Health Organization[65] recommends an application of an emulsion concentrate (benzyl benzoate 68 per cent, DDT 6 per cent, benzocaine 12 per cent and Tween-80 14 per cent) diluted 1:6 in water for the killing of head lice and nits.

It is important to remember that reinfestation may occur through infested headwear, and this therefore should be thoroughly ironed with a hot iron to destroy any lice or nits which may be there.

Pediculus vestimenti. Previously called pediculus corporis, and still often referred to as the body louse. The female lays her eggs in the seams of clothing, but in dirty conditions, and when day clothes are not changed for night clothes (as may happen in wartime), the eggs can be attached to the body hair. They hatch out in one week, become mature in two, and live for four to five weeks. They are similar to, but a little larger than, the head louse. They too live on human blood, and their crawling and biting cause intense irritation which results in scratching, carrying with it the possibility of secondary infection and involvement of the nearby glands. Irritability, restlessness, sleeplessness and loss of appetite are characteristic of infestation. Body lice are known to be capable of transmitting typhus (jail), trench and relapsing fever.

Treatment. The personal clothes are removed into a bag, in which they are generously dusted with DDT powder. The bag is securely tied and left for at least two hours, after which the clothes are washed or dry-cleaned as is appropriate.

The infested person should partake of a cleansing bath, thoroughly dry himself and sprinkle the whole body generously with DDT powder.

Some local authorities deal with persons and clothing infested with body lice at a disinfecting station.

Pediculus pubis. The proper name for this creature is phthirus pubis, but it is most frequently known as the crab or pubic louse. Its life history is like that of the pediculus vestimenti, but it mainly infests the pubic and axillary hair. The signs and symptoms produced by infestation are exactly the same.

Treatment. This can be the same as for pediculus vestimenti plus shaving of the axillary and pubic hair. When applicatio dicophani (National Formulary) is applied to affected hair and skin, shaving is not necessary. The lotion is allowed to dry and is washed off after 12 hours. Treatment should be repeated after seven days.

Other external animal parasites that can infest the human skin are: **Sarcoptes scabiei,** causing the condition of scabies. It has been discussed in relation to its occurrence on the feet and hands, and here we need only mention the other areas of the body where the skin is thin, viz. under the breasts and on the front of the wrists. Infested people are nearly always treated at the local authority clinic.

With some of the current hair styles that are merely 'tidied' daily between 'sets' there is an increasing incidence of infestation with Sarcoptes scabiei, many of which are resistant to benzyl benzoate.

Common flea (Pulex irritans). The female lays her eggs in bedding, dust, rubbish, cracks in floor and furniture, etc. The fleas do not actually transmit any disease, but leave a channel for entrance of infection. The irritation caused by flea bites induces scratching and herein lies the danger of fleas.

Treatment. Meticulous measures must be taken with regard to external cleanliness and personal hygiene, with the addition of DDT powder in cracks and crevices and on bedding.

Rat flea (Zenopsylla cheopis). The rat flea is a carrier of plague, a disease which attacks rats. The flea is a flat, wingless insect. When infection arises amongst rats, killing large numbers of them, the fleas living on them migrate and may bite human beings, at the same time regurgitating the plague germ into them. The flea excretes the germ in faeces, and if a human being with a wound comes in contact with such faeces infection can arise from inoculation.

Strict measures are taken by the Port Health Authorities so that infested rats do not get into this country.

Bed bug (Cimex lectularius). This is a flat, reddish-brown parasite with characteristic pungent odour. It lives in cracks in walls, floors, furniture, etc., most often around beds. It is nocturnal in habit and bites the bed occupants, sucking their blood. It has been known to exist as long as a year without food, when it is as thin as tissue paper, and can easily pass through wall crevices into an adjoining house.

Disinfestation of houses and their contents is by liberal use of a DDT lotion in spray form. One big factor in slum clearance is getting rid of the bugs. Sometimes the furniture, etc., has to be disinfested in a van *en route*, to prevent carrying bugs to new property.

EXTERNAL VEGETABLE PARASITES

Ringworm (Tinea). We have already written about this external, vegetable parasite, a fungus, in its infestation of the hands and feet. There are several varieties of these fungi, and they can infest the scalp and the skin of the body. Diagnosis is confirmed by examining the area under Wood's glass with the aid of a mercury vapour lamp. Affected hairs fluoresce; this examination can be repeated to confirm the termination of infestation.

The condition is highly contagious, so that all articles and clothing must be kept and washed separately; clean clothing must be given daily, and that which is removed must be boiled immediately. Any articles which cannot be disinfected must be burned. Fungicidal ointment, e.g. Whitfield's, undecylenate cream or Tolnaftate cream can be applied locally.

Recently local treatment to the infested area has lessened in popularity, and an antifungal antibiotic has been administered orally, e.g. griseofulvin, 1 to 2 grams daily, gradually reducing as clinical response occurs, and continuing this reduced dose for four weeks.

If local treatment is used, then the applier must remember to protect himself during application by wearing rubber gloves and discarding them into a bactericidal solution.

SUMMARY OF THE SKIN

External covering of body.

THE EPIDERMIS

Compound, stratified epithelium.

Scales constantly shed from surface.

Germinal or 'growing' layer at base. Nutriments brought to it, and waste products carried away from it by lymph—no blood vessels.

PAPILLARY LINE OF UNION

The line of union between epidermis and dermis has the following functions:

1. Increases area for distribution of dermal blood vessels and nerve endings.

2. Gives finger-prints.

3. Contains ergosterol which, when irradiated by ultra-violet rays, produces vitamin D.

4. Contains pigment to protect from sun's rays.

5. Gives origin to the arrectores pilorum muscles.

132

The dermis supports the following structures:

1. Sweat glands.

2. Sebaceous glands, pouring their secretion into hair follicles, each containing a hair, and giving insertion to arrectores pilorum muscles.

3. Blood vessels.

4. Nerve endings: touch, temperature, pain, pressure.

Can you give the composition of sweat?

Can you define: pore, latent heat, good conductor of heat, bad conductor of heat, conduction, convection, radiation, evaporation, the functions of the skin, blackheads, acne?

EXTERNAL ANIMAL PARASITES

Pediculus capitis, head louse.

Pediculus vestimenti, body louse.

Pediculus pubis, crab louse.

Sarcoptes scabiei.

Common flea.

Rat flea.

Bed bug.

EXTERNAL VEGETABLE PARASITES

Many different types of fungi causing ringworm (tinea).

CLOTHING

As well as being kept clean, the skin is usually clothed from birth onwards. What one wears varies tremendously (p. 21), and is affected by climate, tradition and current fashion. This therefore brings us to the choice and cleanliness of clothing. We have already dealt with clothing for the hands (p. 88) and clothing for the feet (p. 71). Revision of these is essential so that you do not get your knowledge fixed in watertight compartments in your brain.

At birth, in most cultures, a baby is wrapped in a cotton blanket[66,67] and placed safely in a cot whilst his mother is attended to. At a suitable time he is taken from the cot, and his skin is swabbed free of the vernix caseosa with which it is covered at birth. He is clothed in a loose-fitting vest and gown, neither article being drawn over his head, for even the momentary confinement and darkening are frightening to a very young baby. Also the blinking reflex and lacrimal apparatus do not work efficiently in early life. A soft, smooth napkin[68,69] is placed in position, avoiding excessive thickness between the legs as the bones at this stage are mainly organic matter (p. 43); he is then wrapped in a clean cotton blanket. This routine, combined with feeding, will take place many thousands of times before any degree of independence can be established by the child bathing and clothing himself. It therefore is each mother's responsibility to provide loose-fitting vests, day and night gowns, and cotton blankets in which to wrap the baby. They should be made of material that will withstand daily laundering without shrinking or becoming hard and matted. The napkin must be absorbent and it is usually made of a soft white terry towelling that can be boiled and hung out in the fresh air to dry. Disposable napkins may be preferred, and they certainly are very useful for holiday times when there may be restricted facilities for washing. Plastic or rubber knickers should not be worn constantly, but their use reserved for taking baby out or when he is being nursed. A towelling bib will be required at feeding time; plastic bibs have been in vogue but they can be dangerous, as they can be thrown up over the face and the baby suffocated. Woollen bootees, hood or cap, jacket and mitts[70] are added when the baby is taken out in the pram. Sooner or later everything finds its way to a baby's mouth, so that decoration with buttons and ribbons should be thoughtfully done. As the baby learns to kick off the pram clothes, socks, shoes and the traditional pram suits are introduced.

At the commencement of crawling, romper suits are the most suitable day-time wear, for they prevent the feet being caught in the hem of a gown which often impedes the first attempts at standing alone. As toddling is achieved, dungarees become useful as they afford some protection against the inevitable grazed knees. As control over walking and running is achieved, clothes characteristic of the culture are introduced, the arms, legs and head being covered in the colder weather, and gradually increased areas exposed in the warmer weather. Special precautions need to be taken with very fair children, as they do not stand excessive exposure to sunshine. All children's clothes should fasten at the front, and children should be able to dress themselves by the time they go to school at five years old. Road safety authorities are currently concerned about small children wearing anoraks or duffle coats with hoods which can restrict hearing and vision as the wearer steps out on to the road. An acceptable explanation must be given to the younger child, who for economic reasons needs to wear 'handed down' garments. Occasionally he must be given absolutely new clothing to boost his morale.

The current fashion of 'shortie' nightdresses favoured by the younger adult population is welcome from a fire precaution point of view. A regulation made by the Home Secretary under the Consumer Protection Act of 1961 states that inflammable nightdresses over 1 m 15 cm (45 in) long must bear a stitched-on label bearing the words 'Warning—keep away from fire',—but labels can be torn off or become illegible with laundering, and many immigrants cannot read English! There is still a tendency to put little girls in ankle length nightdresses in spite of advice that pyjamas carry less fire risk than nightdresses. Many people believe that hospitals and local

134

authorities should set an example by prohibiting nightdresses and insisting on the use of pyjamas in the children's wards, homes and nurseries under their control.

As soon as is reasonable the child should be allowed to select what he is going to wear on different occasions, and then he should learn to buy his own clothes and keep them clean and in good repair, so that he always appears well groomed, in this way increasing his self-esteem, an essential feature in confident and happy living.

Tolerance and good humour towards adolescents who are experimenting with way-out fashions will reap better rewards all round, than continual derisive remarks. A lot has gone into the making of the teenager that is beyond his control. He does not become a *different person* because he dons 'mod gear'—he might *feel different*—presumably *more comfortable* in it—psychologically if not physically, or he would not don it. Maxis are dangerous in lifts and on escalators, and when manipulating stairs and revolving doorways; when stepping on and getting off buses and trains, especially with no free hand to lift the skirt. When getting out of a car, it is easy to stand on the hem and be pitched forward on rising.

The physically disabled[71] are denied some activities by which normal people find satisfaction. It is specially important therefore that careful thought is given to the clothes available for them so that they can get maximum pleasure from their clothes.

With an aging population[72] it becomes imperative for all to think about clothing for senior citizens. With failing eyes and shaking hands it becomes increasingly difficult for them to retain their independence with conventional clothing. Garments with a back fastening can no longer be managed; front fastenings are preferable to side fastenings; zips and velcro tape are preferable to small buttons, hooks and eyes. Reflective arm bands are useful if they are out at dusk. White also shows up in poor light.

Protective clothing is provided wherever there is a known health risk to the workers, or danger of spreading disease. In spite of this provision, and education about the consequences of not wearing protective clothing, some workers fail to take the advice offered and thus fail to 'care' for their own health, and/or the health of others. Encouragement from his family may help a worker to submit to the necessary discipline of wearing protective clothing.

Ergonomics is the study of man in his working environment and it increasingly results in purpose-built fashion for work-clothes.

REASONS FOR WEARING CLOTHING

1. To reduce strain on the heat-regulating mechanism.
2. To give protection from the elements—rain, wind, cold, heat, sun.
3. To give protection from injury, e.g. crash helmets. Extra protective clothing has to be worn in many occupations.
4. To conform to the social pattern into which the individual is born.

In this respect several questions need to be asked.

1. Is it suitable for the occasions on which it is to be worn?

2. If made of an inflammable material, has it been treated to render it non-inflammable? This is often shown by a written statement along the selvedge. In 1971 the Safety Foundation suggested fireproofing of *all* materials, from childrens' night attire to carpets and curtains.

3. Does it fit and is the weight suspended from the shoulders? Tightness at the neck, waist or legs is to be avoided.

4. Does it restrict movement?

5. Is it non-irritant? Dyes and badly cured skins can cause trouble on sensitive skins.

6. Is it a good or bad absorber of moisture?

7. When it does become wet, does it allow slow or rapid evaporation?

8. Is it a good or bad conductor of heat?

9. Does it absorb or reflect light rays?

10. Is it waterproof?

11. Will it be sufficiently warm or cool for the purpose for which it is intended? Heavy clothing wastes energy.

12. Is it hard wearing?

13. Will it shrink in washing or cleaning? Many fabrics are now sanforized to prevent shrinking.

14. Is it easily and inexpensively kept clean?

15. Can it be afforded without causing financial stress?

16. Does the colour fit in with the (a) existent, or (b) anticipated wardrobe?

To answer some of these questions we need to know a little bit about the different fabrics that can be used for clothing.

Cotton. This is obtained from fibres attached to the seeds of the cotton plant. It is light, strong, durable and can be boiled. It will take starch which gives it a glossy surface to which dirt and germs do not so easily adhere. It readily absorbs moisture, allows rapid evaporation of same, and is a good conductor of heat. All these factors make it unsuitable for underclothing, unless the fibres are woven in a cellular fashion, when air is retained in the spaces, thus making it suitable. It is popular and good for summer outer garments such as dresses, shirts and shorts, and many of these are treated so that they are easily laundered and require the minimum of ironing, which is a great asset.

Wool. This is obtained mainly from sheep, and when woven it is capable of absorbing moisture and allows slow evaporation of same. Because of the air retained in its interstices it is a bad conductor of heat and this makes it suitable for underclothes. Many people, however, find its rough surface too irritating next to the skin. Its advantages for this purpose are outweighed by the fact that it shrinks with washing, becomes hard and matted, and therefore loses its absorbent power.

Its greatest usefulness in these days is for outer garments in the form of closely woven tweeds for winter wear, loosely woven tweeds for spring wear, and machine and hand-knitted garments, which often have the added advantage of being self-made, thus boosting the morale.

Linen. This is made from flax and forms a smooth, strong, durable fabric; it takes starch well and has a glossy surface when ironed which discourages the adherence of dirt and germs. It has to be carefully washed and takes considerable ironing, which makes it less favourable for summer dresses and suits. It sbsorbs moisture well, allows rapid evaporation of same, is a good conductor of heat and is therefore unsuitable for underclothes.

Silk. The silkworm secretes a fluid, which it then makes into fine threads, and from these silk is woven. It is smooth, light in weight, does not absorb moisture as well as wool, allows considerable evaporation of that moisture, and is a bad conductor of heat. This makes it suitable for underclothes, but it is the most expensive.

Artificial silk or rayon. This is manufactured from cellulose. It does not absorb moisture well, but it is a good conductor of heat. Its main use in these days is for dresses and blouses.

Rubber. The gum from the rubber tree can be used to make garments waterproof. Rubberized cotton or silk mackintoshes are light and easily transportable, but as they are impervious to air also, their continuous use will lead to heat stagnation; to offset this they need to be well ventilated in the axillary region.

Plastic. This substance is synthesized from coal tar and now has many and varied uses in the clothing kingdom, from shower caps, vanity capes, cuffs for the hostess washing up in her finery, gloves for doing her housework, boots to draw over her more slender outdoor shoes, to mackintoshes which have become popularly known as 'pacamacs'. This impervious material needs to be ventilated when used for outdoor garments. We have already mentioned baby's bibs and knickers, and spoken of the dangers inherent in these articles. We cannot leave this subject without reminding ourselves of the tremendous hazard of the plastic covers with which the dry-cleaning firms are now returning cleaned clothing. They have been grabbed by children and in their imaginative play used as space helmets with several fatalities. It is important that we, as health teachers, should urge the public to avoid leaving these articles lying about, to dispose of them by burning if possible or if not, to tie them into several knots before putting them into the ashbin.

Nylon, terylene, etc. These substances are also synthesized from coal tar, and many are their uses for clothing. They are presented in a variety of undergarments, night attire and outdoor wearing apparel. Whatever disadvantages these fabrics may have as poor absorbers of moisture are offset by the ease with which they can be washed each night as they are removed, and the fact that when drip-dried they do not require ironing.

Protex is a new non-inflammable fabric in which the fibre itself repels burning and in contact with the flame is said to shrivel up and retreat from the flame. Unlike all synthetics it is claimed that it does

137

not melt or drip. The fabric has a soft wool appearance and is reputed to be crease resistant, quick drying and non-iron.

Leather and fur. These are the warmest clothing and are much favoured by those who have to endure snow and ice for part of the year. Leather is completely windproof and, lined with thick, fleecy wool, it forms a cosy and durable garment. If left with the fur on the outside it is more elegant and a further layer of air is entrapped within the hairs, this advantage being outweighed by the fact that fur looks very bedraggled in the rain. In 1971 the International Fur Trade Federation recommended that trading in the skins of the five endangered species should cease.

Flannelette. This is a cotton substance manufactured with a fluffy surface in an attempt to overcome some of the disadvantages. Air is retained within the fibres thus making it conduct heat less readily. It is, however, highly inflammable, but has to be made non-inflammable[73] before it is offered for sale. With repeated washings this becomes less effective, and it is a pity that the cheapness and durability of this material still tempts many people to buy it for night attire for children and the elderly. Dr William Farr, the great medical statistician advocated flame-proof clothing for children and elderly women in 1861! After the 1964 regulations, Oxfam destroyed 500 new infants' nightdresses made of highly inflammable winceyette, which were given to the organization by stores no longer able to sell them in this country.

Paper. Various articles for human attire, including some for hospital staff, are now made of paper. People who wear paper clothing near an open flame, risk burning themselves if the garments have been laundered, dry-cleaned or worn in soaking rain. The manufacturers acknowledge that after washing, many of these paper dresses, etc. lose whatever flame-retardent finish they had.

Electric cloth. Garments are made of nylon containing an inner lining of terylene, coated with a rubber polymer impregnated with carbon, enclosing the 'electric cloth' with fine copper wires attached. A flex and small plug connect to the power source. It is claimed that garments can be sat on, crumpled, cut and even soaked with safety. Heated waistcoats were tested by police motor-cyclists.

A word must be said about the choice of colour of clothing. It is well known that white *reflects* all the light rays; it therefore affords considerable protection against heat. Conversely dark colours *absorb* light rays and therefore afford considerable protection against the cold.

Colour schemes are all important, and every encouragement should be given to the artistic blending of colours, without which 'smartness' cannot be achieved.

It is reasonable that in the declining years, as physical activity and heat production are lessened, the elderly feel the need for more clothing. Two thin layers (separated by a film of air) are warmer than one thick layer. Small buttons, fine zips and 'step-in' clothes are increasingly difficult to manage for stiffened, fumbling fingers. Many of the ideas of clothing for the disabled are also useful for frail, aged people.

You will now be more capable of answering the list of questions on the choice of clothing (p. 136), and you will realize that no one material is suitable for all garments, for all weather conditions and for all individuals, and that there is great latitude in individual choice.

CLEANLINESS OF CLOTHING

Washing. Soft water that forms a good lather with soap is best for this purpose, so that there is no resultant 'curd' to block the interstices of the material, rendering it less clean in appearance and less effective as an absorber of moisture and a conductor of heat. The temperature of the water varies for the different materials, instructions usually being given by the makers. Thorough rinsing is essential, and drying in the fresh air has bleaching properties.

Underclothing, being in contact with the skin, is also in contact with the sweat, sebum, scales and bacteria thereon. The last named have ideal conditions for rapid multiplication, and it has been estimated by a bacteriologist that after one day's wear each square inch contains four thousand bacteria, and after six days' wear 10 million bacteria! Clean underclothing daily would therefore seem to be the goal at which to strive. Underclothing worn during the day should never be worn at night.

Boiling. For all fabrics that will stand this process it is the best way of ensuring that the article becomes germ-free. It also keeps the colour of white garments, though again modern methods have produced white materials that will stay absolutely white without boiling.

Starching. Can you remember the advantage of this process? (p. 136). Do you consider the extra ironing involved a disadvantage?

Dry-cleaning. As parts of the under surface of our dresses and suits are always in contact with the skin, and parts of the outer surface with the atmosphere, frequent cleaning of non-washable garments is necessary. Many firms offer a 24 hour service, so that even those with a limited wardrobe can comply with the rules of cleanliness. There are several preparations available for more frequent sponging of collars and cuffs.

One golden rule is that clothes must never be stored (e.g. when removing winter clothes from the wardrobe) in a dirty state as stale perspiration has an unpleasant odour, rots fabric and encourages moths, which cause further destruction; and don't forget the multiplication of those bacteria.

Summary of Clothing

APPLY THE FOLLOWING QUESTIONS TO GARMENTS MADE OF THE
FOLLOWING FABRICS:

Is it suitable for the occasions on which it is
to be worn?

Is it inflammable?

Does it fit, and is the weight suspended from
the shoulder?

Does it restrict movement?

Is it non-irritant?

Is it a good or bad absorber of moisture?

If it does become wet, does it allow slow or
rapid evaporation?

Is it a good or bad conductor of heat?

Does it absorb or reflect light rays?

Is it waterproof?

Will it be sufficiently warm or cool for the purpose for which it is intended?

Is it hard-wearing?

Will it shrink in washing or cleaning?

Is it easily and inexpensively kept clean?

Can it be afforded without causing financial
stress?

Does the colour fit in with (a) existent, or (b)
anticipated wardrobe?

Cotton
Wool
Linen
Silk
Artificial silk
Rubber
Plastic
Nylon
Terylene
Leather
Fur
Flannelette
Paper
Electric cloth

The Circulatory or Cardiovascular System

Jimmy is now respectable with his skeleton clothed with muscle tissue and covered with skin. We have acquired the knowledge that all tissues require blood, so that we are ready to find out more about the blood that keeps the body's cells bathed in fluid—extracellular fluid.

Blood is the transport medium of the body and is *interdependent* on the other systems so that it is not in danger from depletion of oxygen and nutrients, nor from too drastic a change in its temperature, chemical reaction and composition due to the waste produced by the millions of body cells. The lungs serve it with oxygen and remove its carbon dioxide. The digestive system serves it with nutrients. After a meal, the liver and muscles store extra glucose as glycogen, and extra fat goes into the fat stores. Between meals these are released as needed to keep the level of nutrients in the blood constant. The constant movement of blood ensures that heat from those organs of tremendous metabolic activity is quickly spread round the whole body, thus maintaining a constant temperature. The waste products are filtered off by the kidneys. All these activities are controlled by the nervous system, enzymes and hormones! Now you can begin to glimpse the meaning of the word *interdependence*.

THE BLOOD

Blood is a red, alkaline (pH 7·4), viscous fluid, salty to taste, which escapes from the tissues when they are injured, and then clots. There are about 6 litres (10 pints) in the body, of which 4·8 litres (8 pints) is water. A person can safely lose 0·6 litre (1 pint), which is the quantity taken from a blood donor at each session. Blood so taken, can be stored as such under specific conditions for 21 days, after which the cells disintegrate. The cells are then removed and the liquid part of blood—plasma—can be used. Blood (whole) can now be frozen, but it is an extremely expensive process—especially the defreezing. It will be very useful for storing blood of rare groups. It is convenient to describe the blood in two parts, a liquid called plasma, and a solid part comprising red and white blood cells, and blood platelets. The solids and liquid are present in roughly equal proportions by volume.

PLASMA

This is a clear, straw-coloured fluid composed of 90 per cent water in which is dissolved all the food factors being taken to the cells,

141

and all the waste products of metabolism being taken from the cells, plus bile pigments, oxygen, hormones, enzymes, antibodies and anti-coagulant substances—a formidable list, but since you need to know the food factors and a little bit about metabolism, it is as well to become acquainted with the names here and learn more about them later.

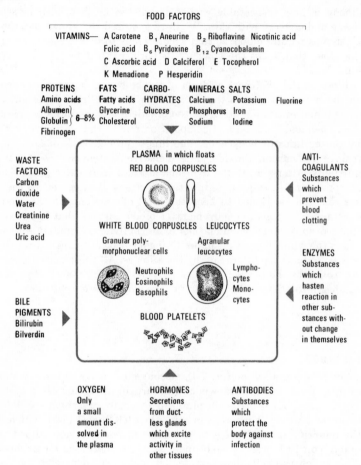

FOOD FACTORS

VITAMINS— A Carotene B₁ Aneurine B₂ Riboflavine Nicotinic acid
Folic acid B₆ Pyridoxine B₁₂ Cyanocobalamin
C Ascorbic acid D Calciferol E Tocopherol
K Menadione P Hesperidin

PROTEINS	FATS	CARBO-	MINERALS	SALTS	
Amino acids	Fatty acids	HYDRATES	Calcium	Potassium	Fluorine
Albumen	Glycerine	Glucose	Phosphorus	Iron	
Globulin } 6–8%	Cholesterol		Sodium	Iodine	
Fibrinogen					

PLASMA in which floats
RED BLOOD CORPUSCLES

WASTE
FACTORS
Carbon
dioxide
Water
Creatinine
Urea
Uric acid

ANTI-
COAGULANTS
Substances
which
prevent
blood
clotting

WHITE BLOOD CORPUSCLES LEUCOCYTES

Granular poly-
morphonuclear cells

Agranular
leucocytes

Neutrophils
Eosinophils
Basophils

Lympho-
cytes
Mono-
cytes

ENZYMES
Substances
which
hasten
reaction in
other sub-
stances with-
out change
in themselves

BILE
PIGMENTS
Bilirubin
Bilverdin

BLOOD PLATELETS

OXYGEN
Only
a small
amount dis-
solved in
the plasma

HORMONES
Secretions
from duct-
less glands
which excite
activity in
other tissues

ANTIBODIES
Substances
which
protect the
body against
infection

Fig. 81   Diagrammatic representation of the blood.

PROTEINS

The body cannot utilize atmospheric nitrogen and it has to rely on its protein intake for this element. As a house is built of bricks proteins are built of amino acids, and however complex the protein as it is taken by mouth, it is broken down into its constituent amino acids in the digestive tract. As well as these amino acids that are being transported round the body, the plasma has to maintain 6 to 8 per cent built-up protein within its own substance, so that it has a sufficient osmotic pressure to suck back tissue fluid at the venous end of a capillary network (p. 158). The main plasma protein is albumin, the others being globulins, fibrinogen and prothrombin. The latter two are necessary for the clotting of blood. Of the globulins, the immunoglobulins (Ig) G, M, A, D and E take part in various immune responses of the body to antigens, which can be bacteria—when we speak of immunity;[74] a foreign substance, when we speak of allergy; a substance produced by the body, when we speak of auto-immunity. When the substance produced by the body is in response to transposed tissue, e.g. organ transplant, we speak of rejection.

Table III.   Levels of food factors present in the blood.

FOOD FACTOR	STATE IN WHICH PRESENT IN BLOOD	BLOOD LEVEL PER CENT
Protein	(a) Amino acids for distribution	...
	(b) Albumin	
	Globulin	6–8
	Fibrinogen	
Fat	(a) Fatty acids and glycerine	0·125–0·3
	(b) Cholesterol	
Carbohydrate	Glucose	0·08–0·12
Vitamins—		
A Carotene (anti-infective, prevents infection)		
B₁ Aneurine (antiberiberi, prevents beriberi)		
B₂ Riboflavine		
Nicotinic acid (antipellagra, prevents pellagra)		
B complex Folic acid (anti-anaemia, prevents anaemia)		
B₆ Pyridoxine		
B₁₂ Cyanocobalamin (antipernicious anaemia, prevents pernicious anaemia)		
C Ascorbic acid (antiscorbutic, prevents scurvy)		
D Calciferol (antirachitic, prevents rickets)		
E Tocopherol (anti-abortive, prevents abortion)		
K Menadione (antihaemorrhagic, prevents haemorrhage)		
P Hesperidin (antipurpuric, prevents purpura)		
Mineral salts—		
Calcium		0·009–0·011
Phosphorus		0·0026–0·004
Sodium		0·3–0·35
Potassium		0·014–0·021
Iron		
Iodine		
Fluorine		
Chlorides (serum) as Na Cl		585 to 630 mg/100 ml
Chlorides (serum) as Cl		355 to 383 mg/100 ml
Bicarbonate (alkali reserve)		54 to 72 vol per cent CO₂ 24 to 32 mEq/litre

Plasma proteins can be lost from the body in some kidney diseases, and you will not be surprised to learn that such patients are fed with a high-protein diet, otherwise they will develop oedema, as the blood will not have sufficient sucking power to get the tissue fluid back into the vascular bed (p. 158).

FATS

These are broken down into fatty acids and glycerine in the digestive tract. In this form they are transported in the blood. A little is built into other complex substances. Cholesterol is a crystalline substance of a fatty nature. A higher than normal blood cholesterol is thought to predispose to the laying down of roughened yellow plaques along the artery linings, especially those of the coronary arteries. Blood clots on a roughened surface and so thrombosis arises. An increased blood cholesterol is also present in those people who develop gall-stones and myxoedema.

CARBOHYDRATES

The blood sugar is able to remain constant, even after consumption of a sweet meal, and in spite of the fact that we only eat four-hourly and not at all whilst we sleep. As the blood sugar is rising after a meal, the excess is withdrawn into the liver and muscle tissue where the hormone insulin, normally secreted by the pancreas, converts glucose into insoluble glycogen, in which form it stays until the blood sugar is beginning to fall. In direct response to a falling blood sugar the glycogen is converted into glucose. By this fascinating mechanism the blood sugar remains remarkably constant between 80 and 120 mg per 10 ml of blood. Seventy per cent of 1,000 atherosclerotic males with 170 mg of glucose per 100 ml of blood developed fatal or non-fatal coronary infarction within three to four years.[75] Things can go wrong with the glucose→glycogen→glucose conversion, and you will learn how to look after patients with diabetes mellitus, and how to teach them to look after themselves.

ELEMENTS

In solution these substances are now frequently referred to as electrolytes, and their correct balance in the blood and cells is essential for health; electrolyte imbalance has to be vigorously treated to preserve life. Electrolytes maintain the slight alkalinity of blood. In the wards you will find that many specimens of blood are taken from the patients for electrolyte estimation, and treatment is adjusted according to the results.

WASTE PRODUCTS

The waste products that are being taken from the tissues are mainly excess carbon dioxide and water from the combusion of all food-stuffs, and creatinine, urea and uric acid from the breakdown of proteins.

BILE PIGMENTS

The pigment bilirubin is formed in the liver from the breakdown of red blood cells. Bilirubin is secreted in bile and gives it its character-istic colour. In conditions of biliary obstruction, bilirubin banks up in the blood stream, and estimation of this is useful in diagnosis.

There is only a small amount of oxygen dissolved in the plasma, as the red blood cells which float in the plasma are the main oxygen carriers.

HORMONES

These are complex chemical substances, made in the ductless glands and absorbed directly into the blood to be carried to the specific tissue in which they excite activity.

ENZYMES

These are organic substances that bring about a chemical change in another substance, without themselves being changed or used up in the process. Many of them have the suffix *ase* combined in their names. Estimation of the amount of enzyme in the blood is a help in diagnosis in some diseases.

ANTIBODIES

These are substances that protect the body against infection, each one being specific, i.e. it can only protect against one species of bacteria; diphtheria antibody can only protect against diphtheria. Some antibodies cause disintegration of the germs and are therefore called bacteriolysins, some prepare the germs so that they are more easily killed by the white blood cells and are therefore called opsonins, and some cause agglutination or clumping together of the germs and are called agglutinins. All antibodies are attached to the globulin fraction of the plasma, and globulin can be injected into a person who has been exposed to an infection and in whom such infection would be serious. The presence of specific antibodies in the blood can be confirmatory evidence of previous infection.

ANTICOAGULANTS

As the name implies these substances prevent the coagulation of blood within the vessel, but anticoagulants are neutralized when there is trauma to, or disease of, the vessel wall. Estimation of anti-coagulant activity in the blood forms a useful guide to treatment in thrombosis.

RED BLOOD CELLS OR ERYTHROCYTES

These are tiny, circular, biconcave, non-nucleated discs about 7 microns or $\frac{1}{2500}$ of an inch in diameter. They are made in the red bone marrow, and amongst other things, protein, iron, copper, ascorbic acid (vitamin C), folic acid and cyanocobalamin (vitamin B_{12}) are required for their formation. Cortisol, secreted by the adrenal cortex, influences production of red blood cells. There are about five million

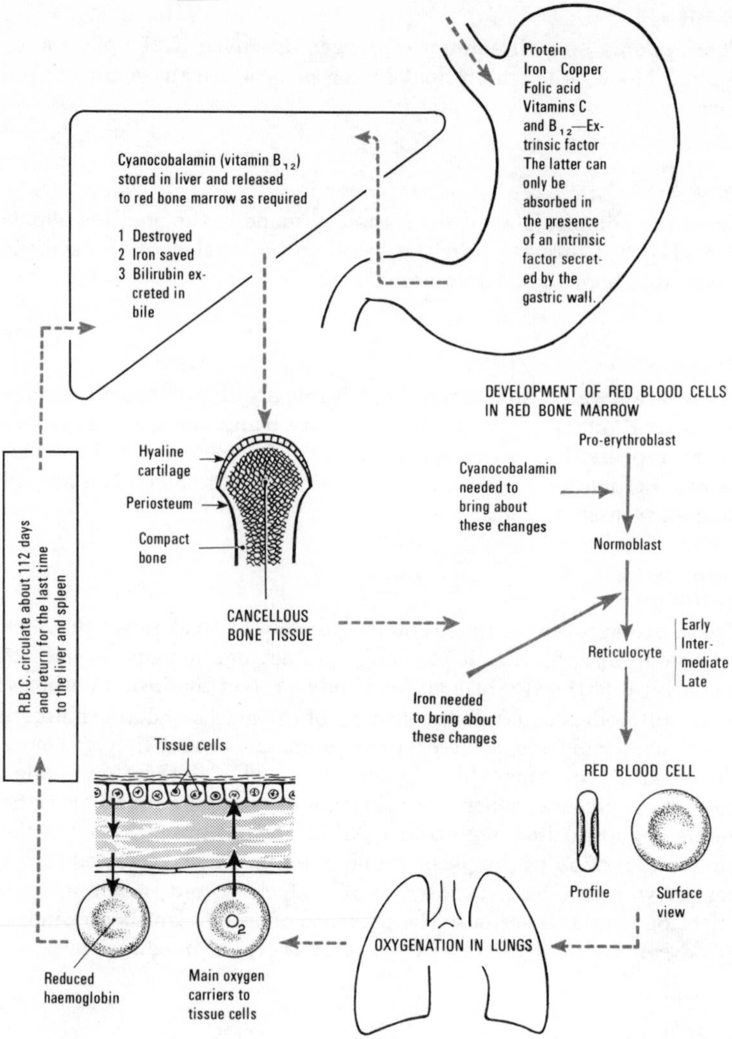

Protein
Iron Copper
Folic acid
Vitamins C
and B$_{12}$—Extrinsic factor
The latter can
only be
absorbed in
the presence
of an intrinsic
factor secreted by the
gastric wall.

Cyanocobalamin (vitamin B$_{12}$)
stored in liver and released
to red bone marrow as required

1 Destroyed
2 Iron saved
3 Bilirubin excreted in
  bile

DEVELOPMENT OF RED BLOOD CELLS
IN RED BONE MARROW

Pro-erythroblast

Cyanocobalamin
needed to
bring about
these changes

Normoblast

Hyaline
cartilage

Periosteum

Compact
bone

CANCELLOUS
BONE TISSUE

Reticulocyte

Early
Intermediate
Late

R.B.C. circulate about 112 days
and return for the last time
to the liver and spleen

Iron needed
to bring about
these changes

RED BLOOD CELL

Tissue cells

Profile   Surface
             view

Reduced
haemoglobin

Main oxygen
carriers to
tissue cells

O$_2$

OXYGENATION IN LUNGS

Fig. 82   Diagrammatic representation of red blood cells.

in each cubic millimetre of blood. They live about 112 days and are
then destroyed by phagocytic action of cells—mainly in the liver and
spleen. We have already learned that the pigment bilirubin released
at their breakdown gives bile its characteristic colour. The iron so
released is saved and used for the formation of more haemoglobin.
A pigment (porphyrin) combines with iron to form a substance
called haem which combines with a protein (globin) to form haemoglobin. Haemoglobin has an affinity for oxygen and in combination
is called oxyhaemoglobin, so that the blood contains most oxyhaemoglobin as it leaves the lungs. As it reaches cells deficient in oxygen,

146

the oxygen dissociates from the haemoglobin and by diffusion enters the cells. At this stage the haemoglobin is said to be reduced and it returns to the lungs, there to combine with some more oxygen; thus the red blood cells are the main oxygen carriers throughout the body.

In the early stages the name pro-erythroblasts is used for the large cells. Cyanocobalamin (vitamin B_{12}) is necessary to convert the pro-erythroblasts into smaller cells called normoblasts; iron is then necessary to convert the normoblasts into non-nucleated red blood cells ready for circulation.

WHITE BLOOD CELLS (LEUCOCYTES)

These are larger than the red blood cells, and are present in less numbers, there being only 6 000 to 8 000 in each cubic millimetre of blood. There is said to be 1 white to every 500 red cells. According to whether they have granules in their cytoplasm, there are two types of white cell—granular and agranular leucocytes.

Granular leucocytes. These are made in the red bone marrow and have an irregular-shaped nucleus which gives rise to the name polymorphonuclear (many-shaped nucleus) cells.

Some of the polymorphonuclear leucocytes can take up different stains and are accordingly called basophils, eosinophils, and neutrophils, the latter making up 70 per cent of the total white cell count. This differentiation is sometimes useful in diagnosis, and you will

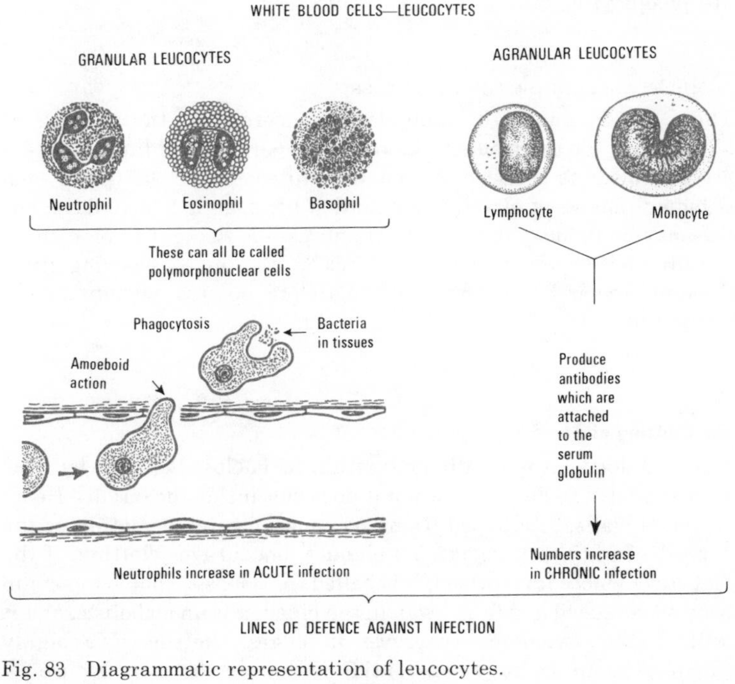

Fig. 83 Diagrammatic representation of leucocytes.

147

need to learn more about the significance of these as you learn to read laboratory reports in the wards.

The neutrophils possess the power of amoeboid movement and can squeeze out between the cells of the capillary walls into the surrounding tissue to engulf bacteria (phagocytosis). Neutrophils are therefore one of the body's lines of defence against infection, and their numbers increase in all *acute* infections. They pour out an enzyme that liquefies dead tissue. The eosinophils are concerned in the allergic state and the basophils in heparin (anticoagulant) formation.

Agranular leucocytes. These are of two types and are so named because there are no granules in the cytoplasm.

1. *Lymphocytes.* These form about 30 per cent of the total white cell count. They are made in the lymphoid or reticulo-endothelial tissue throughout the body, and are numerous in the fluid called lymph. They are concerned in the formation of antibodies and are the chief source of the body's globulins. Their numbers increase in *chronic* infections, and a differential white cell count is sometimes performed to confirm a diagnosis. The thymus gland plays, as yet a little understood part, in the immunological competence of lymphocytes. Lymphocytes are thought to be responsible for transplant organ rejection. Drugs that depress the lymphocyte count are used in an attempt to overcome rejection. An alternative is to use antilymphocytic globulin (ALG) sometimes prepared as antilymphocyte serum (ALS).

2. *Monocytes.* These are the largest cells in the blood, but are few in number. Like the polymorphonuclear leucocytes, the monocytes are phagocytic.

BLOOD PLATELETS OR THROMBOCYTES

These are fragments of protoplasm concerned in the clotting of blood. They too are made in the red bone marrow, and their numbers are intermediate between the red and white cells, i.e. 250 000 in each cubic millimetre of blood. Their normal life span is 8 to 10 days, but this may be reduced in disease. An increase is thought to predispose to thrombosis, and a decrease leads to prolonged bleeding time. Current research suggests that platelets possess immunological properties.

The Clotting of Blood

Blood does not normally clot within its enclosed circuit, but it is very sensitive to damage—when it does clot inside the circuit. Heart or vessel walls, roughened from any cause, e.g. inflammation or the deposit of fatty (atheromatous) plaques, encourages clotting. If the clot stays where it is formed, it is called thrombosis, should a portion become detached and float away in the blood, it is an embolism, and it will produce symptoms wherever it lodges. Clotting is a highly complex business, but the essential events are shown in Figure 84.

148

It is a subject that engages the bio-engineers making heart valves, synthetic vessel transplants, artificial hearts and kidneys, etc.

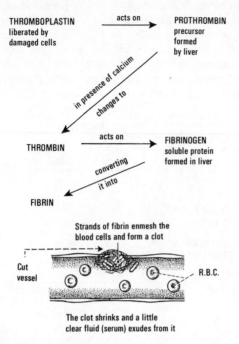

Fig. 84   Diagram of blood clotting.

Blood Groups

These are based on the agglutination or clumping of red blood cells. This is produced by two substances, an agglutinogen in the red blood cells, and an agglutinin in the plasma. The agglutinogens are called A and B and are inherited as Mendelian dominants. The agglutinins are α (alpha) and β (beta) and are inherited as Mendelian recessives. Agglutinin α in the plasma, agglutinates red cells containing agglutinogen A, so that these cannot be present in one person. Similarly, agglutinin β in the plasma, agglutinates red cells containing agglutinogen B, so that they cannot both be present in the same person. Everyone must possess two of the factors—A, B, α, β. This gives rise to the combinations shown in Figure 85.

Blood Group	Agglutinogen in Red Cells	Agglutinin in Plasma
A	A	β
B	B	α
AB	AB	0
0	0	$\alpha\beta$

Fig. 85   Blood groups.

149

In transfusing blood, the serum of the donor is ignored, as it is soon diluted by the larger amount of serum in the recipient. It is the effect of the serum of the recipient on the red blood cells of the donor which is important. The result of transfusion between the various blood groups is shown in Figure 86, x denoting clumping, i.e. incompatibility. Group A can give to groups A and AB; group B can give to groups B and AB; group AB can give only to group AB, and group O can give to any group. Group AB is therefore a universal recipient and group O is a universal donor.

DONOR	RECIPIENT			
	Group A $A\beta$	Group B $B\alpha$	Group AB ABO	Group O $O\alpha\beta$
Group A $A\beta$	√	x	√	x
Group B $B\alpha$	x	√	√	x
Group AB ABO	x	x	√	x
Group O $O\alpha\beta$	√	√	√	√

Fig. 86   Possible result of blood transfusion.

There are many other blood groupings of which the best known is the Rhesus (Rh) group. Eighty-five per cent of human beings are Rh positive, the remaining 15 per cent being Rh negative. The unborn child of an Rh negative mother and an Rh positive father, may inherit Rh positive red cells from the father. In such a situation the mother may become immunized by the periodic escape of fetal red cells across the placenta into her circulation. In response to these, she produces anti-Rh antibodies which cross the placenta into the fetal circulation, where they can break down (haemolyse) the fetal red cells. Such haemolytic disease of the newborn is rarely encountered in a first pregnancy. It is thought that the immune reaction occurs late in pregnancy and during delivery, so that once the mother is immunized, the risk of a further Rh positive fetus may increase with each succeeding pregnancy. Fortunately incompatibility in the ABO groups between mother and fetus helps to minimize the risk of maternal Rh immunization, as it enables the mother to destroy fetal red cells escaping into her circulation *before* the Rh antigen can become effective. Preventive measures can now be taken, and Rh negative mothers at risk, immediately after delivery, are given anti-Rh serum by injection. This combines with the fetal cells in the mother, resulting in their destruction, before they have stimulated antibody formation.

With this elementary introduction to blood we are now ready to discuss the pump that distributes this life-giving fluid.

THE HEART

The heart is likened to a double-sided pump.

A little knowledge has already been gained of the muscle making up this organ (p. 106), and now we must put some more pieces of the jigsaw in position around it.

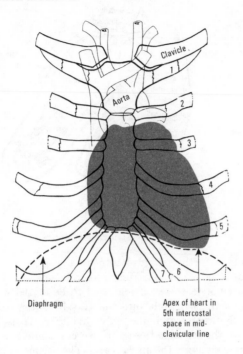

Diaphragm

Apex of heart in
5th intercostal
space in mid-
clavicular line

Fig. 87   Position of heart in relation to the anterior chest wall.

POSITION AND RELATIONSHIPS

The heart occupies the greatest part of the mediastinum in the thoracic cavity. The wider base of this cone-shaped organ is upper-most and projects slightly to the right of the sternum; the narrow apex is below, and projects more to the left of the sternum.

Anteriorly. The sternum and costal cartilages lie in front of the heart.

Posteriorly. The oesophagus, aorta and lungs separate the heart from the vertebral column.

Inferiorly. The heart rests on the diaphragm, its apex striking the chest wall at the fifth intercostal space in the mid-clavicular line. (A nurse learns to take and to chart this apex beat.)

Superiorly. The aorta arches over the top of the heart from right to left, so that in close relationship are the large arteries given off from the aortic arch. The trachea divides into two bronchi.

Laterally. A lung lies on each side of the heart, the cardiac notch of the left lung being much deeper than that of the right lung.

151

The hollow cone is divided into a right and left half by a septum composed of fibrous tissue. In two places there is special tissue called the atrioventricular node and bundle (Fig. 88).

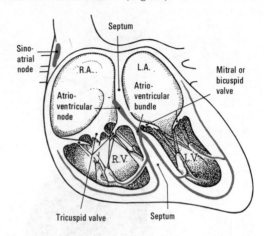

Fig. 88   Cardiac septum showing atrioventricular node and bundle.

Each half is further divided into an upper atrium and a lower ventricle by a valve, so constructed as to be open whilst blood flows from the atrium to the ventricle, and to be closed whilst blood is pumped from the ventricle into the vessel leading from it. On the right side the atrioventricular valve has three cusps and is called tricuspid; on the left side there are two cusps and it is called bicuspid, or more commonly the mitral valve. The perimeter of these valves is attached to the ring of fibrous tissue which separates the atria from the ventricles. The free edges have strong cords (chordae tendinae) attached to them, and these cords blend with papillary muscles projecting slightly from the ventricular wall. The increased pressure before the onset of ventricular contraction forces the valve flaps up into position, so that they form an effective barrier between the atrium and the ventricle, the cords preventing the flaps from going too far.

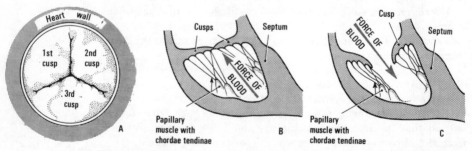

Fig. 89   Diagram of an atrioventricular valve.   A, Seen from above when closed.   B, In section when closed.   C, In section when open.

152

The heart and valves are lined with endocardium to provide a smooth surface so that blood will not clot.

The middle layer of the heart is composed of cardiac, involuntary muscle tissue and can be called the myocardium. It is thinnest in the two atria as they are receiving chambers and merely have to squeeze the blood into the adjacent ventricles. It is thicker in the right ventricle which has to pump blood to the nearby lungs, and it is thickest in the left ventricle which has to pump blood round the whole body. Independent of its nerve supply cardiac muscle can contract automatically and rhythmically at a slow rate which is insufficient to maintain an adequate circulation to the tissues. Cardiac muscle has therefore to be accelerated by the sympathetic and decelerated by the parasympathetic (vagus nerve) portion of the autonomic nervous system to meet the varying needs of the body from moment to moment.

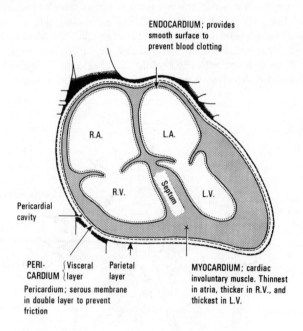

Fig. 90   Diagram of heart to show structure of walls.

The outer layer of the heart is of serous membrane so that all the movement can take place without friction. It is caller pericardium and that layer that closely invests the heart is called the visceral layer; that which is reflected to form a loose sac is called the parietal layer. The potential space between these two is the pericardial cavity. Both layers of pericardium are strengthened by fibrous tissue on the outside.

The superior vena cava returns deoxygenated blood from the upper body to the right atrium, and it enters the upper portion of this chamber. Deoxygenated blood is blood that has circulated through-

153

out tissues, is not devoid of oxygen, but contains diminished oxygen. Some people prefer the term venous blood. The exception to this is that the pulmonary veins return freshly oxygenated blood to the heart. The inferior vena cava returns deoxygenated blood from the lower body to the right atrium, and it enters the lower portion of this chamber—only a very small bit of the inferior vena cava traverses the thoracic cavity. Lying between these two openings the coronary sinus returns deoxygenated blood from the heart's own walls. The right atrium relaxes to allow all this blood to enter (Fig. 92, A); the weight forces open the tricuspid valve, the blood flows into the right ventricle, which is relaxed to receive it (Fig. 92, B). The atrium contracts to squeeze out any remaining blood. The weight of ventricular blood forces up the cusps of the tricuspid valve, so that blood cannot flow back. The right ventricle then goes into contraction, opening the pulmonic semilunar valve as it does so (Fig. 92, C). The blood flows in the pulmonary artery which rises over the *front* of the atria and divides into the right and left pulmonary arteries (Fig. 91). The deoxygenated blood reaches the fine capillaries interlacing the single-celled walls of the alveoli in each lung, and is thus in close contact with the alveolar air (Fig. 186). By diffusion the blood loses 4 to 6 volumes per cent of its carbon dioxide and gains 5

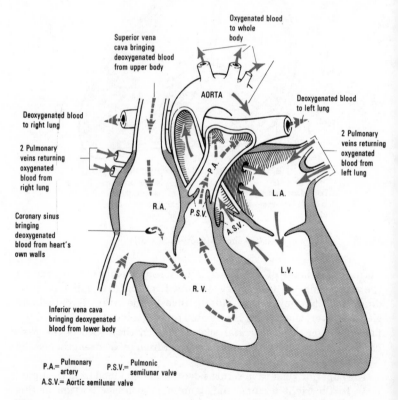

P.A.= Pulmonary artery   P.S.V.= Pulmonic semilunar valve
A.S.V.= Aortic semilunar valve

Fig. 91   Diagram showing the vessels entering and leaving the heart.

to 7 volumes per cent oxygen so that it returns to the left atrium as oxygenated blood via the four pulmonary veins. As there are two of these from each lung they enter the left atrium at opposite sides (Fig. 91). A model of a heart, or a butcher's specimen of an animal's heart is essential to get these points straight. Again the left atrium is relaxed to receive this oxygenated blood, the weight of which forces open the mitral valve. The blood flows into the left ventricle which is relaxed to receive it. The atrium contracts to squeeze out any remaining blood. The weight of ventricular blood forces up the cusps of the mitral valve, so that blood cannot flow back. The left ventricle then goes into contraction, opening the aortic semilunar valve as it does so. The oxygenated blood flows into the aorta to be distributed to all parts of the body. The aorta rises from the upper corner of the left ventricle and passes over the *front* of the atria to arch over the top of the heart.

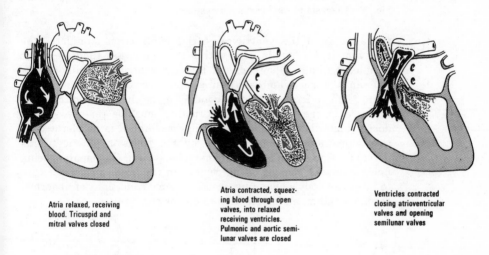

Atria relaxed, receiving blood. Tricuspid and mitral valves closed

Atria contracted, squeezing blood through open valves, into relaxed receiving ventricles. Pulmonic and aortic semilunar valves are closed

Ventricles contracted closing atrioventricular valves and opening semilunar valves

Fig. 92   Diagram to illustrate the flow of blood through the heart.

BLOOD SUPPLY

As the aorta leaves the left ventricle it gives off the coronary arteries which supply the heart's own walls with oxygenated blood.

VENOUS DRAINAGE

Via the coronary sinus entering the right atrium.

NERVE SUPPLY

The sino-atrial node in the right atrium at the base of the superior vena cava is the pacemaker. It initiates stimuli, which then spread over the whole of the atrial walls to produce atrial contraction, after which they are dampened out at the fibrous ring which separates

155

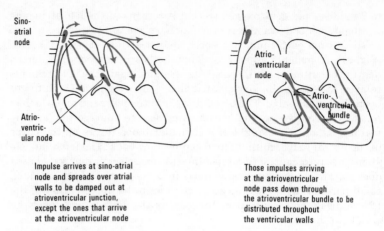

Impulse arrives at sino-atrial
node and spreads over atrial
walls to be damped out at
atrioventricular junction,
except the ones that arrive
at the atrioventricular node

Those impulses arriving
at the atrioventricular
node pass down through
the atrioventricular bundle to be
distributed throughout
the ventricular walls

Fig. 93   Diagram of the heart's nerve supply.

the atria from the ventricles. Atrial contraction stimulates the
atrioventricular bundle lying within the septum. At the base of the
bundle, the impulse spreads out into the ventricular walls (rather
like an upturned umbrella) to produce ventricular contraction.
Things can go wrong with this conducting mechanism and these are
reflected in the pulse. Any abnormality must be reported to the
doctor immediately. This conducting mechanism is accelerated by
the sympathetic and decelerated by the parasympathetic (vagus
nerve) portion of the autonomic nervous system, the impulses being
finely adjusted by controlling centres in the brain and spinal cord
to meet the body's changing needs. In some conditions of abnormal
cardiac rhythm, an 'artificial' pacemaker can be used.

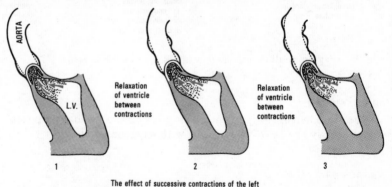

The effect of successive contractions of the left
ventricle on the aorta

Fig. 94   Diagram illustrating the stretch and recoil of aortic elastic fibres.

Arteries

Vessels which carry blood from the heart. They are widest as they
leave the heart and become progressively smaller as they traverse
the tissues, becoming arterioles which gradually shed their coats
until they are single-walled capillaries. Arteries are lined with simple

endothelium for its smooth surface. The middle coat is composed of elastic and plain involuntary muscle fibres, the former being in predominance in the large vessels, for they have to stretch into an egg-like swelling as 70 ml (2 oz) of blood are pumped out into them at each ventricular contraction (Fig. 94). As the ventricles relax to receive more blood, the arterial elastic fibres return to their normal length, and in so doing squeeze blood along the tube so that adjacent elastic fibres become stretched and yet another egg-like swelling appears, and because of this mechanism the rate of the heart beat can be estimated by palpation of an artery conveniently overlying a bone at some distance from the heart. We speak of this as 'taking the pulse'. This wave-like movement together with arteriole resistance causes the tissues to receive a *continuous flow* of blood as opposed to a jet spray at intervals. The outer layer of fibrous collagenous tissue blends with the outer layer of the large veins that are found alongside the large arteries. It is because of this close proximity that a 'pulse' can sometimes be felt in a vein, though a vein does not pulsate in the usual meaning of the word. The lumen of the arteries is constricted and widened by sympathetic and para-sympathetic nerves. Binding arterial elastic and muscle fibres on the outside is a layer of white fibrous tissue.

Arterioles

These are the smaller arteries, and the three coats become thinner, until finally the two outer ones disappear, and at this stage the vessel is called a capillary.

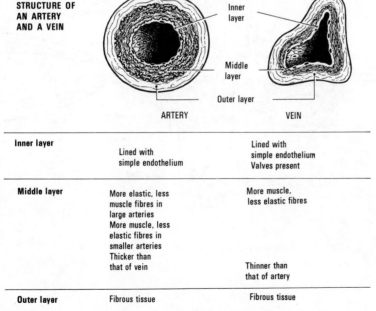

STRUCTURE OF AN ARTERY AND A VEIN

Inner layer

Middle layer

Outer layer

ARTERY   VEIN

	ARTERY	VEIN
Inner layer	Lined with simple endothelium	Lined with simple endothelium Valves present
Middle layer	More elastic, less muscle fibres in large arteries More muscle, less elastic fibres in smaller arteries Thicker than that of vein	More muscle, less elastic fibres Thinner than that of artery
Outer layer	Fibrous tissue	Fibrous tissue

Fig. 95   Chart showing the comparison of an artery with a vein.

Capillaries

Fine, single-walled, hair-like tubes that connect an arteriole to a venule, which eventually becomes a vein. The pressure of the blood in the capillary is sufficient to force the fluid part of the blood through the single layer of cells, to bathe the tissue cells. The plasma proteins (6 to 8 per cent) have too large a molecule to be pushed out, so they, together with all the blood cells, stay inside the capillary lumen and provide the sucking power (osmotic pressure) to get the fluid back into the lumen at the venous end of the capillary network. Meantime this fluid has given nutriments and oxygen to the tissue cells and absorbed their waste products, which are returned to the blood stream to be distributed to the appropriate organ of excretion.

Tissue fluid accounts for 12 litres of the body's water, only about 3·5 litres are contained in the total blood stream. Some tissue fluid is drained away by a nearby lymphatic capillary. Oedema is the name for excess fluid in the tissues. You can now work out some causes of oedema.

Veins

Vessels which carry blood back to the heart, in many instances against the force of gravity. They are smallest at their union with the venules and become increasingly larger as they near the heart.

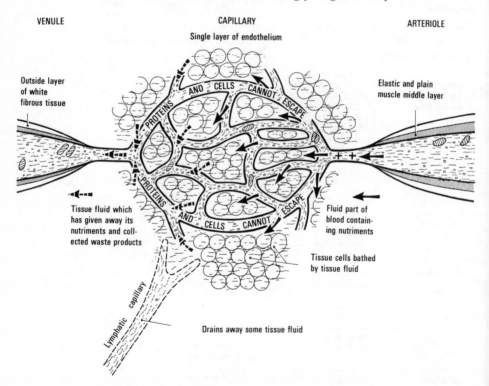

Fig. 96   Illustration of the physiology at a capillary network.

In their endothelial lining they have valves, usually two cusps, to prevent the backward flow of blood. Veins are designated deep or superficial according to whether they are inside or outside the muscle fascia of a limb. The deep veins usually accompany the main arteries and often take the name of the artery.

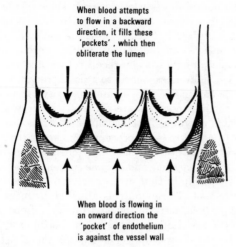

When blood attempts to flow in a backward direction, it fills these 'pockets' , which then obliterate the lumen

When blood is flowing in an onward direction the 'pocket' of endothelium is against the vessel wall

Fig. 97   A valve as seen when the vessel is split and laid open.

The residual force imparted by heart's contraction helps to force blood back to the heart. In the legs the contraction and relaxation of the big powerful muscles also helps to 'milk' blood upwards. Descent of the diaphragm on inspiration not only sucks air down into the lungs but it simultaneously sucks blood up from the lower limbs. Stagnation of blood in the lower limbs is fraught with danger; most people who are up and about manage to avoid this condition, but in those who are confined to bed it is an ever-present threat. All patients should be advised and encouraged to take six deep breaths, and to describe a circle with each big toe six times each hour.

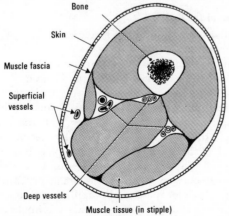

Bone

Skin

Muscle fascia

Superficial vessels

Deep vessels

Muscle tissue (in stipple)

Fig. 98   Cross-section of a limb illustrating 'deep' and 'superficial' vessels.

159

Blood Pressure

(See also p. 374, blood pressure in the kidney.)

This is defined as the pressure exerted by the blood against the vessel walls. This cannot possibly be the same throughout the enclosed cardiovascular system at any given time, because the circumference of the containing structures (two sides of the heart, large arteries, smaller arteries, arterioles, capillaries, smaller veins and large veins) is so different. Furthermore, the walls of these structures contain plain muscle capable of contracting, thereby decreasing the lumen, and elastic fibres that by recoil return a stretched vessel to its normal circumference. Added to this there are slight changes throughout the day in the volume and viscosity of the blood, so one can see that complex controls are necessary to keep blood pressure at an acceptable average. These complex controls are—nerves that are sensitive to *pressure* (pressoreceptors); nerves that are sensitive to the *chemical composition* of the blood (chemoreceptors); cells that are sensitive to the *hormone* and *enzyme* content of the blood (p. 142); and cells in the medulla that are sensitive to *emotion*, particularly fear and excitement. These controls allow adjustments in response to changes in posture, activity and environment. Blood pressure is an indication of the resistance exerted by the peripheral vessels to the blood flow. Increased resistance means an increased pressure to force blood round the body. The changes occur so smoothly that it is difficult to detect them, and for the research worker this presents difficulty in separating the *receptor* and *effector* activity in control of circulation—and thus of blood pressure.

Blood pressure is at its maximum during systole (contraction of the heart). In a healthy adult at rest, the systolic pressure is about 120 mm of mercury. Blood pressure is at its minimum during diastole (resting of the heart), and the diastolic pressure is about 80 mm of mercury. In the process of ageing the arteries lose their elasticity and this raises the blood pressure.

The Main Arteries of the Body

The aorta, the largest artery in the body, leaves the left ventricle of the heart. It immediately gives off a pair of coronary arteries to supply the heart walls. It rises over the front of the heart and arches from right to left, to pass down behind the heart as the descending thoracic aorta. The first branch given off from the arch is the brachiocephalic (innominate) artery, which soon divides into the right subclavian to supply the right arm, and the right common carotid to supply the right neck and head. The second branch given off from the arch of the aorta is the left common carotid to supply the left neck and head. The third branch given off from the arch of the aorta is the left subclavian. Each subclavian artery, after traversing underneath the clavicle, reaches the axillary space and takes the name of the axillary artery. As this reaches the humerus it becomes the brachial artery, and at the bend of the elbow divides to form the

160

radial (lateral) artery, and the ulnar (medial) artery. In the palm these vessels anastomose to form the deep and superficial palmar arches, from which the digital branches are given to each digit.

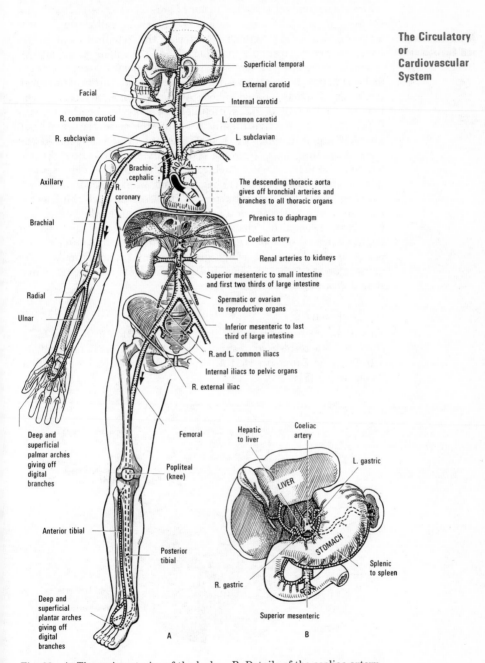

Superficial temporal

External carotid

Facial

Internal carotid

R. common carotid

L. common carotid

R. subclavian

L. subclavian

Axillary

Brachio-cephalic

R. coronary

The descending thoracic aorta gives off bronchial arteries and branches to all thoracic organs

Phrenics to diaphragm

Brachial

Coeliac artery

Renal arteries to kidneys

Superior mesenteric to small intestine and first two thirds of large intestine

Radial

Spermatic or ovarian to reproductive organs

Ulnar

Inferior mesenteric to last third of large intestine

R. and L. common iliacs

Internal iliacs to pelvic organs

R. external iliac

Deep and superficial palmar arches giving off digital branches

Femoral

Hepatic to liver

Coeliac artery

L. gastric

Popliteal (knee)

LIVER

Anterior tibial

STOMACH

Posterior tibial

Splenic to spleen

Deep and superficial plantar arches giving off digital branches

R. gastric

A

Superior mesenteric

B

Fig. 99 A, The main arteries of the body. B, Details of the coeliac artery.

161

After the arch, the aorta descends behind the heart and gives off branches to all the structures contained in the thoracic cavity, those taking oxygenated blood to the lungs being the bronchial arteries.

The aorta pierces the diaphragm, and the first branches to be given off by the abdominal aorta are a pair of phrenic arteries to supply the diaphragm. A wide tube leaving the aorta is called the coeliac artery, and it divides into three arteries, the hepatic to supply the liver, the gastric to supply the stomach and the splenic to supply the spleen. The superior mesenteric artery is the next to leave the aorta and it supplies the small intestine and the first two-thirds of the large intestine with oxygenated blood. A pair of renal arteries to the kidneys, and a pair of spermatic (male) or ovarian (female) arteries supply the reproductive organs. The inferior mesenteric artery is given off and it supplies the last third of the large intestine. At the level of the iliac crest the aorta ends by dividing into the right and left common iliac arteries. Each common iliac artery divides into an internal (to supply the pelvic organs) and an external iliac artery, which becomes the femoral artery as it passes through the femoral ring (see Fig. 62). The femoral artery passes down the thigh, and as it traverses the popliteal space behind the knee it takes the name of popliteal artery. Just below the knee it divides into the anterior and posterior tibial arteries, which give off many branches and finally anastomose to form the deep and superficial plantar arches which send digital branches to the digits.

CIRCLE OF WILLIS

This is the name given to the intricate circulation at base of brain from which arteries rise to supply the brain. The anastomosis of the arteries is so extensive that all the brain tissue receives a double blood supply (Fig. 100).

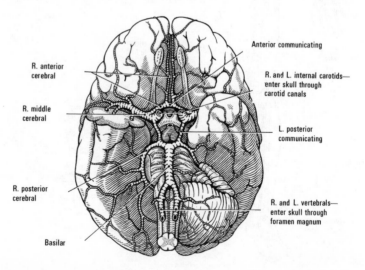

Fig. 100 Arteries forming the circle of Willis and supplying the brain.

Man, his Health
and Environment

162

Drainage of blood from the brain is accomplished by venous channels or sinuses (Fig. 101). These are either extradural as in Figure 101, or those within the brain tissue that have dense connective tissue walls and cannot distend or collapse. They do not possess valves. These channels or sinuses finally link together and pour their contents into the internal jugular vein, which leaves the base of the skull via the jugular foramen, to pass down the neck and join the subclavian vein to form the brachiocephalic (innominate) vein on either side. Union of the right and left brachiocephalic (innominate) veins forms the superior vena cava, which enters the right atrium of the heart (Fig. 102).

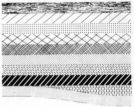

Skin and superficial fascia
Deep fascia
Skeletal muscle
External periosteum of skull
Outer table of compact bone
Marrow-containing cancellous bone
Inner table of compact bone
Internal periosteum of skull
Deoxygenated blood in venous sinus
Dura mater—blends with periosteum, except where there are venous sinuses

Fig. 101   Diagrammatic representation of extradural venous sinuses of skull.

Amongst the venous sinuses, the superior longitudinal or sagittal one is important since the pulse can be taken here in the first 18 months of life. The contained venous blood transmits pulsation from the arteries immediately below. The superior longitudinal sinus can also be used as a site for intravenous infusion (see p. 48).

The cavernous sinuses lie one on either side of the body of the sphenoid bone. They drain blood from the nose and mouth so that sepsis in these areas has an added danger (see Fig. 102).

Venous Drainage from the Upper Limbs

The deep tissues are drained by the radial and ulnar veins which unite just below the elbow to form the deep brachial, which becomes the deep axillary vein.

The lateral superficial tissues are drained by the cephalic vein, which below the elbow gives off the median cubital vein (the most common site for intravenous injection) to join the basilic vein at the medial aspect of the elbow. The median and basilic veins drain the remaining superficial tissues of the forearm, uniting just below the elbow, and continuing as the basilic, which communicates with the deep axillary vein (Fig. 103).

The deep axillary vein becomes the subclavian vein, which receives the jugular vein to become the brachiocephalic (innominate) vein,

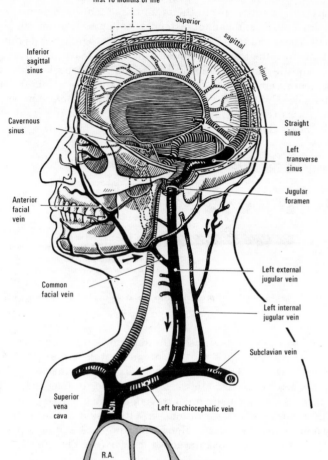

Fig. 102   Diagram illustrating venous drainage from brain, head and neck.

the two brachiocephalic (innominate) veins joining to form the superior vena cava which enters the right atrium of the heart.

Venous Drainage from the Lower Limbs

Most of the blood is returned via the deep anterior and posterior tibial veins, which unite below the knee to form the popliteal, which

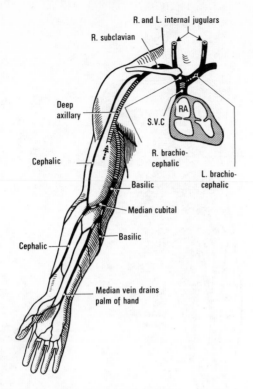

Fig. 103   Venous drainage from the right arm.

traverses up the thigh as the deep femoral vein. The deep popliteal receives the superficial short saphenous vein (Fig. 104), and the deep femoral receives the superficial long saphenous vein, before the deep femoral vein enters the femoral ring, to become the external iliac, which receives the internal iliac to form the common iliac vein. The two common iliac veins unite to form the inferior vena cava, which traverses alongside the aorta and receives many more branches until it finally pierces the diaphragm to enter the right atrium of the heart.

FACTORS AFFECTING THE CARDIOVASCULAR SYSTEM

EXERCISE

Experiments have proved that during exercise there is an increase in the rate of the heart beat, and there is an increase in the amount of blood leaving the heart at each ventricular contraction. Exercising tissues therefore have an improved blood supply and this gives a feeling of wellbeing. In those veins which are surrounded by skeletal muscle, venous stasis is further prevented by the milking action of the muscles against the vein walls. This will help to prevent thrombosis.

165

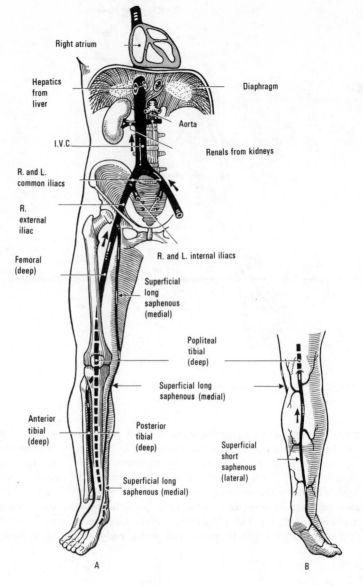

Right atrium

Hepatics from liver

Diaphragm

I.V.C.

Aorta

Renals from kidneys

R. and L. common iliacs

R. external iliac

Femoral (deep)

R. and L. internal iliacs

Superficial long saphenous (medial)

Popliteal tibial (deep)

Superficial long saphenous (medial)

Anterior tibial (deep)

Posterior tibial (deep)

Superficial short saphenous (lateral)

Superficial long saphenous (medial)

A   B

Fig. 104   Venous return from lower limbs. A, Anterior view. B, Posterior view.

REST

Metabolic processes are at a minimum whilst the body rests, consequently the heart can function more slowly. Worn-out cells can be replaced, and the tissues are refreshed. Each individual finds a posture conducive to rest; a reclining position suits most people, and if the feet are raised it assists the return of blood to the heart from the lower limbs.

Prolonged rest can produce such a slowing of the circulation (venous stasis) in the lower half of the body, that the blood clots (venous thrombosis). A portion of blood clot may become detached and travel (as an embolus) via the ever-increasing bore of the veins until the heart is reached. The heart is sufficiently wide for the embolus to pass through, then it leaves the right ventricle in the pulmonary artery, which is of ever-decreasing bore, so that finally the clot sticks as a pulmonary embolism, the area deprived of blood being an infarct. Prolonged rest also encourages constipation, which can cause varicosity of the rectal veins (piles, haemorrhoids).

SLEEP

Sleep gives the maximum rest of which the human being is capable. Since few people sleep more than eight hours daily, there cannot be any ill-effects, as with prolonged rest.

POSTURE

When standing, gravity opposes the return of blood to the heart from the lower half of the body, and when continued over a period of time can result in lack of blood to the brain, with consequent fainting. Deep breaths and moving of the toes when on parade will help to offset this. Occupations that involve long periods of standing predispose to varicose veins.

EMOTION

In response to such emotions as fright, fear and anger, the heart beats more rapidly and more forcibly, so that the owner may become conscious of it pounding against the chest wall (palpitation). It is for this reason that you will be taught to advise cardiac patients and their families to avoid emotional upheavals. Depression and dejection can cause slowing of the heart beat.

SHOCK AND/OR HAEMORRHAGE

Shock and/or haemorrhage produces a fall in blood pressure. Initially the pulse may be bounding, later it increases in rate and decreases in volume. These effects are mainly due to a reduction in the volume of circulating blood. In shock, the increased potassium in the blood (p. 229) can cause sudden heart failure.

CARE OF THE CARDIOVASCULAR SYSTEM IN THE MAINTENANCE OF HEALTH*

Take a diet which will allow optimum production of blood, especially iron as that in food is much more readily absorbed than medicinal iron. Red meat, eggs and green vegetables all provide iron. Other

* Care for Your Heart. Kenneth Mason Publications Ltd. 10p.

essential food factors are protein, a trace of copper, ascorbic acid (vitamin C), folic acid (part of vitamin B complex), cyanocobalamin (vitamin B_{12}) and menadione (vitamin K) (p. 224). Avoid tight garments which will impede the return of blood to the heart from the legs. 'Pantie-girdle' syndrome is the name given to such impediment. Avoid pressure on the posterior surface of the thighs when sitting or lying, as this contributes to poor venous return, and the resultant stasis encourages thrombosis. Take some deep breaths and be sure to have sufficient leg movement throughout the day. Avoid sitting with the knees crossed for long periods. When standing still for long periods, e.g. on parades, move the toes at frequent intervals to assist return of blood to the heart thereby improving supply to the brain, thus preventing the feeling of faintness and varicose veins. Avoid obesity as it predisposes to varicose veins. Avoid constipation as it can cause haemorrhoids (piles: varicose veins around the anus). Avoid smoking as it thickens vessel walls and raises the blood pressure. Take some exercise in the fresh air daily. Avoid stress. Don't take office work home. Take proper holidays. The Office of Health Economics suggests that men at risk of high blood pressure should be screened, so that necessary preventive treatment can be applied.

Avoid walking in bare feet: (1) in areas known to be infested with hookworm for infestation in the human being can cause gross anaemia: (2) in snail-infested areas, for the snail is the intermediate host of Schistosoma and infestation of the human being causes severe anaemia.

Accidents frequently cause shock and/or haemorrhage, therefore it behoves each citizen to take precautions against accident, especially in the home, at work and on the roads. Since rapid loss of blood can kill a healthy person quickly, the ability to stop bleeding (first aid) helps to prevent severe circulatory disturbances; external cardiac massage can re-start beating of a heart that has suddenly failed.

With our present state of knowledge it would seem reasonable to avoid excessive refined sugar (thought but not proved to be atherogenic) and animal fat in the diet, and where possible to substitute vegetable fat.

SUMMARY OF THE CARDIOVASCULAR SYSTEM

THE BLOOD

There are 6 litres (10 pints) in the body and 600 ml (1 pint) can be lost without discomfort.

PLASMA

Contains 90 per cent water in which is dissolved amino acids for distribution, 6 to 8 per cent plasma proteins—albumin, globulins and fibrinogen; fatty acids, glycerine, cholesterol, glucose, vitamins,

168

calcium, phosphorus, sodium, potassium, iron, iodine, chlorides and bicarbonate. The elements are referred to as electrolytes when in solution. Plasma also contains the waste products carbon dioxide, creatinine, urea and uric acid. Bile pigments, oxygen, hormones, enzymes, antibodies and anticoagulants are found in the plasma.

RED BLOOD CELLS

Erythrocytes. These are tiny, circular, biconcave discs about 7 microns in diameter. They have no nucleus and therefore cannot reproduce. They live 112 days and are then destroyed in the blood vessels, liver and spleen, the resultant bilirubin is excreted in bile and the iron is saved for further red blood cells. They need protein, iron, copper, folic acid, vitamins C and B_{12} for their production which is in the red bone marrow. They contain haemoglobin which has a great affinity for oxygen, with which it readily combines to form oxyhaemoglobin. This change occurs in the lungs, and as the red cells go round the body they release their oxygen and return to the lungs to collect more. There are five million in each cubic millimetre of blood.

WHITE BLOOD CELLS

Leucocytes. There are 6 000 to 8 000 in each cubic millimetre of blood.

Granular leucocytes. These are made in the bone marrow and have a many-shaped nucleus—polymorphonuclear cells. Neutrophils possess the power of amoeboid movement and can thus escape from the vessel to engulf surrounding bacteria. Their numbers increase in *acute* infections. They pour out an enzyme that liquefies dead tissue. Basophils form heparin, an anticoagulant. Eosinophils are concerned in the allergic state.

Agranular leucocytes. Lymphocytes make up the bulk of these cells, and are made in the reticulo-endothelial tissue throughout the body. They are numerous in the fluid called lymph. They are concerned in the formation of antibodies and are the chief source of the body's globulins. Their numbers increase in *chronic* infections.

BLOOD PLATELETS

Thrombocytes. These are protoplasmic fragments concerned in the clotting of blood. There are 250 000 per cubic millimetre of blood. An increase is thought to predispose to thrombosis, while a decrease leads to prolonged bleeding.

THE HEART

This lies behind the sternum, occupying a large part of the mediastinum in the thoracic cavity. It is divided into a right and left side by the septum. The right atrium receives deoxygenated blood from the venae cavae and the coronary sinus, and forces it into the right ventricle via the tricuspid valve. From the right ventricle the blood is forced through the pulmonic semilunar valve into the pulmonary artery for conveyance to the lungs for oxygenation. This freshly oxygenated blood returns to the left atrium by way of the four pulmonary veins. It is forced through the mitral valve into the left ventricle, from which it is forced through the aortic semilunar valve into the aorta for distribution around the body.

Endocardium lines the heart and valves; myocardium forms the middle layer and is of cardiac involuntary muscle fibres, which increase in rate of contraction when stimulated by sympathetic nerves, and decrease in rate of contraction when receiving parasympathetic stimuli. The myocardium is thinnest in the two atria, thicker in the right ventricle and thickest in the left ventricle. It has the special property of automatic, rhythmical contraction at a slow rate independent of its nerve supply. The serous membrane—pericardium—is found on the outside of the heart as a double sac, that closely investing the organ being the visceral, and that reflected to form the sac the parietal layer. The potential space between these two is the pericardial cavity.

ARTERIES

These are vessels which carry blood from the heart, being widest as they leave this organ, and becoming progressively smaller until they become arterioles, which shed their coats and are finally single-walled capillaries. The arteries have an endothelial lining to prevent blood clotting. Their middle coat is thicker than that of a vein, being composed of elastic and plain involuntary muscle fibres. The elastic fibres predominate in the large arteries and are stretched as 70 ml (2 fl oz) of blood are pumped out of the heart at each ventricular contraction. Whilst the ventricles are relaxing to receive more blood, the elastic fibres are returning to their normal length, so that blood is continually being squeezed along the artery and the tissues receive a constant flow of blood instead of an intermittent jet spray. The plain involuntary muscle fibres are innervated by grey neurones from the autononic nervous system.

CAPILLARIES

These are fine, hair-like vessels connecting arterioles to venules. Capillaries have a single layer of endothelial cells so that the liquid part of blood, minus the plasma proteins, can be pushed out to nourish the tissue cells. The plasma proteins and the blood cells stay inside the lumen and create the osmotic pressure whereby the tissue fluid is sucked back into the venous end of a capillary network.

VEINS

These are vessels which carry blood back to the heart, having the same three coats as an artery. The endothelial layer has valves to assist the flow of blood against gravity. The middle layer is not as thick as that of the artery, and has some plain muscle fibres to support the column of blood.

Figures 99 to 104 summarize the blood supply to, and the venous return from, the various parts of the body.

THE EFFECT OF EXERCISE ON THE CARDIOVASCULAR SYSTEM

1. Exercise increases the rate of the heart beat.

2. It increases the amount of blood leaving the ventricles at each contraction.

3. The improved blood supply to muscles gives a sense of well-being.

4. It prevents venous stasis, with a possible consequent thrombosis and embolism.

THE EFFECT OF REST ON THE CARDIOVASCULAR SYSTEM

1. Metabolic processes are at a minimum so that the heart beats more slowly.

2. Worn-out cells are replaced and the tissues are refreshed.

3. Prolonged rest can cause thrombosis in the lower half of the body, with the danger of consequent pulmonary embolism.

4. Prolonged rest can also cause constipation which can lead to piles.

THE EFFECT OF SLEEP ON THE CARDIOVASCULAR SYSTEM

The greatest degree of rest is achieved in sleep so that the heart and vessels have the least possible work to do.

THE EFFECT OF POSTURE ON THE CARDIOVASCULAR SYSTEM

1. Gravity opposes the return of blood from the lower half of the body.

2. After prolonged standing this can result in lack of blood to the brain with subsequent fainting.

THE EFFECT OF EMOTION ON THE CARDIOVASCULAR SYSTEM

The heart beats more rapidly and more forcibly, making the person conscious of its beating.

THE EFFECT OF SHOCK AND/OR HAEMORRHAGE ON THE CARDIOVASCULAR SYSTEM

Low blood pressure; pulse may be bounding at first, then increases in rate and decreases in volume. Sudden heart failure from increased potassium in blood.

171

CARE OF THE CARDIOVASCULAR SYSTEM IN THE MAINTENANCE OF HEALTH

Take a diet which will allow optimum production of blood, especially iron contained in red meat, eggs and green vegetables. Protein, a trace of copper and vitamins C, B and K should also be included. Excessive refined sugar and animal fat should be avoided.

Avoid any clothing or posture which will impede the return of blood to the heart, obesity, constipation, smoking and excessive stress.

The Lymphatic System

The last decade has brought increased knowledge of the highly complex lymphatic system. Lymphatic cells are the commonest in the body, probably because of their fundamental role in dealing with *immunity processes*—to maintain the integrity of the body when it is invaded by organisms, or foreign protein, including that of transplanted tissue or organ. Two components of the lymphatic system are now recognized:

1. **Central lymphatic system.** Includes the ring of lymphatic tissue guarding the entrances to the respiratory and alimentary tracts and the thymus gland. It is now believed that proper development of the lymphatic system depends on the thymus, without which the secondary component, i.e. the peripheral lymphatic system cannot develop adequately. In the absence of the thymus, gamma-globulins are not produced, the tonsils and adenoids do not develop, and death results from lack of normal immunological processes. This condition is called agammaglobulinaemia. A less severe form is called hypogammaglobulinaemia.

2. **Peripheral lymphatic system.** Consists of all the other lymph nodes, the spleen and the bone marrow. This triad is now called the lymphoreticular (or reticulo-endothelial) system.

The lymphatic system is composed of a network of vessels and glands, but whereas the cardiovascular system is a 'closed circuit' in which blood is transmitted from the heart, circulates round the body and back to the heart, the fine lymphatic capillaries start out in the tissue cells, draining away any excess tissue fluid, and pouring it

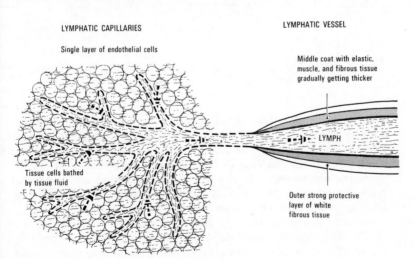

LYMPHATIC CAPILLARIES

Single layer of endothelial cells

Tissue cells bathed by tissue fluid

LYMPHATIC VESSEL

Middle coat with elastic, muscle, and fibrous tissue gradually getting thicker

LYMPH

Outer strong protective layer of white fibrous tissue

Fig. 105   Lymphatic capillaries and vessel.

173

into vessels of ever-increasing bore, until finally it is poured into the blood stream. In other words, this is a 'one-way traffic' system.

Lymphatic capillaries. Fine hair-like vessels composed of a single layer of endothelial cells. Lymphatic capillaries are arranged like the branches of a tree amongst the tissue cells, and connect up with the 'trunk' in that area. As already mentioned they drain away the tissue fluid, which has not been sucked back at the venous end of a blood capillary network.

Lymphatic vessels. These are formed from the union of several lymphatic capillaries, and in structure are similar to veins, even to having valves at intervals throughout their length. They connect with lymphatic nodes, the afferent vessel transmitting fluid *to* the node, and the efferent vessel transmitting fluid *from* the node. The specialized vessels in the small intestine are called lacteals into which digested fat is passed from the intestinal lumen.

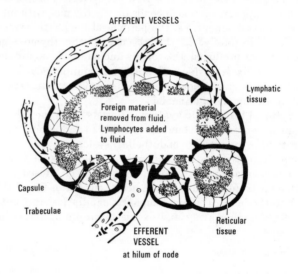

Fig. 106   Lymphatic node.

Lymphatic nodes. These have a convex and a concave border, and vary in size from that of a hemp seed to a bean. They have a fibrous capsule which is pierced by the afferent vessels entering on the convex border. Inside the capsule is lymphatic pulp which is capable of removing any foreign matter from the entering fluid, and of adding fresh lymphocytes to the fluid before it leaves the node via the efferent vessel at the hilum on the concave border.

In many places in the body lymphatic nodes are aggregated, so that each lymphatic vessel must pass through at least one node or gland. (These glands do not make a secretion and so some people prefer the term node.) A group in the groin filter lymph from the

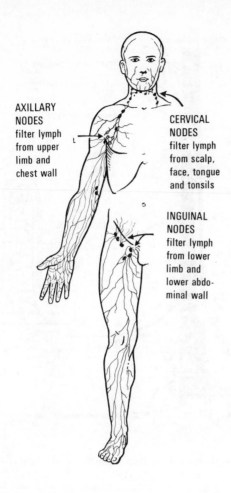

AXILLARY
NODES
filter lymph
from upper
limb and
chest wall

CERVICAL
NODES
filter lymph
from scalp,
face, tongue
and tonsils

INGUINAL
NODES
filter lymph
from lower
limb and
lower abdo-
minal wall

Fig. 107   Superficial lymph nodes.

lower limb, the pelvic cavity and the lower abdominal wall. There are many nodes throughout the intestinal mesentery, and these can be the site of tuberculous infection, if they drain the germ from the gut. Another cluster of nodes is found in the axilla, and these filter lymph from the upper limb and chest wall. Other groups are found in the neck and these filter lymph from the scalp, face, tongue and tonsil. The adenoids, faucial and lingual tonsils form a ring of lymphatic tissue guarding the entrances to the respiratory and alimentary tracts.

If the foreign material in the lymph brought to any of these nodes happens to be germs or cancer cells, the nodes themselves may become involved, and be the site of inflammation or cancer deposits. The lymph nodes form part of the lymphoreticular (reticulo-endothelial) system which is concerned with blood formation, and also with anti-infection and repair mechanisms.

175

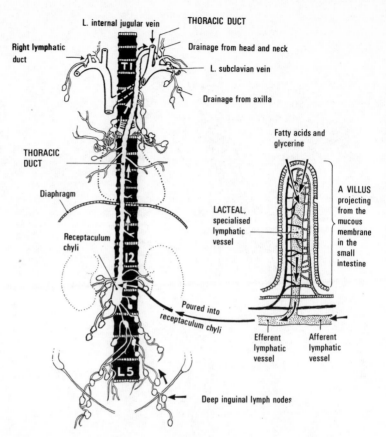

Fig. 108   A lacteal, thoracic duct, deep lymph nodes and right lymphatic duct.

LYMPHATIC DUCTS

The thoracic duct. This starts in front of the lumbar vertebrae as a dilated pouch into which several lesser dilations pour their contents. This more elaborate arrangement is necessary because digested fat, in the form of fatty acids and glycerine, is absorbed into the lacteals in the villi of the small intestine, and as this is a milky fluid (chyle), the dilated pouch is called the receptaculum chyli. The thoracic duct emerges from its upper border and rises to pass through the diaphragm at the same opening as the aorta. The duct continues up the thoracic cavity behind the heart, then swings to the left to empty its contents into the union of the left jugular with the left subclavian vein, there to be mixed with deoxygenated blood which is on its way through the right side of the heart to the lungs, before being returned to the left side of the heart for distribution round the body. The thoracic duct also drains all lymph from the lower limbs, the pelvic and abdominal cavities, the left side of the thorax, the left upper limb, head and neck.

176

The right lymphatic duct. This vessel is only about 3 to 5 cm (1½ to 2 in) long and pours its contents into the union of the right jugular with the right subclavian veins. It drains lymph from the right upper limb, chest, head and neck.

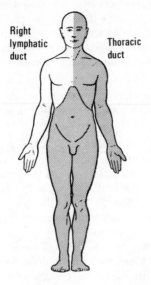

Right lymphatic duct

Thoracic duct

Fig. 109   Areas of body drained by the right lymphatic and thoracic ducts.

FUNCTIONS OF THE LYMPHATIC SYSTEM

1. To provide a constant stream of new lymphocytes which are essential in the formation of antibodies and are the chief source of globulin, both of which are important in immunity reactions.

2. To pass absorbed fat from the intestine to the blood stream.

3. To filter foreign matter from the fluid passing through the nodes.

4. By its filtration it helps to localize infection and prevent it reaching the blood stream.

5. To drain away excess tissue fluid into the blood stream.

6. Lymph nourishes avascular structures, e.g. the lower layers of the epidermis and the cornea.

7. The nodes form part of the lymphoreticular (reticulo-endothelial) system.

CARE OF THE LYMPHATIC SYSTEM IN THE MAINTENANCE OF HEALTH

Since proteins and electrolytes enter into the formation of all cells, and enzymes and vitamins are necessary to trigger off many metabolic processes, then an adequate, well-balanced diet is a prerequisite to a healthy lymphatic system. Never drink warm milk straight from the cow. Avoid injury to the exterior of the body, and do not leave things in the environment that can cause such injury, e.g. a glass bottle on the beach. Where injury has occurred avoid infection by covering immediately with an adhesive dressing. Always take adequate muscular exercise as this massages lymph along the vessels, and this is especially necessary in the legs where the lymph is being returned against gravity. Do some deep-breathing exercises daily to assist in sucking up lymph in the thoracic duct. In tropical countries it is wise to protect against mosquitoes which besides transmitting malaria can inject filariae (worms) which, in turn, can block lymphatic vessels, causing elephantiasis.

CONDITIONS WHICH CAN INTERFERE WITH THE HEALTH OF THE LYMPHATIC SYSTEM

Infestation of the head with lice (p. 129) gives rise to itching, and scratching leaves a route via which infection can enter. The occipital and cervical nodes drain this area, and they can be overcome by the germs which they are valiantly trying to filter, and so they succumb to abscess formation. Thanks to the health visitors, school nurses and health education, this condition is much less frequently found than 20 years ago.

Milk in the past has caused much ill-health and many deaths by infecting lymphatic nodes with tuberculosis from the cow. The nodes draining the tonsil and those in the mesentery were commonly the site of such infection.

Cancer cells can be spread from the primary growth by the lymphatic system, and this holds good for cancer of the lung which is of such topical interest. The nodes that become involved are those in the mediastinum, and their enlargement further embarrasses breathing. Every effort should be made to discourage young people from taking up smoking (p. 363).

The Spleen

This organ is as well discussed after the lymphatic system, for there are some similarities.

POSITION AND RELATIONSHIPS
The spleen[75] is contained in the upper, posterior abdomen on the left side. Its outer surface is domed to fit underneath the diaphragm.

Anteriorly. The lateral stomach lies in front of the upper spleen, and the 'tail' of pancreas lies in front of the lower spleen.

Posteriorly, Laterally and Superiorly. The spleen lies on the muscle fibres forming the diaphragm.

Medially. The left kidney is in contact with the medial surface of the spleen.

Inferiorly. The left colic flexure is in contact with the lowest part of the spleen.

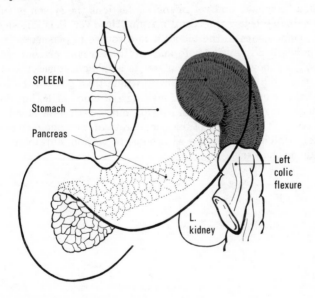

Fig. 110   Diagram showing position and relationships of the spleen.

STRUCTURE

The spleen is about the same size as the kidney, and it is a spongy organ with the lateral border in contact with the diaphragm. It is bound by a capsule of white fibrous tissue, from which strands containing plain, involuntary muscle fibres, pass into and divide up the splenic pulp. There is a great deal of reticulo-endothelial tissue

179

scattered throughout these sections and the blood comes into intimate contact with all the cells making up the spleen. The vessels enter and leave at the concave medial border.

BLOOD SUPPLY

The splenic artery, which is a branch from the coeliac artery of the abdominal aorta.

VENOUS DRAINAGE

Via the splenic vein, which unites with the gastric, superior and inferior mesenteric veins to form the portal vein going to the liver.

NERVE SUPPLY

Grey motor neurones from the autonomic nervous system.

FUNCTIONS

These are somewhat obscure, for a person can live perfectly well without a spleen, and yet in some infections the spleen is enlarged and palpable below the costal margin. However, it is known to add new lymphocytes to the blood; it is a source of plasma cells which are concerned in antibody formation; it destroys worn-out red blood cells, lymphocytes and blood platelets by phagocytosis. It conserves the iron from red cell breakdown—which then goes to the liver. It can take over the function of making red blood cells, but does not normally make them. It can squeeze out a reserve of blood in cats and goats. It forms part of the lymphoreticular (reticulo-endothelial) system which is concerned with blood formation, anti-infection and repair mechanisms.

The Nervous System

Jimmy is now respectably presenting a skin-covered exterior, but even with his skeleton, muscles, joints and skin all complete with their blood supply, he is still inert, for he has no mechanism whereby he can move his muscles. Combustion or metabolism within the muscle is necessary to produce movement. So far we are aware of the complex *interplay* of enzymes, hormones, vitamins and food in the process of metabolism. We have hinted at nerve supply to various tissues and organs, and nervous control of various functions, so that we are ready to learn more about this.

Turning back to page 13 you will revise the fact that there are four main groups of tissues making up the human body. A glance at page 15 will ensure revision of epithelial tissue, page 15 connective tissue, page 92 muscle tissue, and now we are ready to complete this list by adding nerve tissue.

NERVE TISSUE

The basic unit making up this tissue is called a neurone, which is a nerve cell together with the fibres called dendrons or dendrites, for transmission of a stimulus *to* the cell, and a fibre called an axon, for transmission of a stimulus *from* the cell. This is called the law of forward conduction (Fig. 111) and rarely is it reversed. The nervous system is made up of millions of these neurones. A nerve cell is even more complicated than the cell described at the beginning of this book. The nerve cells are surrounded by special supporting cells called neuroglia. The fibres of each neurone, that is the dendrons

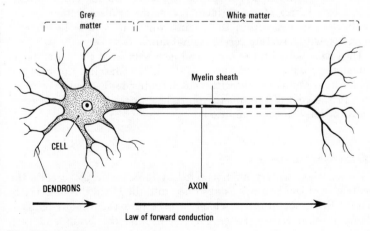

Fig. 111 A neurone.

and the axons, need to be insulated for transmission of stimuli, and this is afforded by a fatty sheath, the myelin sheath, with an outer protective sheath of connective tissue, the neurilemma. In the nervous system *white* matter is a collection of nerve *fibres*, as opposed to *grey* matter, which is a collection of nerve *cells*.

ORGANS FORMING THE CENTRAL NERVOUS SYSTEM

1. Cerebrum or large brain
2. Cerebellum or small brain
3. The midbrain
4. Pons varolii
5. Medulla oblongata or spinal bulb
6. Spinal cord
7. Twelve pairs of cranial nerves given off from the brain
8. Thirty-one pairs of spinal nerves given off from the spinal cord.

COVERINGS OF THESE ORGANS (MENINGES)

The first six of the organs mentioned above, all have the same coverings.

Dura Mater. This is the outermost covering and it is strong and durable. Not only does it *cover* the brain and cord, but it also forms a *lining* for the cranial bones and the neural canal. In places it forms a *double layer,* between which the venous blood collects, and these areas are spoken of as the *venous sinuses* (p. 163).

Arachoid Membrane. This is so named because its microscopic appearance is likened to a spider's web. It hangs like a loose bag between the dura and the innermost covering. It encloses cerebrospinal fluid, but does not secrete it.

Pia Mater. This is a delicate membrane, richly supplied with blood vessels and closely investing the brain, dipping down into all the irregularities of its surface. Pia mater is also present in the ventricles (spaces within the brain), where it covers the choroid plexuses, which secrete cerebrospinal fluid.

The space between the arachnoid membrane and the pia mater is called the *subarachnoid space* or *theca*. It contains the cerebrospinal fluid and connects with ventricles inside the brain via openings in the roof of the fourth ventricle. In certain areas injections (intrathecal) can be given into it.

CEREBROSPINAL FLUID

This is a clear watery fluid secreted by the choroid plexuses in the ventricles of the brain. It is absorbed into the blood through arachnoid villi, vascular tufts which project into the venous sinuses. Being present like a water cushion around the brain and cord cerebrospinal fluid acts as a shock absorber. It prevents the delicate nerve tissue being damaged against the bony containing walls. It nourishes, cleanses and removes waste products from the tissues with which it comes into contact. It has a normal pressure of 50 to 150 mm of water, and this can be greatly increased in disease. You will later learn to assist at the procedure for estimation and reduction of an increased pressure.

Cerebrum

Large brain. Can you remember the cranial bones and the formation of the three fossae along the base of the cranium (p. 50)?

POSITION AND RELATIONSHIPS

Having done this revision you will now be capable of deciding what lies above the cerebrum—the two parietal bones; below, the anterior and middle fossae house the cerebrum, but posteriorly and inferiorly is the cerebellum, small brain. Anteriorly is the frontal bone; posteriorly is the occipital bone, and on either side a temporal bone.

STRUCTURE

The cerebrum is so constructed that its average weight is 1·4 kg

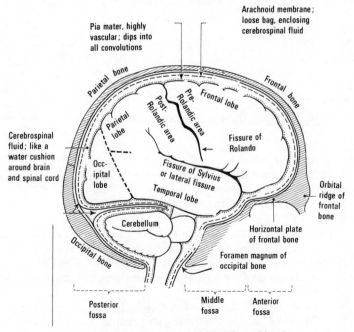

Fig. 112   The cerebrum or large brain.

183

(3 lb). The nerve cells are arranged on the outside as grey matter. To increase the surface area for these, there are many infoldings known as convolutions, sulci and fissures. The deepest one of these, the sagittal or longitudinal fissure, runs medially from front to back, and divides the cerebrum into two halves, each controlling the opposite side of the body. Each half is further divided into two by the fissure of Rolando.

Each cell making up this convoluted cortex is attached to a nerve fibre, and these are all gathered together in one 'stalk' which leaves the cranial cavity as the spinal cord. The fibres collected together form the white matter, and from previous statements you will realize that it is found on the inside of the cerebrum.

Many of these fibres are transmitting impulses *in* for the brain cells to *interpret*, so that we become aware of our environment, e.g. we hear, see, smell, taste and touch things in our proximity, and only thus can become aware of them. The remaining fibres are transmitting impulses *out*, and these result in *action,,* which is our reaction to our environment, e.g. we see a pencil on the floor and we pick it up. It is reasonable therefore to call the fibres transmitting impulses *in* to the brain, *sensory*, and those transmitting impulses *out* from the brain, *motor*, fibres.

There is a *space* in each cerebral hemisphere called a *lateral ventricle* in which the pia mater covers the choroid plexuses from which the cerebrospinal fluid is secreted.

BLOOD SUPPLY
Circle of Willis.

VENOUS DRAINAGE
Via venous sinuses into the jugular veins.

FUNCTIONS (Fig. 113)

1. Cells in pre-Rolandic area *initiate* impulses to all skeletal muscles.

2. Cells in the base of the frontal area *initiate* impulses to the eye muscles.

3. Cells in the frontal area *initiate* impulses to the muscles concerned in speech. This centre is best developed on the left side in right-handed people and vice versa. Not only will injury to the left side of the brain produce a right-sided paralysis, but it can interfere with speech.

4. Cells in the occipital area *interpret* sight.

5. Cells in the temporal area *interpret* sound.

6. Cells in the higher temporal area *interpret* taste and smell.

7. Cells in the post-Rolandic area *interpret* sensation from the whole body.

8. Cells in the remaining frontal area deal with the most abstract

184

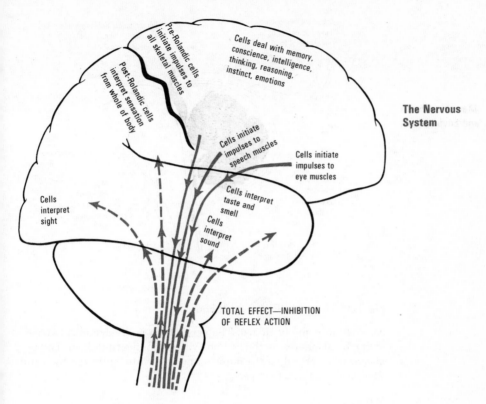

Fig. 113   The functions of the cerebrum.

qualities which make man superior to the animals. These are memory, intelligence, thinking, reasoning, conscience, instincts and emotions.

9. The cerebrum *inhibits* reflex action.

Cerebellum

The cerebellum is also known as the small brain.

POSITION AND RELATIONSHIPS

The cerebellum lies below the posterior portion of the cerebrum, separated from it by a special shelf of dura mater, and the inferior cerebellar surface is supported on the posterior cranial fossa formed by the occipital bone. Anteriorly lies the fourth ventricle, in front of which lies the pons varolii, which inferiorly becomes the medulla oblongata.

STRUCTURE

Whereas the surface of the cerebrum is deeply convoluted, that of the

185

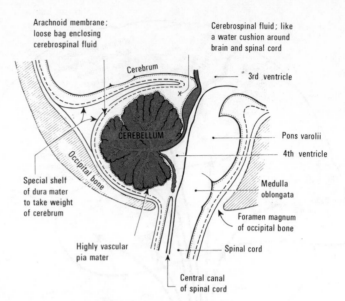

Arachnoid membrane;
loose bag enclosing
cerebrospinal fluid

Cerebrospinal fluid; like
a water cushion around
brain and spinal cord

Cerebrum

3rd ventricle

CEREBELLUM

Occipital bone

Special shelf
of dura mater
to take weight
of cerebrum

Pons varolii

4th ventricle

Medulla
oblongata

Foramen magnum
of occipital bone

Highly vascular
pia mater

Spinal cord

Central canal
of spinal cord

Fig. 114   The cerebellum or small brain.

cerebellum is finely ridged. The cerebellum is divided into two cerebellar hemispheres, with the pons varolii connecting them. The grey
matter is arranged on the outside and the white matter on the inside
like the branches of a tree.

BLOOD SUPPLY

The circle of Willis.

VENOUS DRAINAGE

Via venous sinuses into the jugular veins.

FUNCTIONS

These are complex and as yet not fully understood, but we do know
that many of them are performed at a subconscious level.

1. The cerebellum takes care of our sense of *balance* and
equilibrium.

2. It is responsible for *muscle co-ordination*, i.e. muscles working
together and in immediate sequence to produce smooth, steady and
graceful movement. Jerky movements and unsteady gait result from
disease of the cerebellum.

3. Maintenance of *muscle tone*. Do you remember about one-third
of each muscle's fibres being 'on duty' to maintain muscle tone
(p. 93)? Here is an instance of a function being performed at a subconscious level.

4. *Nutritional* or *trophic impulses* are supplied to all tissue, and
these are constantly streaming from the cerebellum, down the spinal
cord, to be relayed out to the tissues in the lower motor neurone.

When you learn about disease of the lower motor neurone you

will be told that it produces a flaccid paralysis, and you will realize that is because no muscle tone impulses are reaching the muscles to keep one-third of the fibres contracted.

You will also be told when you are learning about diseases of the lower motor neurone that there is wasting of the limb. You will understand that it is because no nutritional impulses can get out to the tissues. The Bible term of withering palsy is very descriptive!

5. *Postural reflexes* occur at cerebellar level, so that we have not consciously to be willing ourselves to maintain good posture. How important therefore for the cerebellum to be given the pattern of *good* posture to take over at an automatic level.

So one could add to the list, for many of the complicated muscular movements that we learn in a lifetime are practised with conscious effort, and are eventually taken over at an automatic level by the cerebellum. You have only to watch a small child learning to knit and compare this with the adult knitting and watching television! As you think about this you will recognize many more patterns that are taken over at cerebellar level.

The Midbrain

This resembles two stalks leaving the base of the cerebrum and sinking distally into the pons varolii. The two stalks are sometimes called the cerebral peduncles; they are mainly *white matter* of '*in-going*' and '*outgoing*' *fibres*. As they merge together they contain a canal which communicates the third ventricle at the base of the cerebrum with the fourth ventricle which lies in front of the cerebellum.

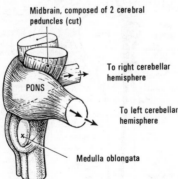

Midbrain, composed of 2 cerebral peduncles (cut)

To right cerebellar hemisphere

PONS

To left cerebellar hemisphere

Medulla oblongata

Fig. 115 The midbrain and pons varolii.

The Pons Varolii

Part of this structure is a continuation of the midbrain, and these fibres pass directly into the medulla below. Running at right angles to these (or transversely) is another set of fibres which give the functional name of pons, which means bridge, for they act as a bridge connecting the two cerebellar hemispheres.

187

The Medulla Oblongata

The medulla oblongata is also known as the *spinal bulb*. The fibres in this communicate all structures above, i.e. cerebrum, cerebellum, midbrain and pons, with the spinal cord below. It lies just within the foramen magnum of the occipital bone. Here we find the *grey* matter on the *inside* and the *white* matter on the *outside*. It is here that the nerve fibres *cross over*. We said earlier that the right side of the brain controlled the left side of the body, so this is another piece of the jigsaw puzzle that has been put in place! The grey matter in this region controls many vital activities such as respiration, body temperature and heart rate at an 'unconscious' level, for these activities must continue whilst the 'conscious' brain sleeps. Injury here will cause instantaneous death. Do you remember the injury inflicted by the hangman's rope (p. 56)?

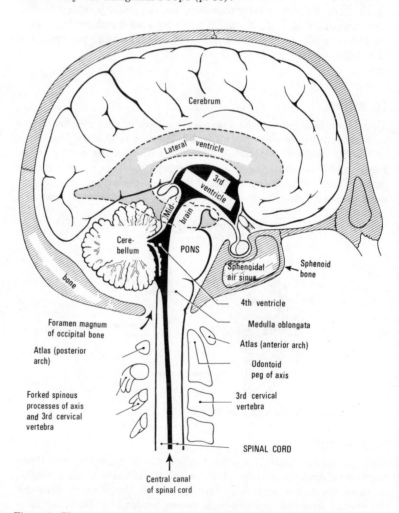

Fig. 116   The ventricles, midbrain, pons and medulla oblongata.

This grey matter is also essential for the reflex actions of swallowing, vomiting, coughing and sneezing. There are dire results when these reflexes do not work efficiently, for the saliva which should be swallowed trickles down into the lungs, there to drown the patient or set up pneumonia or lung abscess. You can foresee being asked to nurse some of your patients in the semiprone position with the foot of their beds raised.

Spinal Cord

POSITION AND RELATIONSHIPS

This structure is contained within the vertebral or neural canal. It is continuous with the medulla oblongata above and is thickest in the cervical and lumbar regions. It ends at the level of the second lumbar vertebra.

STRUCTURE

It is a cylindrical structure containing a tiny cavity, the central canal, which is filled with cerebrospinal fluid, and is continuous with the fourth, third and lateral ventricles above. You can see now that not only are the brain and cord surrounded by fluid, but they also have a column of fluid within them.

The spinal cord is safely anchored to the base of the sacral canal by a cord of strong white fibrous tissue, the *filum terminale*. This helps to prevent it being knocked against the bony vertebral canal. It is further kept in place by 31 pairs of spinal nerves which are given off from its substance to communicate with all tissues of the body. Each pair of nerves has its own pair of intervertebral foramina, via which it leaves the canal. The cord ends at the level of the second lumbar vertebra, but the lower lumbar and the sacral nerves have to travel within the bag of cerebrospinal fluid until they reach their foramina of exit (Fig. 117). This arrangement of nerves is called the *cauda equina* because of its resemblance to a horse's tail. Senior nurses are shown how to set for, and assist in, the procedure of lumbar puncture. You will realize that this can be safely performed, as there is little possibility of piercing a structure similar to a white thread floating in fluid. From Figure 117 you will also see that the pia mater closely invests the cord, whereas the arachnoid membrane hangs loosely down to the end of the sacral canal.

A cross-section of the spinal cord will confirm the presence of the central canal filled with cerebrospinal fluid. It will also show the cord partially divided into two by a deep cleft anteriorly and a lesser cleft posteriorly. The grey matter continues to be on the inside as in the medulla, but is arranged in the form of the letter H. The four ends of the H are called horns, roots or cornua. The two *anterior* roots contain the cells of *motor* neurones, which receive impulses *from* the brain. The two *posterior* roots contain the cells of *sensory* neurones, which relay impulses *to* the higher centres in the brain,

189

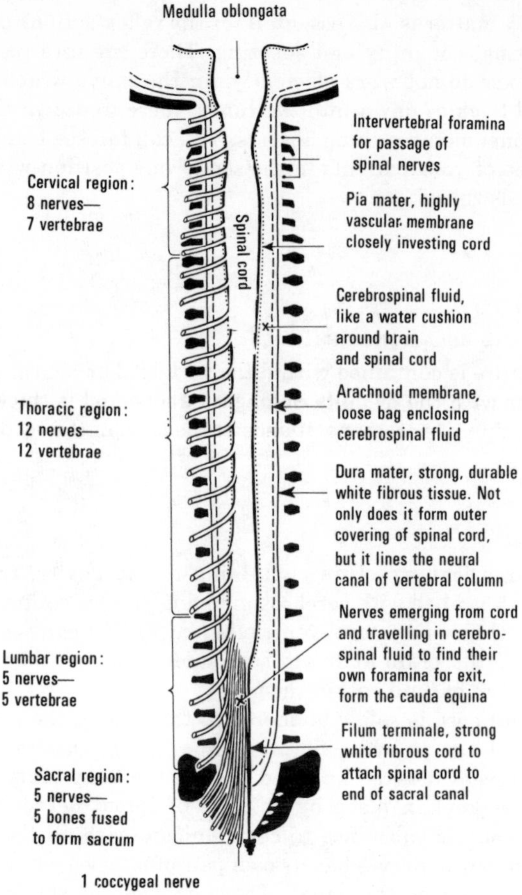

Fig. 117   The spinal cord, its coverings and nerves.

received from the posterior sensory afferent neurone seen in Figure 118. The cells of these posterior sensory afferent neurones are gathered together in a ganglion, marked the posterior root ganglion (Fig. 118). There are also connector neurones between the posterior and anterior roots as shown.

The white matter in the cord is conveniently divided by the grey matter into anterior, posterior and two lateral tracts. These convey sensory impulses *into* the higher centres, and impulses for muscle tone, muscle co-ordination, nutrition of tissues, and motor impulses for movement *out* from the brain to the periphery.

BLOOD SUPPLY

Branches from the thoracic and abdominal aorta.

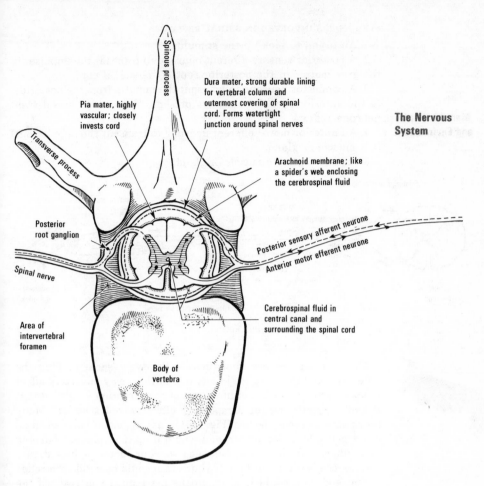

Fig. 118   Cross-section of spinal cord within its bony canal.

VENOUS DRAINAGE

Branches into the superior and inferior vena cava.

FUNCTIONS

   1. Connection between brain and all other parts of body.
   2. Centre for reflex action.

Reflex Action

An automatic or involuntary motor or secretory response to a
sensory stimulus.

1. An organ to pick up the stimulus.

2. A posterior sensory afferent neurone to transmit the impulse to the grey matter in the posterior root of the spinal cord.

3. A connector neurone to transmit the impulse from the posterior to the anterior root. (Not always present. The knee jerk is a two-neurone reflex.)

4. An anterior motor efferent neurone to transmit the impulse out to a muscle or gland.

5. An end organ—a muscle or gland.

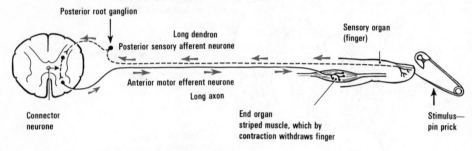

Posterior root ganglion

Long dendron
Posterior sensory afferent neurone

Sensory organ (finger)

Connector neurone

Anterior motor efferent neurone

Long axon

End organ striped muscle, which by contraction withdraws finger

Stimulus— pin prick

Fig. 119 A protective reflex action.

These structures are sometimes collectively spoken of as the reflex arc, and interference in any one of them may adversely affect reflex action. In our list of functions of the cerebrum we stated that it had an inhibiting (or dampening) effect on reflex action. Many reflex actions are *protective* (Fig. 119) in nature, e.g. if you picked up a hot plate, by reflex action you would drop it to prevent burning your fingers. If this plate belonged to your mother's best dinner service, the 'willing' portion of your brain could over-ride the reflex action and you would hold on to the plate until you reached the nearest safe place on which to deposit it! Other reflex actions are tested as an aid to *diagnosis* (Fig. 120). Some reflex actions have a *secretory* response (Fig. 121).

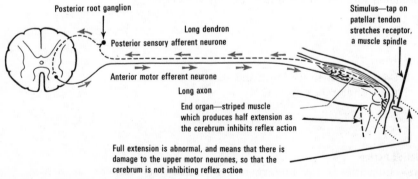

Posterior root ganglion

Long dendron
Posterior sensory afferent neurone

Stimulus—tap on patellar tendon stretches receptor, a muscle spindle

Anterior motor efferent neurone

Long axon

End organ—striped muscle which produces half extension as the cerebrum inhibits reflex action

Full extension is abnormal, and means that there is damage to the upper motor neurones, so that the cerebrum is not inhibiting reflex action

Fig. 120 A diagnostic reflex action.

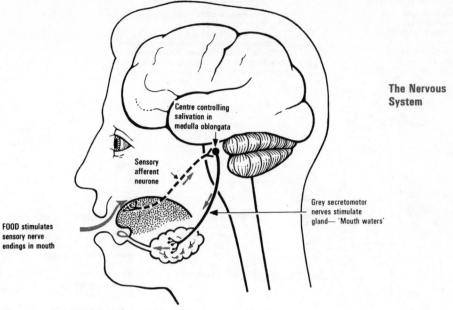

Labels in figure:

Centre controlling
salivation in
medulla oblongata

Sensory
afferent
neurone

Grey secretomotor
nerves stimulate
gland— 'Mouth waters'

FOOD stimulates
sensory nerve
endings in mouth

Fig. 121   A secretory reflex action.

THE AUTONOMIC NERVOUS SYSTEM

Having learned that much of the activity within our bodies goes on
at a subconscious level, and that our internal, plain, involuntary
muscle is supplied with grey neurones from the autonomic nervous
system, it is now time for us to try and master the rudiments of this
highly complex organization.

The autonomic nervous system derives in the first instance from
the brain and its 12 pairs of cranial nerves composed of white
neurones, and the spinal cord and its 31 pairs of spinal nerves which
are also white neurones belonging to the central nervous system.
Somewhere then, these white neurones must end and give rise to grey
neurones belonging to the autonomic nervous system, and this is
exactly what does happen in ganglia (collections of nerve cells
outside the brain and cord). The white neurones from the upper
cranial and lower sacral nerves arborize with grey neurones going
to form the *parasympathetic* portion of the autonomic nervous system,
the ganglia where this happens being *within the organ* concerned.
The white neurones from the remaining cranial and spinal nerves
arborize with grey neurones going to form the *sympathetic* portion of
the autonomic nervous system, the ganglia where this happens
being arranged in *two chains*, one on either side of the vertebral
column.

All the internal organs have a *double nerve supply*, i.e. they are
supplied with *grey neurones* from the *sympathetic* portion, and with
grey neurones from the *parasympathetic* portion of the autonomic
nervous system. These neurones have *opposite effects*; if one stimu-

193

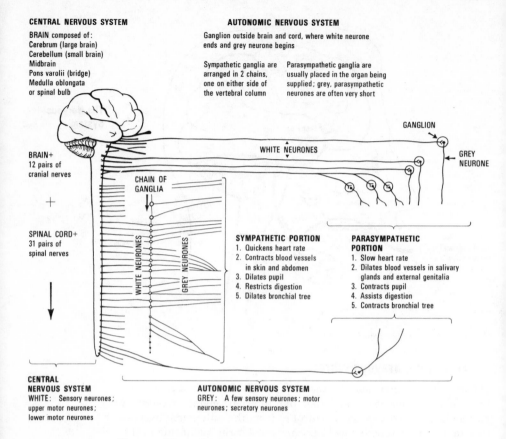

CENTRAL NERVOUS SYSTEM

BRAIN composed of:
Cerebrum (large brain)
Cerebellum (small brain)
Midbrain
Pons varolii (bridge)
Medulla oblongata
or spinal bulb

BRAIN+
12 pairs of
cranial nerves

SPINAL CORD+
31 pairs of
spinal nerves

CENTRAL
NERVOUS SYSTEM
WHITE: Sensory neurones;
upper motor neurones;
lower motor neurones

AUTONOMIC NERVOUS SYSTEM

Ganglion outside brain and cord, where white neurone
ends and grey neurone begins

Sympathetic ganglia are   Parasympathetic ganglia are
arranged in 2 chains,   usually placed in the organ being
one on either side of   supplied; grey, parasympathetic
the vertebral column   neurones are often very short

GANGLION

WHITE NEURONES

GREY
NEURONE

CHAIN OF
GANGLIA

WHITE NEURONES   GREY NEURONES

SYMPATHETIC PORTION
1. Quickens heart rate
2. Contracts blood vessels
   in skin and abdomen
3. Dilates pupil
4. Restricts digestion
5. Dilates bronchial tree

PARASYMPATHETIC
PORTION
1. Slow heart rate
2. Dilates blood vessels in salivary
   glands and external genitalia
3. Contracts pupil
4. Assists digestion
5. Contracts bronchial tree

AUTONOMIC NERVOUS SYSTEM
GREY: A few sensory neurones; motor
neurones; secretory neurones

Fig. 122   The autonomic nervous system.

lates then the other checks, and it is in this way that a 'normal' activity for each organ is brought about. There are, fortunately, only a few of these specific actions that a nurse needs to know, and they are shown in Figure 122

The motor and secretory fibres are not influenced by the will but they are influenced by the *emotions*. The *sympathetic* fibres are stimulated by *fear* when they clamp down on activity in the abdominal organs, dilate the bronchial tree to get more oxygen, and give the skeletal muscles maximum blood supply, so that the body is ready for the instinctive reaction of *running away* from the source of fear, facing up to it or fighting it. A similar pattern of activity occurs in response to *anger*, and then the body is ready for the instinctive reaction of *fight*.

The *parasympathetic* nerves are stimulated by *pleasant emotions* in response to which the digestive juices are secreted, and the bowel stimulated to deal with the food taken. The hoteliers who provide low lights and sweet music in their dining-rooms are assisting in the production of pleasant emotions!

194

From the above you will be able to recognize the highly emotional people with frequent digestive upsets as those with an over-playing sympathetic portion of their autonomic nervous system, and the extremely placid people as those with an over-playing parasympathetic portion of their nervous system. There are always extremes, but most people strike a happy balance between the two portions of their autonomic nervous system.

The nervous system can be likened to a computer into which information is received from all the sensory organs, i.e. from the skin, mouth, nose, eye and ear. We have talked briefly about information received via the skin (p. 126), the emotions aroused in response to that information, and the building of attitudes to objects and persons causing the sensations. Later we will speak of the information received via the other sensory organs and of the learning and adaptation that is necessary in the sexual field for the develooment and maintenance of mental health. As a general introduction to Normal Development[224], chapter III of the *Report of the Committee on Maladjusted Children* is quoted below.

NORMAL DEVELOPMENT
'DEVELOPMENT AND MATURITY

Education is deeply concerned with the process of maturing: indeed, it is in essence the means by which the immature are enabled to become mature. In this sense it takes place not only at school; the whole environment, both human and material, in which the child grows up is the true educative medium. Modern research suggests that the most formative influences are those which the child experiences before he comes to school at all, and that certain attitudes have then taken shape which may affect decisively the whole of his subsequent development.

No human being is fully mature, nor does the degree of his maturity remain constant. Under stress of violent emotion anyone can regress temporarily to a childish form of functioning—as when a man kicks and abuses the door which 'will not' open. The level of a person's maturity will vary with his state of health, the ease or difficulty with which his basic needs are being met at the time, or the company in which he finds himself. In essence it may be said that the mature person is one who accepts the responsibility of ordering his own life and making his own decisions, and who does not act simply on the impulse of the moment.

A feature of maturity is that conduct becomes expressive and characteristic of the person himself. Principles and values are integrated into a coherent system which gives shape and stability to the personality. Inner conflict and indecision are thus reduced, and it becomes possible for an individual to exercise control and persistence, and to pursue remote ends.

195

Normal development, therefore, is development towards independence, stability and control, and the gradual drawing together and realisation of all man's capacities. One very important point at the outset is the meaning to be attached to the term 'normal'. A criterion of normality is peculiarly difficult to obtain, for the following reasons:

Man, his Health and Environment

1. What is normal for one child may not be normal for another. Every child is unique, and his personality is a complex blend of hereditary traits and environmental influences—the latter including not only the people and objects round a child, but also the attitudes, feelings and events which affect him or to which he may respond. On a child's make-up will depend what is normal for him. A child of introverted temperament will, for example, normally be cautious in making friends with other children, but an extrovert will not. The child of high intelligence will normally not be slow in learning to read, but the dull child will.

2. Behaviour of a certain kind may be normal at one stage but not at another. It is natural, for example, for a very young child to be completely dependent on his mother, but it would be abnormal if this continued until he was much older.

3. Development takes place in many directions, and not all of a child's powers will mature at the same rate, though for good adjustment and healthy growth there should be some degree of harmony between them. Concentration on one aspect of growth may temporarily retard the growth of another, as when a baby becomes less vocal while he is perfecting manual dexterity, or a six-year-old more demanding and dependent while he is struggling with formal work at school. On the other hand, the development of one power often assists the development of another, as when a child of twelve months becomes more tractable as he begins to crawl, or a two-year-old as he learns to speak or becomes steadier on his feet.

4. Normal behaviour is not always 'good' behaviour. All children will from time to time display behaviour problems, but, if development is taking place normally, so far from these holding up the maturing process they will promote it. Compare, for example, the frantic compulsive destructiveness of a maladjusted ten-year-old, which teaches him nothing and may so stifle his curiosity that he is unable to learn at all, with the dispassionate destructiveness of a normal two-year-old, which helps him to understand the material world better and so to control it.

From what has been said it is clear that the normal must be thought of as a group which includes wide variations rather than as a single type. This notion is generally accepted in the sphere of intelligence, where the normal group is thought of as comprising the central 50 per cent of the population. Progress towards maturity is even more difficult to measure than intelligence, and it is not possible to chart clearly either the range within the normal at any age or the line of progress from one stage to the next. All that can be attempted is a description of the manner in which a child progresses and the series of experiences and satisfactions which naturally come his way.

As a child grows, he should increasingly become independent and at the same time capable of forming satisfactory and lasting relationships with other people. This is in essence an affair of the feelings, but the feelings do not mature in isolation and there is constant interplay between all the aspects of the child's growing self.

The Nervous
System

Each phase in development has its own appropriate emotional satisfactions. The normal course of development seems to be to experience these satisfactions in an unhurried, confident fashion, gaining something from them and either leaving them behind or building them into the next stage. Deprivation, curtailment or perversion of these natural satisfactions may lead to regression or to a general disinclination to go forward; and advance to the next stage is only possible if previous stages have been satisfactorily accomplished. Even on the physical plane this seems true; it may be said, for example, that an infant learns to walk from the neck downwards, progressively coordinating the muscles of the neck, back and limbs, as he learns first to hold up his head, then to heave a shoulder off the pillow, then to sit up, crawl, stand and finally stagger forwards.

Advance towards maturity is helped at each stage because the child's mind, body and feelings mature together. He is constantly discovering, often by accident, substitute satisfactions for those it is time to leave behind, and these new satisfactions in their turn help on the maturing of his feelings. When he begins to run about, for example, he can occupy himself better in play and find new interests for himself, so that he has alternative pleasures to fall back on when his mother is too busy to give him her attention. Painful contacts with table corners or doorsteps help him to distinguish fact from fantasy, and to adapt his responses to it. He also plays his way into a dim realisation of what it will be like to be a grown-up person, and may find it both possible and likeable. In this way he is helped by seeing older children, as well as his parents and other adults, obviously enjoying a more mature way of living which, because of its manifest controls and disciplines, may to the natural man in him seem impossibly difficult and distasteful. In this way he is led on towards maturity, finding at each step that he gains more than he loses and that he is equal to the increasingly complex demands which are made on him.

INFANCY AND EARLY CHILDHOOD

The first act of the drama in a child's struggle for independence is normally played out within the family, and in it the mother normally takes the leading role, since it is she who makes his first essay at independence possible and it is from her that he must first detach himself. But before the child becomes the man, the struggle has to be repeated more than once, each time on a more complex and conscious plane and accompanied at each repetition by the possibility of stress and breakdown. Roughly speaking, the first cycle takes seven years and may be described as the period of infancy and early childhood, in which the support and approval of adults are the most potent influences on a child's development.

To enjoy satisfactory relationships with people is always vital for a child's development, but his capacity for making them will largely depend on the quality of his emotional and physical experience in the first years of life. An infant has all his feelings in his body; from the start he feels the attitudes of other people through their care of his body—for example, through the way in which his mother picks him up, holds him to her and feeds him. The world is presented to him right from the beginning as predominantly good or bad according to the quality of the mothering he receives. He needs one person constantly with him, not only to feed, care for and love him, but also to allow him to enjoy this relationship. In this way he builds up the sense of security which he needs if he is to reach out or respond to other people sufficiently to commit himself and run the risk of getting hurt; and it is not until a person has the confidence to take this risk that he can fully give or receive affection. The same early experiences appear to affect a child's moral sensitiveness, his curiosity and liveliness of mind, and his ability to learn up to the limits of his native capacities.

Even for an infant who is loved and wanted, life soon presents problems. He quickly realises that the giver of good things can also refuse them. The first great crisis of separation occurs when he is weaned and has to learn to take solid food. Even at this early stage the principle of substitute satisfactions is at work, for by this time an infant is usually learning to sit up, curiosity is developing and he is beginning to play.

If a child is assured of his mother's love he can bear to be away from her, and by about the age of two has advanced sufficiently both in knowledge and bodily skill to want to do so, although at first only for short periods. The good mother, while maintaining a steady, secure intimacy with the child, is at the same time directing his interest away from herself, helping him to improve his speech and the use of his body and encouraging him to make friendly advances to other children. It is, however, during this period, somewhere between six months and three years, when a child is detaching himself from his mother through play and exploration, that he needs her most and that maternal deprivation is most damaging. This is seen in the way he will sometimes not play with a new toy until his mother gives it to him; and in the way he rushes back to her for protection as soon as anything goes wrong with his attempts at friendliness with other children or with animals.

The question whether the world is fundamentally a friendly or a hostile place is continually before a child in his first years. Somewhere about the third year this uncertainty culminates in a period of

198

conflict with authority, when he is often wilful, aggressive and difficult to control. On the handling he receives at this time will largely depend his ability to accept discipline and frustration in later years.

At this period, a child's relationship with his father becomes increasingly important. The father takes his place as the embodiment of authority in the family, not only protecting and supporting the mother, but by firm though affectionate control, buttressing the young child against his own aggressive feelings and designs. It is natural for a child to want to be grown up, and the father, by giving a satisfactory example of grown-up life and by allowing the child to watch and share some of his masculine activities, provides a further means of stimulating development. Through this second and significantly different relationship the child emerges out of a world centred on his mother into one of wider human relationships.

If the period of conflict with authority is brought to a satisfactory conclusion, a child normally enters on a much calmer phase, greatly helped in his progress towards independence by the development of a genuine desire for friendship with other children. This stage is often reached during the fifth year and makes the second great separation from the mother, when he goes to school, much easier to endure.

From this point the teacher takes over some of the functions of the mother, and it falls to the teacher to give the child in school the warmth of affection needed if he is to learn satisfactorily. The teacher also makes it possible for him to continue his emotional education. To this end she encourages the formation of small groups, both for play and work, and, while seeing that children do not harm each other, allows scope for feelings to be expressed naturally, including feelings of hostility and aggression. Like the good mother, the teacher has to accept a child's dependence and need for protection and at the same time encourage him in every way to become independent of her. In this, skilful teaching in the narrower sense is a great help; for the more a child knows, the more confidence he has in managing his own affairs. Growth in understanding the use of numbers, for example, may encourage him to lay out his pocket money for himself, and being able to read may make him less dependent on adults for his pleasures.

The young school child will, however, transfer to his teachers the attitudes he has taken over from his parents; he will welcome the chance to learn new things if he has a mind 'innocent and quiet' and has been encouraged at home to do things for himself. As he progresses through the school, he has gradually to forsake real things for symbols and learn to deal with the abstract; this will be a very difficult and painful process if growth is not continuing satisfactorily on the emotional side, as, for example, if he has been unable to tolerate the separation from his mother. Many children go through a period of nervous tension at about the time when the formal business of learning to read and write commonly begins, but this is normally quickly over; most children enjoy going to school and probably find membership of a large group a relief from the close-knit, intimate atmosphere of the family.

The fact that important steps forward are being made on the intellectual level at this stage also helps on emotional adjustments in other directions. The child of six or seven can usually distinguish between fact and fantasy, although the ability to do this is likely to break down under emotional stress (as when he is badgered to 'own up'). Having a better understanding of the world around him, he can accept it more easily; he realises, for instance, that if he wants parts for his Meccano set he must wait until he has saved up enough for them. At the same time, he may still protest violently if he does not get his own way. He also begins to grasp that the rules and restrictions of life apply to other children as well as to himself. As he grows up physically he gains more confidence in his relationships with other children. Fortified in all these ways he enters, somewhere about the eighth year, on the period of later childhood.

LATER CHILDHOOD

It used to be held that the years between the ages of seven and twelve constituted a kind of golden age, when a child gave little trouble and lived happily and thoughtlessly, reaching an almost mature state of stability and reasonableness towards the end of the period. It has, however, recently been remarked that the eighth and ninth years constitute one of the peak periods for references to child guidance clinics, and educational backwardness also becomes a pressing problem at about the same age. There is some truth in both points of view. Whether a child is happy and stable in this period, or unhappy and out of step with society or with his lessons, largely depends on one thing, the adequacy of his early nurture.

The average child of eight has developed a capacity for stepping outside himself, as it were, and viewing himself as he appears to others. He needs the approval of other children and has become keenly aware of any differences between himself and them which are likely to diminish their approval. This may have either good or bad effects on the maturing process. For example, he may first begin to suffer consciously at this age from failure, whether at work or play. On the other hand, the result may be an improvement in his standards on both the social and intellectual planes. He likes to look at the work of other children and compare it with his own, he considers their skill in relation to his own, he notices how they behave in the face of frustration, rebuke or victory. He wants to know as much as they do and to be as well thought of, and so he strives to emulate them. Above all, he wants to be accepted by them, and this often stimulates him to control his feelings and adapt his behaviour to the requirements of the group. If he is accepted, he can be very generous in his praise of others who are more successful than he is.

In this way he identifies himself with a larger group and can enjoy vicariously its success, so that the quality of his own individual performance matters less. As a result of his identification with the group, he thinks less about himself. Like the child of two or three years, he becomes absorbed in outside interests, and these are no longer confined to what is happening in his immediate environment. His actual world is likely to be rather restricted; he does not travel much, does not meet many people and cannot exercise much control over the kind of experience which comes his way. But his imagination, though realistic and practical, is well developed and active, and in poetry and story or through the cinema and wireless he can enter into the lives and feelings of people still more remote and unfamiliar. He turns from one interest to another without much persistence, but all the while he is comparing, noting, and widening his experience. In this way he lives in a perpetually expanding universe, and his mind and feelings expand in harmony with it.

Although the influence of other children is very great at this stage, adults still play a most important part in his development. At school what some of his teachers say and do will have a lasting effect on him. The most formative influence in his life will, however, still be his home, and the security of a good home is vital for his sound growth.

One of the most important factors in fostering this sense of security is a harmonious relationship between the child's parents. Another is the existence, in the parents and the other adults with whom children come into contact, of firmly held moral principles and standards of conduct. These will not necessarily be based on religious beliefs; many parents without religious beliefs bring up their children satisfactorily, just as others with such beliefs fail to do so. But there are many people who feel that children who are brought up believing in a loving and merciful God are thereby helped to develop harmoniously and without constraint; and that a child can more readily get rid of a sense of guilt after wrong-doing, and can more readily forgive other people, if he knows that he himself is loved and forgiven.

In this period a child's life and interests are no longer entirely centred on his home and his school. He may join other groups where he will meet adults in a rather different guise, as leaders of co-operative enterprises in a freer and more friendly atmosphere than is usually possible in school. His passion for experimentation, his desire for independence or the influence of other children may occasionally lead him to lie, cheat or pilfer, but by the end of the period he has normally developed considerable resistance to temptation and he is reliable, cheerful and co-operative.

ADOLESCENCE

Adolescence is often thought of as an unsettled period between two relatively settled ones. The junior school child should have adjusted himself to the world of childhood, and he can usually disregard the problems of the adult world. His body and mind serve him so well and his feelings trouble him so little that he is not tempted to introspection or self-mistrust. When maturity has been finally reached he will normally return to a like state of stability and serenity, and will feel himself confident to deal with the problems which in childhood he disregarded. Whether the transition from childhood to adult life is made easily or with difficulty will to a considerable extent depend on whether the early nurture of the child has been good or bad, and on the degree of strain and pressure which his environment is exercising at the time.

The far-reaching physical changes of puberty are matched by equally far-reaching emotional and intellectual changes. These may be described as the breaking down of an established pattern and its forming again on a more adult and complex level. The adolescent has to learn to manage a more adult body and to deal with unfamiliar and stronger feelings. This is made even more difficult, because the system of values which sufficed to control feeling during the previous period is itself being broken down and re-formed. The adolescent will have a wider interest in the world around him—in its day-to-day problems as they most immediately affect him, in broader political and social issues, and perhaps in the nature of the universe and the purpose and meaning of his and its existence. There will at this time often be an awakening or deepening of his religious feelings and a quickening of his aesthetic sensibility. He has to re-mould and make his own the principles and beliefs with which he has been growing up in infancy and childhood.

The period of adolescence is one of such instability that it is particularly difficult to determine what constitute the normal limits of behaviour and adjustment. An adolescent may, for example, give way to uncontrollable bouts of temper or crying, may wander off, pilfer or romance, and may swing from exultation to depression, now flinging into society, now withdrawing into solitariness. All this may merely mean that, like the infant, he is adjusting himself to an unfamiliar and complex world by means of a personality which is itself unfamiliar and fluid, and that, also like the infant, he has to experiment in order to understand and control.

Adolescence has been described as having more in common with infancy than with the intervening period of childhood. The main need of the adolescent, as of the infant, is emancipation. Whereas, however, the infant has to emerge from a life centred on his mother into the family, the adolescent has to emerge from the family into the world at large. He has to free himself finally from dependence on adults, and to accept the responsibility of ordering his own life, even though he may go on living at home. In a way, therefore, he repeats the infant's struggle with authority and may experience again something of the infant's hostility and antagonism to his parents. But his struggle is more conscious, and is often accompanied by an ability to formulate and express criticism which may make it far more bitter and distressing both to himself and to the adults around him. Out of it, however, normally comes full understanding, an increased tolerance and a new relation of friendly equality. He may experience again something of the conflict which occupied him as an infant, in that he desires both protection and freedom, both fears independence and is irresistibly attracted to it. But between him and his infancy lies the period of childhood in which he has experienced the pleasure of comradeship with his peers, and the natural course of his maturing feelings is to drive him still further beyond the family, to form lasting friendships and to find new objects for his affections.

In the early part of adolescence there is generally a greater emotional interest in the same sex. This is a normal phase, which ordinarily gives place in the later teens to a greater emotional interest in the opposite sex. If a person has had ample opportunity to experience the progressive emotional stages of infancy, childhood and adolescence, he will as a young adult make a satisfactory adjustment more easily in the sphere of personal affections.

The adolescent needs also to come to terms with his work and to develop ambitions which are both possible of achievement and satisfying to his feelings of self-respect and self-interest. If he has to leave school at fifteen or sixteen he may not have much choice in what he does and may have to look elsewhere for his main interests, but at least he has the satisfaction of earning his own living and of having money to spend as he pleases. Absence of this obvious stage of independence may make the period more difficult for those who continue full-time education, but they have compensating opportunities for pursuing the career of their choice and for achieving intellectual emancipation, and have more leisure for reflecting on the world in general. In either event, the adolescent as he matures identifies himself to some extent with both his work and the community at large; he is prepared to accept the requirements and restrictions of both and to contribute what he can to their successful functioning.

SUMMARY

In reviewing this process of development from infancy to maturity one can pick out various pointers to normality or the reverse. At every stage the child who is progressing satisfactorily is able to profit from the experiences which can normally be assimilated by someone of his age and innate equipment. A child will, for example, learn to talk or to read at about the time when children of his intelligence commonly do. On the emotional side also, he becomes increasingly sensitive to the feelings and wishes of others and is able to profit to the normal degree from the opportunities which life offers him of learning to control his feelings. One might even say that physically he profits in the same way; his food does him good and he is built up by fresh air and exercise. The concept becomes clearer if one considers instances where something has gone wrong with the process of development: the child of average intelligence who, though present at all the lessons, does not learn to read; the nervous child who may eat voraciously and who fails to thrive for no obvious physical reason; the clever child who remains pig-headed and babyish in dealing with frustrations; or the ten-year-old delinquent whose face still looks chubby and infantile, as if experience had washed over it without leaving its customary mark.

All through the process of development the normal child keeps in reasonable touch with the world around him (though with many sidesteps into fantasy and evasion), respects it and tries to adapt himself to it. In this way he gradually learns to control fantasy, accept discipline and persevere in the pursuit of more distant goals, and his prevailing mood is one of serenity and optimism. Year by year he is building up his own style of life, developing characteristic ways of meeting situations and dealing with problems. The older he gets, the more difficult it becomes to alter his life pattern and to modify a faulty style. His way through life from birth to maturity may be likened to the progress of a Channel swimmer, necessarily affected by waves and current and exulting in the resistance they offer him, but shaping a course to the opposite shore in spite of them.'

CARE OF THE NERVOUS SYSTEM IN THE MAINTENANCE OF HEALTH

As we have seen, much of the care that goes into the structuring of a healthy nervous system is not given by oneself, but has been administered by others! Earlier in this book we looked at exercise and the challenge that it can offer to mental, as well as to physical health (p. 108). We also spoke of fatigue (p. 110) and acknowledged that it has a mental component. For mental health[78,79] it is thought that most days should be balanced and divided into hours of recreation, sleep and work, so that we need to know something about these.

LEISURE/RECREATION

Recreation can be defined as what one does with one's leisure time. It is for the purpose of 're-creating' one's mind and body, so that one will return to work refreshed, enriched and invigorated. It is therefore very important that one's leisure time is spent in pursuit of those things which give one the greatest satisfaction. One man's work may be another man's hobby, e.g. gardening. By inclination most people choose for their hobby that which is entirely different from their work, e.g. the sedentary worker chooses tennis, swimming, etc.; the person who has worked alone all day chooses to go into company for recreation, and the one who has spent the day with many other people may well enjoy his own company in the evening.

With increasing automation there will be a shorter and shorter working week. Society needs to provide more interesting and challenging, strenuous and less strenuous recreational facilities for its members. It may well be that society will have to think about offering the recreational facilities in schools to the adults in the evenings. Whatever money is spent on achieving adequate recreational facilities is well spent, because a bored society can so easily become a delinquent society.

Adequate care of the aged is an increasing commitment throughout the world as people live longer due to improving standards of

204

living and advances in medical science. Many voluntary organizations in this country help in providing recreation for pensioners, arranging all sorts of activities to suit local tastes during the year, and by arranging annual holidays for them. It is important for ageing people to retain their independence for as long as possible. It is equally important for them to recognize when they need help and to accept such help. There is a range of independence varying from the stolid type that cannot accept even necessary help, to the other extreme when dependence allows advantage to be taken of the goodwill of others. As in so many things—to strike a happy medium seems to be the formula for success.

REST

According to the *Concise Oxford Dictionary*, the verb *to rest* means to be still, cease or abstain or be relieved from exertion or action or movement or employment; lie in sleep or death, be tranquil, be let alone. We know that without periods of rest the body would quickly become worn out. The best rest is gained in sleep, but a considerable degree of rest can be achieved in some forms of recreation, e.g. lying in a hammock, lounging in a chair on deck, etc. Recreation, rest and sleep can therefore be used as interchangeable terms, but this is not always so, and each has something special about it.

Rest incorporates the art of relaxation, contentment and the achievement of a peaceful frame of mind. With the pace of modern living these arts are slowly being lost, but their resurrection and practice may well lead to a lessening of the number of people visiting doctors' surgeries for 'sleeping pills'.

Rest in the form of complete relaxation can be practised in a comfortable armchair with the feet supported on a footstool. Sir Winston Churchill said that 20 minutes spent in this way after lunch put two hours on to his day!

SLEEP

Though much has been discovered about sleep in the last decade, there is still a lot to be discovered[80,81,82]. Sleep is essential for human beings; we can last longer without food than we can without sleep. Two types of sleep are now identified. *Deep sleep* is characterized by a slower heart rate; deep, regular breathing; complete muscle relaxation, and a distinctive electrical tracing pattern from the brain. *Paradoxical (dreaming) sleep* is characterized by an irregular heart rate and breathing; a period of rapid eye movements (r.e.m.) when the electrical activity of the brain approximates to that of the waking state. It accounts for a quarter of the night's sleep. When wakened during the r.e.m. period a person can remember what he is dreaming. If wakened each time he has rapid eye movements he becomes mentally disturbed after a few nights. It looks as if the purpose of sleep is to enable us to dream! It would seem that dreaming plays an essential part in maintaining the brain in a functional state.

United States scientists have shown that the peaks of growth hormone production always occur during the phases of dreamless sleep, of which there are usually two or three in a good night's rest. In adults it allows repair of worn out tissues. People who sleep well, heal well! Toddlers' bones with their large amount of animal matter become compressed during the day from weight bearing; sleep allows these 'shrunken' bones to regain their length.

The young baby sleeps most of the day with increasing periods of wakefulness until, by the time he is adult, his needs are fixed at an average of eight hours' sleep daily. Some babies learn to sleep in a crib in the living room, some learn to sleep in a hammock, and some learn to sleep alone in a separate room, having been in the presence of the family throughout the day. In some cultures it is normal for a baby to sleep in the same bed with the parents, but from a physical safety point of view—the danger of suffocation from overlaying, this is discouraged in the west. To avoid association of unpleasant ideas, confinement to bed should never be used as punishment for a child. Children should not be told that a dead person 'has gone to sleep' as it can induce fear of sleep.

Housing enters into any consideration of where adults sleep. Some are faced with sleeping in the room in which they have lived and cooked during the day. A house or home can mean different things to different people in different parts of the world. Each country has a quota of people who choose a vagrant way of living and this may include sleeping rough. Wherever one sleeps the first and foremost thing which encourages sleep is a healthily tired mind and body. A warm bath withdraws blood into the skin; the brain with its consequent lessened blood supply is less active, which is conducive to the onset of sleep. Many people find fresh air, and a darkened, pleasing room a requisite for sleep. Most people would list a comfortable bed as a requisite, though some people sleep on a mat on the floor of a mud hut. Commerce can play a part in the changing needs for a nation's health. The average height of man is increasing. In 1971, the National Bedding Federation announced that the industry is changing to king size beds, i.e. double beds are to be 5 ft wide by 6 ft 6 in long. The previous size was 4 ft wide by 6 ft 3 in long. Many people require loose, warm bedclothes and personal attire, but you may visit homes that do not have sufficient clothing to keep the inhabitants warm at night, and you will learn of the agencies that can help these people to acquire clothing. Some people need quietness for getting off to sleep and others learn to sleep with noise in the background and miss it when they have to sleep in a quiet place. Florence Nightingale said, 'Unnecessary noise (however slight) injures a sick person much more than necessary noise'. Final year nurses were asked to discuss this statement and suggest ways of avoiding unnecessary noise in the wards.

Most people spend an average of eight hours a day for five days of each week at work. Fostering of a healthy attitude to work is important in a nation's health and economy. For the achievement and maintenance of mental health, work should bring a feeling of satisfaction to the workers. Many jobs that *have* to be done, can be seen as either dirty and/or boring, or as contributing to the general good of society. To mention a few: roads have to be cleaned, refuse and sewage has to be collected and rendered harmless, so that the nation is not threatened by epidemics of infectious disease. Vehicle maintenance can be thought of as a greasy, dirty job to be done with as little concentration as possible, or worthy of being done excellently as a contribution to road safety. Filling milk bottles can either be considered a soul-destroying job, or it can be thought of as providing the public with a clean milk supply—and thus as a contribution to the public health. And so one could go on and on, the point being that all work can be seen to be contributing to the social good, if it is looked at within a national framework.

The minimum age at which the majority of a nations' young people leave school and start work varies, and is usually laid down by law. At the other end of the working-life-span, countries also vary about the age at which its people retire from active gainful employment. In Britain it is 60 for women and 65 for men. Our working population pay National Insurance contributions, so that each member will have an income (in Britain called the Old Age Pension) on retirement. The numbers of workers, pensioners and unemployed persons (together with the number of children) are important factors in a nation's economic policy.

The pattern of work life up to the last decade was that a person was trained to do a particular job and thereafter spent a life time doing that job. At the current rate of technical advance it is thought that workers will probably need to be trained to do several different jobs in a lifetime. Much research has been done into the best methods of training the older worker to do a different job. We have several Government sponsored re-training schemes. These will probably be instituted and increased in all nations in the next decades as work patterns change, so that insecurity which is anathema to mental health, will be minimized.

Anyone who finds that he is a round peg in a square hole at work, can seek advice from the Occupational Guidance Units, sponsored by the Department of Employment. They are specialist centres, staffed by trained interviewers who, after investigations, provide a list of options of work, of which that person will be capable. An appointment can be arranged through the local Labour Exchange.

Addiction/Dependence[83]

Now we come to the addictions and though they have physical effects in other parts of the body, it is in behaviour that they produce their greatest effect. If, in the pursuit of mental health, one should

The Nervous System

207

avoid dependence, then one needs to know something about dependence. The word addiction has a punitive ring and the connotation centres round the addict. The World Health Organization prefers the word dependence and in order to recognize the complex set of circumstances that precede such a state, the World Health Organization recognizes three types of dependence. *Social dependence* is when a person depends on a drug in order to conform to the behavioural standards of his particular community. *Psychological dependence* is when a person depends on drugs to provide enjoyment, and/or suppress or come to terms with mental or emotional conflicts. *Physical dependence* is when the body becomes dependent on drugs for normal functioning. If you look at the booklet published by the British Medical Association[84] you will find that it lists dependence on *tobacco* as 20 times more prevalent than dependence on *alcohol*, which is twice as prevalent as dependence on the classes of *drugs* listed as hypnotics, stimulants, analgesics and fantasticants and the dependence in the latter group (of drugs) occurs in that order.

DEPENDENCE ON TOBACCO

Elsewhere, *disease* in other parts of the body produced by smoking is discussed (p. 363). It would appear that dependence on tobacco grows insidiously over the years. There is little doubt that it starts as social dependence and this can be at an early age while at school. There must be other factors that make it attractive, for some children when removed into a non-smoking group, e.g. at work, still continue to smoke. It may be that there is oral gratification in having a cigarette between the lips; it may be that the act of indrawing gives special pleasure; or the feeling of relaxation may be associated with the blowing out portion of the act, or it may be the greatest attraction lies in having something to do with the hands that otherwise would fidget. Sooner or later this is physical dependence, and many people, in spite of determined effort to give up smoking, find that they have lost their independence in this action and they have become dependent on tobacco. This is the stage of physical dependence on the nicotine content of their blood.

DEPENDENCE ON ALCOHOL[85]

Next on the list is dependence on alcohol. The use of alcohol, still the favourite tranquillizing drug of our society, is of great antiquity. In sites occupied by our remote ancestors, archaeologists found evidence that very soon after cultivating grain and using it to make bread, people learned to ferment it and make beer from it. For the majority of people alcohol is a pleasant thing in their lives. They are able to enjoy it on the occasions when they choose to drink it, but they experience no compulsion towards drinking it. The law in some countries states the minimum age of those to whom alcohol can be sold. However, in other countries, especially where the water supply is suspect, wine drinking is a normal part of meal times.

Drunkenness often leads to irresponsible action, and disorderly behaviour, prior to an irresistible desire to go to sleep. It can cause a person to drive a vehicle recklessly[86], or to assault another person, or to indulge in petty crime, or to indulge in casual sex. A Home Office Working Party found that in 1969 there were 80 500 convictions for drunkenness. Many alcoholic offenders are charged with some offence *other* than drunkenness, so that figures for drunkenness are likely to be higher than the number of convictions. The Working Party estimated that about a third of the weekly discharges from the big London prisons were 'regular' drunks who underwent numerous, and almost consecutive prison sentences. This resulted in the innovation in 1970 of five experimental hostels, that are run by the Special After-Care Trust formed for the purpose, with financial backing from the Carnegie United Kingdom Trust. The response of those living at the hostels is most encouraging. Other people do not experience a desire to commit disorderly behaviour or to go to sleep in reaction to alcohol, but find that they need to imbibe at increasingly frequent intervals during the day, to get through the day's work. Many of these people do not come near the courts, for they are functioning adequately in society. They may however cause great social disruption within a family, as money goes on satisfying their craving for drink, that should go on family maintenance. The 1971 Report of the National Council for Alcoholism says that there are 175 000 chronic alcoholics in Britain and some 225 000 problem drinkers who may become alcoholics, unless helped. In an attempt to do this, the Council is experimenting with nine Information Centres. Its statistics show that, in every 1000 people engaged in industry, between 7 and 10 (men and women) drink heavily, and on average lose between 40 and 60 days' work a year. Dr A. B. Sclare, a consultant psychiatrist addressing an International Conference in Glasgow, said that the increasingly younger age of alcoholics was causing concern. The London Borough of Hammersmith's Health Education Service has published a report (which is available from the Service) concerning the attitudes, habits and impressions of young people on the social drinking scene.

Alcoholism is when a person has no control over the drinking of alcohol. In many countries alcoholism is being recognized as a health, rather than a penal problem, and court procedures are recommended for dealing with alcoholics in a similar way to those already relating to the mentally ill, and to allow persistent drunkards to be detained for up to two years without conviction.

There are important *voluntary associations* working in conjunction with the National Council for Alcoholism. One is for ex-alcoholics and is called Alcoholics Anonymous (AA), a world-wide association. Members of local branches meet frequently to give each other moral support in the long and continuous process of refraining from drinking alcohol. The other is called Al-Anon Family Group which is for the wives and husbands, friends and relatives of the alcoholic. They help each other to learn to cope with the problem of alcoholism as part and parcel of their lives.

The Group does not offer any easy answers, any sentimental feather-bedding of difficulties. It demands of members a ruthless self-searching, and a determination to grow in order to meet the challenge. In return it offers comfort, reassurance, loyal and steadfast support.

DRUGS[87,88]

The word drug is now emotive, but the medical profession use the word for *any medicine* that is *taken by mouth*. Many medicines have their effect through the central nervous system. For instance, pain killers do not act at the site of pain. They deal with any type of pain from any part of the body by their action in the brain. The most commonly taken pain killer (analgesic) is aspirin, and almost every household has a bottle of these tablets. These and any other medicaments should not be available to children to prevent accidental poisoning that can effect the nervous system. The last two decades have brought an increasing number of drugs that have a direct effect on the mind—the tranquillizers, antidepressants, pep pills and sleeping tablets. It is estimated that there are over 40 000 million prescriptions for these each year. This led Professor Carstairs to say at a Conference, 'It is not only our adolescents that are playing dangerous games with drugs, but our whole society shows signs of having been misled by the delusion that drugs can solve all our problems. But there are circumstances in life that have to be coped with'. The member of the family for whom tablets are prescribed must accept the responsibility that it is a contract between himself and the doctor, and see that these tablets are not available to any other person. Health educationists advise parents and adults not to take drugs in the presence of children. Not only can this lead to accidental poisoning by imitation, but it breeds an attitude of acceptance without discrimination to the whole cult of taking tablets.

Marijuana/Pot/Cannabis. The Wootton Report in 1969 recommended lighter punishment for those caught with this drug, until such time as more was known about the long term effects. The picture in this, and many other countries, is that cannabis is illegal and possession of same is a punishable offence. Many countries are making determined efforts to prevent the growing of marijuana. Some blood banks in the United States do not accept blood from those who have smoked pot. Smokers of pot claim that they experience excitement and a heightened awareness. In March 1971, two psychiatrists published a five year survey, 1965 to 1970. They excluded from their survey patients who had used other drugs, such as LSD and amphetamines, and those who had demonstrated symptoms of emotional disorders prior to smoking marijuana. The doctors state that, 'the patients consistently showed symptoms including poor social judgement, poor attention span, poor concentration, confusion, anxiety, depression, apathy, passivity and often slowed and slurred speech'. In the group 'marijuana use seemingly accentuated the very aspects of disturbing bodily development and psychological

210

conflicts which the adolescent had been trying to master . . . the adolescent may reach chronological adulthood without achieving adult mental functioning or emotional responsiveness'.

Professor Paton of Oxford, in 1971, found that the main physically active constituent of cannabis, is tetrahydrocannabinol (THC). It is liable to be absorbed by body fat and can accumulate progressively in the body like DDT. This means that regular smokers of cannabis are progressively building up high, possibly dangerous concentrations of THC in their bodies. Professor Paton says, 'If one were to view cannabis simply as a new drug which might be introduced into medicine, the evidence we already have of health hazards would rule it out. If people are going to reject DDT, cyclamates and dieldrin, I certainly think they ought to reject cannabis'.

Amphetamines. These are the much talked about 'pep pills' and are in tablet or capsule form called 'black bombers', 'French blues' or 'purple hearts'. Another type—methedrine—is called 'speed'. The tablets or capsules are taken by mouth, to keep awake or to feel more alert. About 65 per cent of chronic amphetamine users can expect paranoid psychosis in some form. Stimulation is often followed by anxiety feelings and tension, and the next tablet is taken to avoid these sensations of 'crashing'. Amphetamines are a social drug, usually taken away from home. The pupils dilate, users become boisterous, jerky, mobile and talkative. The laugh gets louder. When the effect wears off, there is usually lethargy, irritability and unsociability. This psychological see-saw is a clue to those who might be using amphetamines.

Barbiturates. These are very useful sleeping tablets, and though they may give rise to psychological dependence, they are usually taken each night only in the dose recommended by the doctor. However a cult is growing in which the tablets are mixed with water and injected. Barbiturate does not dissolve properly and the veins become inflamed, blood clot forms and can cause sudden death. Ulcers can occur at the injection sites and gangrene can ensue. The Department of Health and Society Security has a special committee currently investigating this problem and its report is awaited. There is said to be a flourishing black market in stolen barbiturates. Barbiturates are not illegal. They are only obtainable on a doctor's prescription.

LSD/Lysergic acid. This is classified as a hallucinogen, i.e. a substance that produces hallucinations. It is a transparent liquid, used in minute quantities. It can be dropped on sugar cubes or the back of postage stamps which are common means of distribution. It can also be made in tablet form. It is nearly always taken orally and is rarely injected. It is not legally on the market, and to possess it is a legal offence. Illicit supplies have to be made and the people capable of making them are chemists and laboratory technicians. It increases the heart rate and raises the blood pressure. As stated, it produces hallucination, which is a false perception occurring without any true sensory stimulus. Evidence of chromosome damage in 1969 cut demand sharply, and though this is considered not yet proved,

211

coping with the torrent of psychological experience released by the drug can prove difficult and dangerous without medical supervision. **Heroin, cocaine and morphine.** These all have a useful though small part to play in the medical world. They are not legally available and to possess them is a punishable offence. 'Hard drugs' is the term used when illicit use is being referred to. The inescapable fact seems to be that there is an interaction between social setting, personality and the drug. In other words because of the social setting, one person might decide to experiment and try it for himself. The social setting and the personal introduction, lead some people to liken the spread of drug-taking to that of the spread of a 'contagious' disease. For some personalities, even after just one injection, abstinence is not possible. Such people find themselves on the slow road of decline—to an addiction that rules their lives. Obtaining the next dose ('fix') takes precedence over any other activity. Eventually there has to be injection into their veins up to six times daily. The doses have to be higher and higher to satisfy the craving. There is ever-present danger from infected syringes—jaundice as well as sepsis. Interest in food is lost, malnutrition is a common accompaniment of drug taking. There is a consequent lowered resistance to infection.

World therapy. It is natural that each country wants to be rid of its drug problem. But economics and politics can be the enemy of enlightened social policy. Poppy and cannabis are not grown for fun and the farmers that grow them are going to export them and get the biggest price that they can for them. Drug abuse is spreading to countries and social groups, where it was unknown only a few years ago. The International Narcotics Control Board of the United Nations follows four lines of simultaneous action: education to change the social attitude to drugs in general; raising the peasants' standard of living in countries where opium and cannabis, both requiring much labour, are cultivated; strengthening of measures to suppress illicit traffic; treatment and social reintegration of addicts.

An international crime syndicate appears to wait and watch until the black market prices have risen sufficiently in a country and a sizeable market has been created. Then it moves in, in a big way, and the country has more than a drug problem, it has a drug explosion. It is the men who gain a fortune from selling drugs that should receive society's strictest censure.

CONDITIONS THAT CAN INTERFERE WITH THE NERVOUS SYSTEM

For health of nervous tissue, the genes need to provide the correct *enzyme* and *hormone* organization, e.g. *thyroxin*, the hormone secreted by the thyroid gland, is essential for the proper growth of mind and body. When this gland is deficient at birth, not only is the body stunted in growth, but the mind fails to develop, so that the resulting cretin is incapable of even learning to take care of himself,

212

and often has to be cared for in an institution. However, if the health visitor notices that a child is slow in passing the normal milestones of development, she will assist in the diagnosis of this condition, and the doctor will order thyroid extract to be given by mouth; if this is taken throughout life the child has a good chance of developing its own potential. Several conditions of brain retardation are now known to be due to *enzyme* and metabolic abnormality, and if these can be seen sufficiently early, and dietetic or replacement treatment instituted, that brain can develop its potential. One such condition is diagnosed by using Phenistix to detect the presence of phenylpyruvic acid in the urine. This acid is derived from phenylalanine in the diet and in normal people is further broken down into a harmless product. When it is omitted from the diet of a person with an abnormal enzyme pattern the acid is no longer formed in the body, and the nerve tissue in the brain can develop normally. Another such condition is diagnosed by finding galactose in the urine. In many areas of the country young babies' urine is tested by the health visitors, so you are learning something of their work.

In considering the *interdependence* of the many factors that play a part in the production of an *adequately functioning human being,* I would like to quote Sir Julian Huxley (Fawley Foundation Lecture, Southampton, 1962). He . . .'considers it vital that a person educated in any country should appreciate the role of ecology in a study of man's triple tier of environments—material, social and psychological. Ecology in the customary sense deals with man's relations with the forces and resources of external nature. Social ecology deals with man's social problems, within and between human societies. Psychological ecology deals with man's relations with the forces and resources of his inner nature, and the environment of ideas and beliefs with which he has surrounded himself. Education must have a unitary pattern, reflecting the unity of knowledge and the wholeness of experience; and it must give growing minds a coherent picture of nature and themselves'.

SUMMARY OF THE NERVOUS SYSTEM

The nervous system is composed of millions of neurones each one having a nucleated cell, branching fibres called dendrons that transmit the impulse to the cell, and another fibre called an axon that transmits the impulse from the cell. Where the nerve cells are found together they constitute grey matter; where the nerve fibres are found together they constitute white matter.

The Central Nervous System

CEREBRUM

The cerebrum or large brain occupies the major portion of the cranial cavity, and it gives off 12 pairs of cranial nerves. The surface is convoluted and is made of grey matter, the inside is mainly white

213

matter. It is divided by the longitudinal fissure into right and left cerebral hemispheres, each controlling the opposite side of the body. Each hemisphere also contains a lateral ventricle which contains specialized blood-vessels—the choroid plexuses, from which the cerebrospinal fluid is made. This fluid cleanses, nourishes and protects the structures with which it comes into contact.

Each cerebral hemisphere is further divided into an anterior and a posterior portion by the fissure of Rolando. The pre-Rolandic area is motor and initiates impulses to all skeletal muscles. The post-Rolandic area is sensory and interprets sensation from the whole of the body. Other areas initiate motor impulses to the eyes and to the organs of speech, and other areas interpret sight, sound, taste and smell. In the frontal lobe the cells deal with memory, intelligence, thinking, reasoning, conscience, instinct and emotion. The cerebrum inhibits reflex action.

CEREBELLUM

The cerebellum or small brain occupies the posterior fossa of the cranial cavity and lies under the posterior cerebrum, from which it is separated by a special shelf of dura mater. The surface is ridged and is made of grey matter, the inside is made of white matter arranged like the branches of a tree. It is divided into two cerebellar hemispheres which are joined by the pons varolii. The fourth ventricle lies in front of the cerebellum. At a subconscious level it takes care of balance and equilibrium, muscle co-ordination, muscle tone, nutritional or trophic impulses, and postural reflexes.

MENINGES

The meninges are three coverings of the brain and cord. The outer one is strong and durable and is called the dura mater; the middle one is like a spider's web and hangs loosely forming a bag which contains cerebrospinal fluid; the inner one is a fine, highly vascular membrane that dips in between all the convolutions and ridges.

THE MIDBRAIN

Two stalks mainly of white matter, continuous with the cerebrum above and pons varolii below. They contain a canal joining the third ventricle at the base of the cerebrum to the fourth ventricle in front of the cerebellum below.

PONS VAROLII

The pons varolii is made in two parts, one continuing from the midbrain above to the medulla below. The other fibres are at right angles and connect the two cerebellar hemispheres.

Communicates the whole of the brain above with the spinal cord below. White matter is now on the outside with the grey matter collecting in the middle; the fibres actually cross here so that the right side of the brain controls the left side of the body and vice versa. At a subconscious level the medulla controls the vital activities such as respiration, body temperature and heart rate. It is necessary in the reflex actions of swallowing, vomiting, coughing and sneezing.

The Nervous System

THE SPINAL CORD

This cylindrical structure begins at the foramen magnum of the occipital bone and ends at the level of the second lumbar vertebra. The grey matter inside is now in the form of the letter H, the projections of which are called horns, roots or cornua. It is anchored to the base of the sacral canal by the filum terminale. It gives off 31 pairs of spinal nerves and as the lower ones travel to their foramina of exit they form the cauda equina. Sensory, afferent neurones enter the posterior root, and motor, efferent neurones leave the anterior root, the two roots being in communication via connector neurones. The spinal cord connects the brain with all other parts of the body and is the centre for reflex action.

REFLEX ACTION

An automatic or involuntary motor or secretory response to a sensory stimulus. The stimulus is carried via a sensory, afferent neurone to the posterior root of the spinal cord, where it passes via a connector neurone into an anterior, motor, efferent neurone to reach the muscle or gland and produce a response.

The Autonomic Nervous System

Composed mainly of grey motor and secretory neurones. It is divided into two parts and each organ has a supply from each part; these have opposite effects—one stimulates and one checks activity within the organ, the two working together normally striking a happy balance.

PARASYMPATHETIC PORTION

This is derived from upper cranial and lower sacral white neurones, and the ganglia where these white neurones change to grey neurones is usually within the organ concerned. These fibres slow the heart beat, dilate the blood vessels in salivary glands and external genitalia, constrict the pupils, assist digestion and contract the bronchial tree.

215

SYMPATHETIC PORTION

This is derived from the remaining white neurones emerging from the brain and cord, the ganglia where these change to grey neurones being arranged in two chains, one on either side of the vertebral column. These grey fibres quicken the heart beat, contract the blood vessels in the skin and abdomen, dilate the pupils, restrict digestion and dilate the bronchial tree.

Care of the Nervous System

1. Feed it properly by taking a well-balanced diet (vitamin B is antineuritic).

2. Give it an adequate amount of relaxation, rest and sleep.

3. Avoid fatigue that is not dispelled by a good night's sleep.

4. Avoid over-anxiety.

5. Avoid tension, and take steps to deal with it, should it occur.

6. Stimulate it by trying something new.

7. Do not abuse it by imbibing excessive alcohol or drugs.

Nutrition in Relation to Health

Jimmy's skeletal and articular systems are now complete; he has been clothed in muscle tissue, covered with skin, and his whole body supplied with blood and nerves, but before the muscle tissue can produce movement it must have *nourishment* brought to it and stored in it.

Jimmy's *attitude to food* will be related to the feelings that he experienced during feeding as a baby. If these were associated with warmth, the security of being held in firm arms, and pleasure in sucking as well as dispelling the feeling of hunger, then he is very likely to look forward to meal times. He might in adversity return to the thing that gave him pleasure, and find that he is 'nibbling' between meals and putting on weight. The concept of *meal times* is different for different peoples throughout the world. Some people gather out of doors on a shaded veranda or under a tree, some eat in the room in which they live, others gather in a room set aside for the purpose of eating, but in all cultures eating is a *social occasion* and for some families is the only time when all members of the family are present. *Style of eating* varies considerably, from all members using their fingers or an implement to partake from the same bowl, to having separate plates or bowls and using fingers, chopsticks; or knives, forks and spoons as is the custom in the west. What is eaten is equally varied. Flesh eaters are called carnivores, plant eaters—herbivores, and those who partake of both—omnivores. These adherences can be from choice or can be religiously determined. Every culture has its traditional dishes, hence we have restaurants representing many cultures in most developed countries. Some of these dishes are served in sauces, others are of a drier consistency. Preparation of food is often culturally determined and again some religions have definite rules about the preparation of food. Some meats can be taboo in some seasons and so on. The giving and receiving of food is an important *social expression*. Do we not spontaneously invite our relatives and friends to share a meal in our homes, and do we not equally enjoy partaking of meals in our relatives' and friends' homes?

This is always supposing that we have a home, and we will be talking about housing problems later. And it is always supposing that we have food to eat. Even in developed countries there is need of such things as Poverty Action Groups! And in the world of today even with all its marvellous inventions, there are famine areas in which millions of people die a ghastly death from starvation. And this leads us to the world population crisis about which we will be talking later. This only serves to show the inter-relatedness of problems.

Nutrition is one of the most important physiological needs affect-

217

ing our personal well-being. *Good food selection has to be learned* because even with plenty of food, or plenty of money with which to buy food, the uninformed are likely to develop food habits incompatible with good health. In countries where food is available, the two extremes of obesity and malnutrition can be the result of such bad habit, and from a national point of view, these conditions have vast social and economic implications.

Insurance companies are wary of insuring *grossly over-weight* people, as statistics show that they die earlier than their leaner brethren, from coronary disease, other heart conditions, bronchitis and diabetes (p. 398). It is estimated by the British Medical Association in their booklet *Overweight Children—Victims of a Cruel Kindness* published in 1968, that there are two or three children in every 100 who are overweight. It is assumed that this figure will increase as pocket money increases. Adults should refrain from 'buying' childrens' love by giving them titbits and sweets. Once a child is fat, the proper remedy is less food and more exercise, along with all the moral support, love and patience that the parents can give. Throughout the process the child needs to feel that he is lovable and desirable whatever his shape. If the child is old enough to benefit from instruction the situation can be used to good advantage by teaching sound dietary habits based on accurate dietetic principles. It is useful if the parents and the school can co-operate in this matter. Over the last few years several clubs have been established to help the overweight adults. Some of the local authority schemes are limited to those who need to lose weight for health reasons. Most of the other clubs function on group therapy lines,[89] i.e. the members discuss their weight problems with others who are in the same boat; there is re-education of eating habits; a diet is prepared for each person, that can be maintained all the year round, and that is acceptable financially, socially and psychologically; recipes for non-fattening snacks are shared; experts discuss the principles of regulation of body size by regulating the food intake to meet the energy requirements of each individual. And here is the rub—members of a family can have different energy needs, one member being able to eat relatively large meals without gaining weight, while another member eating an equivalent amount becomes obese. Most of the time individuals rightly revel in maintaining their individuality, but for some individuals one of the things that they have to accept about their individuality is that their bodies need less food than that of the average person to maintain an average body weight.

Gout is now labelled a disease of civilization. Before the Second World War, perhaps 2 in 1 000 people suffered from gout. Professor Gries says that today the figure is 10 times higher.[90] Gout sufferers have an increased tendency to heart attacks and strokes.

In areas where there is an inadequate food supply leading to *malnutrition*, the infant mortality rate and the death rate for the 1 to 4 year olds is high. Those who do survive to go to school cannot get the maximum benefit from education. It is ironic that for the maintenance of good health, protein is an essential constituent of each *meal,* and yet the protein of high biological value is the most expensive item in the diet. In the process of industrialization, as incomes rise, more people can afford meat, fish and dairy produce and they increasingly neglect cheaper sources of vegetable protein. Over the last 10 to 20 years, there has been decreased consumption of peas and beans in Italy and France, and of bread in the United Kingdom and the United States, and yet all these countries continue to have a part of their population in an undernourished state! A survey[91] done in the United Kingdom stated that 25 per cent of boys and girls go without breakfast and may suffer impaired learning as a result. Dr Lynch said that the children were not victims of poverty, but of a lack of communication between school and home, and continued ignorance about nutrition which could only be remedied by intensive health education. Vagrants, young people living away from home for the first time, and people living alone—often in the older age group are at risk of being undernourished, sometimes from a financial point of view and sometimes because they do not cook, but exist on cups of tea and sandwiches, their diet lacking the minerals and vitamins to be found in fruit and vegetables. There is a limit to the distance that perishable food can be transported to an area of need. The distribution of food and its storage over longer periods are real problems of world-wide interest. The World Health Organization's policy is that chronic shortage of food can only be met by enabling people to increase their own *regional* and *local* yields. They cannot be hostages to the policies of other countries. In the developing countries there is a shortage of protein food and an abundance of children whose needs for protein are relatively greater than those of adults. Almost all children in these countries are breast fed, but they are usually weaned on to bulky, starchy foods, often due to local custom and lack of knowledge of nutrition. Kwashiorkor means 'the disease a child gets when the next baby is born'. Previously vast areas of land in the developing countries was unavailable for crop growing because of high soil salinity. This problem can now be overcome by the application of calcium salts. Fungi, algae, yeast and bacteria—'single cell protein sources'— can now be used to produce protein-rich powder and 'dough', as can fish, commercial oil[92] and coal. In the plant world, while fats and carbohydrates are found concentrated in seeds, grains and tubers, the leaves—from which protein can be extracted—are usually left to rot or are lost to insects and pests. Now the leaves from soya bean, cotton seed, pea, tare, clover, lucerne, cassova, corn and sorghum plants, together with other cellular waste can be made into a tasteless, colourless protein-rich powder and 'dough' on an industrial scale. It is the policy of the World Health Organization to encourage the industrial production of these supplemental proteins regionally and locally at prices that can be afforded by the poorest people in the district. Coupled with this, the World Health Organization supports educational programmes trying to reach even the people who still live in remote villages.

The protein rich powder can be made into drinks of almost any flavour suiting local custom.[93] The 'dough' can be spun to resemble any meat in texture and taste, so that it fits in with the existing food culture in any given community. One of the biggest problems before this scientific breakthrough was getting protein that was palatable to the malnourished communities. Furthermore the protein powder and 'dough' does not involve matters of conscience regarding methods of rearing (intensive animal husbandry) and slaughtering (some religions have special rules about this), or questions of health connected with antibiotics, hormones, fertilizers, herbicides, pesticides, etc., nor do they need inspection for salmonellosis and other animal diseases. In an emergency they are of less bulk to be transferred to an area of need, such as after a natural disaster, e.g. hurricane, earthquake.

Food intake is normally controlled by *appetite*, neither too much, nor too little being taken. The nervous system is supplied with information from receptors for the level of glucose and other nutrients in the blood, and the body water content.

Metabolism is the sum of all chemical reactions occurring in the body at a given time. These reactions include the *breaking down* of larger into smaller molecules (catabolism) with the release of ATP (adenosine triphosphate, Figs 6 and 7), and the *building up* of smaller into larger molecules (anabolism) which uses up ATP. You need to remember about metabolism because occasionally patients have their basal metabolic rate (b.m.r.) estimated as this can be below or above average in different diseases. The heat/ energy released when food is oxidized in the process of catabolism is measured in calories. The calorie is the amount of heat required to raise 1 cubic centimetre of water through 1 degree Centigrade. It is such a small unit of heat that the *kilocalorie*, which is 1 000 calories, is used in relation to food. A capital C is sometimes used for the kilocalorie, but it has led to so much confusion that some people use the word kilocalorie or its abbreviation kcal. There are others who think that we should forget the word calorie and in the interests of science use the correct Standard International (SI) unit, the *joule*. Baldwin[94] states that the change would be 'especially beneficial, since the concept of *energy* would become paramount, and thermometry could be made by considering the *temperature* change as the *external sign of internal energy*, and introducing all specific heats in the correct SI units—joules per kilogram per degree Kelvin'.

There are seven food groups from which we should normally partake daily in order that our tissues remain well nourished and healthy throughout life. These groups are proteins, fats, carbohydrates, water, vitamins, mineral salts and roughage.

PROTEIN AND NUCLEIC ACIDS[95,96,97,98,99]

Protein is an organic substance required throughout life to maintain the blood protein at 6 to 8 per cent and to build and repair the body tissues. It is a constituent of RNA, DNA and enzymes which are essential for metabolism in each cell. Since the human body cannot utilize the atmospheric nitrogen, it must rely upon protein for its nitrogen. Protein is a constituent of each cell and the day's supply needs to be shared out and a portion taken at each meal. Protein is a complex substance compounded of amino acids, of which there are 21, 10 of them being designated 'essential', since they cannot be elaborated within the body. Animal proteins contain the essential amino acids. Examples are meat, fish, eggs, milk, butter and cheese. Vegetable proteins contain a selection of amino acids, but not all the essential ones. Examples are the cereals—wheat, oats, rye and the legumes—peas, beans and lentils.

A baby's diet usually contains all animal protein in the first few months, and it is thought that the protein intake should be 50 per cent animal protein until full stature is reached. Adults can maintain health if one-third of the protein intake is of animal origin.

The metabolism of 1 gram of protein yields 4 kcal. In the adult 75 to 105 grams ($2\frac{1}{2}$ to $3\frac{1}{2}$ oz) is the daily requirement. In the digestive tract it is broken down into its constituent amino acids, which are absorbed from the gut into the blood capillaries in the intestinal wall. The amino acids are transported via the superior mesenteric vein into the portal vein and thus to the liver, where those needed are incorporated into liver cells under the influence of insulin (p. 392). Other amino acids circulate in the blood and are synthesized by the body cells into the actual proteins required, again under the influence of insulin. The amino acids not required for building and repairing the tissues return to the liver for deamination—the *non-nitrogenous* part is changed into a substance closely resembling fat, which is then metabolized with the production of heat and energy; the *nitrogenous* part is converted via an intermediary stage of ammonia, into urea (the main waste product from protein metabolism) which is transported in the circulation to the kidney for excretion. It follows therefore that the amount of urea in the urine bears a direct relationship to the amount of protein in the diet (Fig. 150).

Whenever the body cells are assaulted by injury or disease, healing can only take place if there is adequate protein intake.

Glucagon, a hormone secreted by the islets in the pancreas, as well as raising the blood sugar, facilitates the breakdown of protein into glucose. In starvation, this process is accelerated.

FAT AND FATTY ACIDS

Fat is an organic substance containing more carbon than the other food groups. It is required in the human body for the following purposes:

1. To act as a food store.

2. To protect and help to keep in position delicate organs such as the eye and the kidneys.

3. To provide heat, since each gram metabolized yields 9 kcal.

4. To form a sheath around the nerve fibres belonging to the central nervous system, for increase of their conducting power.

5. To form an insulating layer around the body.

The fats which are of animal origin contain vitamins A and D (calciferol). Recently research workers have propounded a theory that animal (solid) fat has a predominance of saturated fatty acids. The fats (liquid) which are of vegetable origin contain vitamin E (tocopherol), and many of them are marketed in an oil form. They have a predominance of unsaturated fatty acids, some of which are 'essential' because they cannot be manufactured by the body from other substances. A partial deficiency of these may produce defective arterial walls resulting in the deposit of atheromatous plaques in blood vessels, especially those in the coronary circulation. The roughened vessel wall encourages thrombosis. Recently there has been a trend towards greater usage of fats of vegetable origin for cooking.

Approximately the same amount of fat as protein should be consumed daily. The carbon contained therein requires considerable oxygen to break it down into the harmless end products of carbon dioxide and water. This is a complex process and some of the acids (ketone bodies) formed in the intermediate stages are poisonous in excess. To assist in the complete oxidation of fats, at least four times as much carbohydrate as fat should be consumed daily. Lack of *available* carbohydrate as in starvation and diabetes causes excess of ketone bodies in the blood—ketosis.

The metabolism of 1 gram of fat yields 9 kcals. In the adult 75 to 105 grams ($2\frac{1}{2}$ to $3\frac{1}{2}$ oz) is the daily requirement, some people believing that one-third of this should be of animal origin, the remaining two-thirds of vegetable origin. Can you explain the reason for this belief? Name the vitamins in animal and vegetable fat respectively.

In the digestive tract fat is broken down into glycerine and fatty acids. These are absorbed into the lacteals of the villi in the small intestine, via which they are transferred to the receptacle of chyle on the posterior abdominal wall. From here they rise in the thoracic duct, to be poured into the blood stream at the junction of the left subclavian with the left jugular vein. From the blood stream they are deposited in those tissues which need them, any remaining being stored in the fat depots of the body (Fig. 151). Growth hormone breaks down stored fat into fatty acids, this being necessary for the maintenance of metabolism (especially energy production) several hours after a meal.

CARBOHYDRATES

The starches and the sugars comprise this group of foodstuffs, and in the main they provide the energy for the body. They are built from a basic chemical unit—the saccharide—and in the order of complexity can be designated monosaccharides, disaccharides and polysaccharides. Two monosaccharides, ribose and deoxyribose are essential components of the nucleic acids, which play such an important part in cell metabolism and reproduction. Protein synthesis takes place in the ribosomes in cell cytoplasm and they are rich in nucleic acids. Ribonucleic acid (RNA) is found mainly in the cytoplasm. Deoxyribonucleic acid (DNA) is found mainly in cell nuclei and contains the genetic code.

The metabolism of 1 gram of carbohydrate yields 4 kcal. Four times as much carbohydrate as protein and fat, i.e. 300 to 420 grams (10 to 14 oz) is an average daily adult requirement. As previously mentioned carbohydrates help in the complete oxidation of fats.

In the digestive tract carbohydrates are broken down into monosaccharides, mainly glucose, and in this form are absorbed into the blood capillaries of the villi in the small intestine, and transported via the superior mesenteric vein into the portal vein and thus to the liver. There is a normal 'sugar' content of the blood, and in order that this should not rise unduly after a meal, glucose is transported by the blood to the liver and muscles, where the hormone insulin causes it to be converted into insoluble glycogen which is stored in the liver and muscle fibres and fat. A falling blood sugar causes conversion of this insoluble glycogen, by glucagon, back into its soluble form, glucose. Together with the action of growth hormone (p. 392), the blood sugar is maintained within the limits of 80 to 120 milligrams (mg) per 100 millilitres (ml) of blood (Fig. 152).

When carbohydrates are consumed in excess of body needs, more glucose is converted into fat, hence the omission of bread, potatoes and pastries from a slimming diet! Excessive sugar can also cause dental caries, hyperlipidaemia (extra fat in blood) with its associated risk of coronary atheroma, diabetes and such skin disorders as acne and seborrhoea. Brain cells rely entirely on glucose for energy and they can be permanently damaged by a low blood sugar. Most other body cells can use fat or protein for energy when there is a low blood sugar.

In the body, under the influence of the hormone, cortisol, (p. 394) glucose can be made from protein and fat. This process is called gluconeogenesis.

WATER (See p. 255)

In its importance to life this is equal to oxygen, for man can live temporarily without food but not without water. Man is composed of 75 per cent water, divided between intracellular fluid, extracellular fluid and transporting liquids—mainly blood and lymph. There is a considerable water content in most solid foodstuffs. Some water is produced within the body from metabolism. This is estimated at 200 to 300 ml/day and should not be forgotton when a patient is on a measured fluid intake and output. In addition to this, each individual requires to drink 1 to 1·75 litres (2 to 3 pints) daily. Water forms the basis of all secretions and body fluids via which each cell receives food and oxygen and gets rid of its waste products. Water dilutes toxins, assists in their removal from the body, and prevents constipation. According to the earth through which water has seeped it contains a variety of elements; when taken at the completion of a meal water washes food particles from between the teeth. As the main constituent of sweat, water helps to maintain normal body temperature.

An *increased* water intake is needed when there is deficient elimination from the bowel, in fevers and all toxic conditions, in rheumatism and gout because of excess uric acid, after excessive loss of fluid as in diarrhoea, vomiting, sweating and bleeding.

The body's water balance is maintained by a complicated *interaction* of the composition of blood, antidiuretic hormone and cortisol. Thirst gives the warning that the body is short of water.

VITAMINS[100]

These are sometimes called accessory food factors. They are chemical factors contained in foodstuffs. Some of them have been synthesized and are available commercially. Deficiency of them is relatively rare in the west. Only small amounts are needed as the molecules are used over and over again in the body. An increased amount is needed during rapid growth, such as pregnancy, lactation, childhood and repair of diseased tissues. Sterilization of the gut by antibiotics and sulphonamides reduces vitamin production in the gut.

Some vitamins are soluble in oil and are called *fat-soluble*. In any condition where there is lack of bile or pancreatic juice in the digestive tract there is minimal absorption of the fat-soluble vitamins. Others are soluble in water and are called *water-soluble*. Some are less stable than others and are destroyed by heat, by alkaline or acid reaction, and by exposure to air. The unfolding of more knowledge about the vitamins is a thrilling story belonging to this century.

Vitamin A (*Retinol*). This *fat-soluble* vitamin is not destroyed by cooking, but can be destroyed by over-long exposure to air. It is *anti-infective,* i.e. it protects the body from infection as its main function, but it is also necessary for the growth of children, and to the eyes for the perception of light. Shortage of vitamin A, which is normally stored in the visual purple of the eye (as well as the liver and other tissues) can give rise to night blindness (nyctalopia), which is a maladaptation of vision to darkness. Vitamin A is necessary for the health of epidermal tissues, and a glance at Figure 13 on page 15 will refresh your memory about these. Broadly speaking, they form the skin and mucous membranes, so that deficiency can give rise to (*a*) skin diseases (dermatoses); (*b*) drying of the conjunctiva, with consequent ulceration (keratomalacia, which causes blindness in tens of thousands of young children each year, in the developing countries; and (*c*) lack of resistance to infection in the respiratory, digestive and genito-urinary tracts.

Vitamin A can be elaborated in the liver from a pigment (carotene) contained in green and yellow vegetables and fruits. Thyroxine is necessary for this conversion. The pigment is often spoken of as a precursor or previtamin.

Liquid paraffin interferes with the absorption of vitamin A from the intestinal tract.

224

The daily intake for health is thought to be 2 500 international units. Suggestions as to how these can be obtained are contained in *Manual of Nutrition*, published by Her Majesty's Stationery Office. In Britain margarine has had 500 international units of vitamin A added to each ounce. Otherwise meat, fish and dairy produce form our main supply, the concentration in these usually being highest in the summer when green grass is available. Overdosage of vitamin A is rare, but may cause anorexia, loss of weight, fever, enlarged liver and tenderness of the long bones.

Vitamin B (*water-soluble*). This term is now known to cover a group of substances often found together in the same foods. At first numbers were added to the several members, but over the years it has become preferable to refer to them by name. Vitamin B complex is sometimes called *antineuritic*, i.e. it prevents neuritis.

Aneurine or thiamine. Aneurine is the British and thiamine the American name for what was previously called *vitamin B_1*. It is slowly destroyed by cooking and always by intense heat. It is readily destroyed by alkali. The greatest loss in vegetables is due to the aneurine becoming dissolved in the water which is so often discarded. Aneurine is concerned with the steady and continuous release of energy from carbohydrates, and mild deficiency gives rise to loss of appetite, fatigue, indigestion, constipation, irritability and depression, whereas more severe deficiency causes polyneuritis and beriberi, fortunately seldom seen in Britain. (The former *is* seen in conjunction with alcoholism.) This has led to the use of *antiberiberi* as a descriptive term for aneurine. Beriberi is a form of severe neuritis, frequently accompanied by cardiac symptoms and generalized oedema. Aneurine is present in large quantities in wheat germ and bran, but milling, polishing, refining, etc., eliminate it from such foods as rice and white flour. From January 1941 all flour sold in Britain has had aneurine added to it. Unrefined grain products, yeast, peanuts, bacon, meat, green peas, cabbage and milk, all yield aneurine for our daily intake.

Riboflavine (vitamin B_2). This is a yellow substance that possesses a green fluorescence when in solution; the slight green fluorescence of whey is attributable to the riboflavine. Riboflavine is destroyed by bright light, so milk should not be left on the doorstep for long. Some is made in the human intestine by bacteria. This vitamin forms an essential link in the metabolism of carbohydrate and protein. Deficiency causes maceration at the corners of the mouth (cheilosis); the tongue becomes red and sore; there is itching and burning of the eyes, blurred vision, conjunctivitis, and sensitivity to light (photophobia).

Among the everyday foods the best sources of riboflavine are dairy produce, liver and eggs.

Nicotinamide. This is called niacin in America, and it forms yet another link in the chain of carbohydrate metabolism from which the body gets its energy. Deficiency produces changes in the skin and in the digestive and nervous systems; in severe degree this gives rise to the condition of pellagra which has been called the disease of the

225

three 'D's'—dermatitis, diarrhoea and dementia. The descriptive term of *antipellagra* has been given to nicotinamide.

Meat, poultry and fish are among the everyday foods that are the best sources of nicotinamide. Yeast and peanuts contain a considerable amount, but are not consumed in sufficient quantity to rely on them. A little is manufactured in the intestine by its natural bacterial flora but antibiotics and sulpha drugs interfere with this. The body can manufacture nicotinamide from the amino acid tryptophan.

Folic Acid. This vitamin is necessary for blood regeneration and participates in the metabolism of nucleoproteins. Rapidly dividing cells need a lot of folic acid, e.g. (1) the growing fetus, and lack of folic acid can give rise to anaemia in pregnancy; (2) cancer cells, so that drugs that interfere with the action of folic acid can be used as anticancer agents. Deficiency in man is not likely from dietary defect, but in the tropics changes in the intestinal bacteria lead to a lessened production that damages the intestinal wall and gives the condition tropical sprue, characterized by persistent diarrhoea. Liver, kidney, yeast and green leafy vegetables are good sources of folic acid, and many other foods contain a little folic acid.

Pyridoxine (vitamin B_6). This is closely associated with protein metabolism. Yeast, liver, cereals and pulses are the best sources; there are small amounts in milk and green vegetables, so that deficiency is unlikely to occur in this country. The antitubercular drug isoniazid (Rimifon) can inactivate pyridoxine, thus signs of deficiency can occur with a normal diet. Convulsions are a warning sign.

It has been administered therapeutically for nausea and vomiting especially in post-irradiation sickness.

Cyanocobalamin (Cytamen). This is still frequently referred to as vitamin B_{12}. It is also called the *antipernicious anaemia factor,* or Castle's extrinsic factor. Its absorption depends on the presence of acid and Castle's intrinsic factor in the stomach. A little is manufactured in man's intestine by the natural bacterial flora but antibiotics and sulpha drugs interfere with this. Cyanocobalamin is essential for the maturation of red blood cells in the bone marrow, as you learned previously. Deficiency produces pernicious anaemia, so named as it was a fatal disease until cyanocobalamin was extracted as a by-product in fermentations which produce antibiotics, such as penicillin and streptomycin. Unfortunately, cyanocobalamin has to be given by injection, but the condition of pernicious anaemia can be kept under control and the blood picture normal if the necessary injections are given at intervals.

Experimental findings suggest that cyanocobalamin may influence the maturation of human spermatozoa.

Inositol, pantothenic acid, choline, biotin and *paraminobenzoic acid.* These are all components of the vitamin B complex, but as yet little is known about them.

Vitamin C *(Ascorbic Acid).* This is *water-soluble* and easily destroyed by cooking. It is *antiscorbutic*, i.e. it prevents scurvy, characterized by great debility, anaemia, bleeding into the skin and subcutaneous tissues, and from the mucous membrane. Vitamin C is necessary for collagen formation in connective tissue. Ascorbic acid must be taken daily as any excess is excreted in the urine. Fresh vegetables and citrus fruits are good supplies. Previous studies showed that 28 per cent of the nation's dietary intake of vitamin C was obtained from potatoes. The now favoured 'instant mashed' potato contains hardly any vitamin C. A further reduction in vitamin C is caused by 'forcing' fruit and vegetables to be ready at an unnatural time. One cigarette destroys about 25 mg of vitamin C so that those smoking more than 15 cigarettes daily may have a 50 per cent reduction of serum vitamin C.

Ascorbic acid is available commercially in the form of convenient tablets, which can be given orally in conjunction with iron in the treatment of anaemia; it is often given as a pre-operative course of treatment as a good supply assists in the healing of wounds by collagen formation.

As ascorbic acid is so unstable, babies in this country are given supplies which their mothers procure from the Infant and Child Welfare Clinic.

Vitamin D *(Calciferol).* This is *fat-soluble* and not destroyed by cooking. It is *antirachitic,* i.e. it prevents rickets, a disease in which the bones are soft and bent, teeth late and carious, and there is an accompanying anaemia and liability to chest and intestinal infections.

There is a similar substance in the skin (ergosterol), and this elaborates calciferol when stimulated by the ultraviolet rays of the sun.

Calciferol must be present in the blood stream to facilitate the absorption of calcium from the intestinal tract. It is also necessary for normal bone growth. Calciferol is found in irradiated ergosterol, cod and halibut liver oil, liver, fat fish and all dairy produce. From 1957 the calciferol content of baby foods in Britain has been as follows:

Cod liver oil, 100 international units per gram or millilitre.

National Dried Milk, 100 international units per ounce (dry).

Infant cereals, 300 international units per ounce (dry).

Daily Requirements. Two months to one year, 400 international units daily from all sources; 1 to 5 years, 400 international units to supplement the daily diet of 400 units

Examples. When using *National Dried Milk,* 7 to 8 drops of cod liver oil gives 50 international units daily. By the time the baby is weaned a maximum dose of 1 teaspoonful (approx. 4 ml), i.e. 400 international units daily is required. When using *breast* and *cow's milk,* half a teaspoonful twice daily from two months onwards will suffice.

Mild deficiency of vitamin D leads to poor calcification of bones and teeth; a more severe deficiency produces rickets.

Hypervitaminosis D. This word literally means too much vitamin D. The condition leads to too much calcium in the blood stream (hypercalcaemia), which can lead to the formation of stones, e.g. in the kidney.

Vitamin E[101,102] *(Tocopherol).* This substance is *fat-soluble* and *anti-abortive*. It has been proved to be necessary for normal fertility of rats. It is claimed that it helps to prevent abortion in human beings, but as yet this has not been proved. It has some effect on the normal development of the muscular and nervous systems, and is at present on trial in the treatment of such conditions as muscular dystrophy, etc. In 1971 a doctor in New Zealand suggested that a shortage of vitamin E and a chemical called selenium could be the cause of unexpected 'cot deaths' of babies.

It is remarkably stable and has a wide distribution, especially in wheat germ and other seeds, green leafy vegetables, nuts and legumes.

Vitamin K *(Menadione).* This vitamin is *fat-soluble* and easily destroyed by acids, alkalis, light and oxidation. It is *anti-haemorrhagic*, i.e. it prevents excessive bleeding. Bile must be present in the intestinal tract before menadione can be absorbed. This vitamin has been prepared synthetically and is available as injections; more recently an oral preparation has appeared on the market. The intestinal bacterial flora manufacture some menadione. In the liver it is essential for the production of prothrombin, without which blood cannot clot. Menadione is present in green leaves, egg yolk and liver, and deficiency is hardly likely to arise from diet. It can arise from (1) absence of bacteria in intestine, especially during the first few weeks of life, and after administration of sulphonamides and antibiotics; and (2) absence of bile in the intestinal tract.

Vitamin P *(Hesperidin).* This was previously called *citrin*. It is *water-soluble* and has been found to be more effective than ascorbic acid in the treatment of increased capillary permeability; it is therefore sometimes called *antipurpura*. It has also been tried in the treatment of rheumatic fever, to lower the erythrocyte (red blood cell) sedimentation rate.

Hesperidin is found in rose hips, citrus fruits and blackcurrants.

Those vitamins that work in association with enzymes in the process of metabolism are sometimes spoken of as co-enzymes.

ELEMENTS

These are next in our list of essential factors that have to be consumed daily in order that our tissues remain well nourished and healthy throughout life.

Calcium. This is a light metal and a weak alkali, being soluble in an acid medium. (A *slightly* acid medium, e.g. saliva pH 6·8 does not dissolve calcium.) Human beings can take it orally in milk, milk compounds and vegetables, but calciferol (vitamin D) must be present in the intestine to facilitate absorption of calcium into the blood stream. In combination with phosphorus it is essential for hardening our bones, nails and teeth. (If starchy foods are left in between the teeth, the bacteria normally present in the mouth ferment the starch, producing an acid and a gas. The calcium in the teeth dissolves in this more acid medium and thus the first stage of dental caries is accomplished.)

Calcium is also necessary for the conduction of an impulse along a nerve fibre, and where appropriate the subsequent contraction of muscle tissue. This takes place normally when there are 9 to 11 mg of calcium per 100 ml of blood. Should this level fall (hypocalcaemia), there is interference with neuromuscular function and the condition of tetany ensues. Should this level rise (hypercalcaemia), as in childhood when there is an associated acidosis, then anaemia, failure to thrive, vomiting, constipation, and deposition of calcium in the kidneys (renal calcinosis) ensue. Using the bones as a storehouse or depot, various enzymes and hormones (chiefly the secretion of the parathyroid glands) play their part in maintaining a constant calcium content of the blood.

Mentioning blood brings us to the last function of calcium, i.e. it is essential for the clotting of blood. Later you will learn how the laboratory staff utilize this knowledge when they need a specimen of unclotted blood.

Phosphorus. This is a non-metallic solid, present in all body cells, especially in the nuclei. Most of it is found in the bones in conjunction with calcium. It is carried in the blood, 2·6 to 4 mg per 100 ml being normal.

It is contained in milk, egg yolk, brown bread, oatmeal, peas, beans, meat and fish.

Sodium. This is present in most foods in the form of sodium chloride (common salt). It regulates the acid-base balance and maintains the osmotic pressure of body fluids, being present in the blood in the portion of one teaspoonful to one pint of blood (0·9 per cent). The kidneys *normally* excrete 3 to 5 grams of chloride per litre of urine, which is increased or decreased according to any change in the amount of chlorides in the tissues. Salt is lost from the body via the bowel and the skin—we have already mentioned the giving of salted drinks to those whose work causes heavy perspiration. Vomiting, and to a less extent diarrhoea, causes pathological loss of chlorides.

Potassium. Potassium is required in the body for cell building, for the maintenance of the fluid within the cell (intracellular fluid), and for the regulation of neuromuscular activity. The normal amount found in blood is 14 to 21 mg per 100 ml of blood. When this level falls there is nausea, drowsiness, muscle weakness, low blood-pressure with a warm, dry skin. Heart failure can supervene.

Under conditions of stress, e.g. accident or operation, potassium leaves the cells and thus increases the potassium content of the blood plasma. If urinary excretion of potassium does not take place, e.g. in a decreased urinary output (oliguria) which accompanies the condition of shock, the increased potassium in the blood can cause sudden heart failure. The heart is sensitive to the amount of potassium in the blood and can fail from too little or too much.

Iron. Iron is essential for the production of haemoglobin in the red blood cells, via which oxygen is transported round the body. Iron is more soluble in an acid medium, so those people who do not secrete much hydrochloric acid in the stomach may have difficulty in absorbing the iron from their food. Shortage of iron gives rise to one form of anaemia.[103] Some proprietary infant milk foods are fortified with iron to as much as 1·1 mg per oz (1 oz = 26·34 g). When red blood cells are destroyed within the body the iron is saved, when they are lost from the body as in bleeding the iron is lost.

Good sources of iron are lean beef, liver, egg yolk and green and yellow vegetables. Milk does *not* contain iron, which fact must be remembered when using milk for infant feeding.

Iodine. This is a non-metallic element, obtained commercially from seaweed. It is present in most green vegetables, which in turn have derived it from the soil or air. Where the soil is deficient in iodine, as in land far removed from the sea, the vegetables cannot be relied upon to supply sufficient iodine to the population. Under the 1950 Salt Act all table salt must contain 15 to 30 parts of iodine per 10 million parts of salt.

Iodine is essential in the thyroid gland for the production of thyroxine, the hormone that controls the *rate* of metabolism through-the body.

Magnesium. Necessary for normal functioning of neuromuscular tissue. Lack of it causes muscle spasm, irritability and eventually unconsciousness. In severe diarrhoea large amounts of magnesium can be lost in faeces. It has been found that patients maintained on intravenous fluids can become magnesium deficient.

Fluorine. This substance is present in cow's and human milk in a concentration of 0·05 to 0·25 mg per litre, and in some drinking waters—one part per million parts of water. Fluorine forms hard teeth. In excess it 'mottles' the teeth. In deficient areas it can be added in the proportion of one part to one million parts of water.

The remaining elements found in the body are needed in such small amounts that they are unlikely to be absent from the diet favoured by most people in this country.

This brings us to the last item, *roughage*, in our list of essential food groups.

ROUGHAGE

Roughage is the indigestible part of food, which in the human being gives bulk to the intestinal contents, thus encouraging peristalsis and preventing constipation.

Fruit and vegetables contain a high percentage of cellulose which forms the main part of roughage. In cases of increased frequency of bowel evacuation a low residue diet is given.

BABY REQUIREMENTS

The sucking reflex is present at birth and breast feeding is considered by many to be the ideal for any baby, not only because of the advantages that accrue from breast milk, but because in this natural process the first feelings of security are said to be engendered within the baby (p. 217). The advantages of breast milk are:

1. Protein. Lactalbumin predominates and this is more easily digested than casein.

2. Fat. Smaller globules increase the surface area exposed to the digestive enzymes.

3. Breast milk contains iron and more vitamins provided that the mother is taking an adequate diet.

As well as nutritional advantages, breast milk is ready on demand, it is at the correct temperature and because of lack of apparatus involved there is less likelihood of the baby getting gastroenteritis. Breast feeding is recommended in an attempt to prevent sudden unexplained death in infants. It is thought that the breast-fed baby has a comparatively wider palate and better-shaped arch to the roof of the mouth.

If for any reason the baby is not breast fed, then it is important that he is not deprived of being nursed during feed time. (After feeding he should be nursed in the sitting up position, or against the mother's shoulder, so that he brings up any swallowed air.) The dietary requirements of babyhood are 165 ml per kg of expected body weight (2½ fl. oz per lb) which should provide 120 kcal per kg (45 to 50 kcal per lb). Healthy babies are usually fed four hourly, missing the night feed, so this means five feeds—6 a.m., 10 a.m., 2 p.m., 6 p.m., 10 p.m. Cow's milk contains 20 kcal in each 30 ml (one ounce), but it is not sufficiently sweet for a young baby and sugar has to be added. It takes one and a half teaspoonfuls of sugar to yield 20 kcal. Most of the dried milks have the calorie content printed on the tin, as have the evaporated and the condensed milks.

We have already learned of the vitamins that have to be given to all babies. Can you remember them?

Weaning. In most households this is accomplished with very little trouble. The baby is mastering a new set of muscular movements, i.e. he is learning to chew instead of suck, and the attempt may prove somewhat messy at first! The psychological aspect of weaning is mentioned on page 198.

ADULT REQUIREMENTS

These vary according to height, weight, activity, age, climate, etc., but from all that we have said, you will be able to work out the kilocalorie content of a 65, a 70, a 75, an 80, an 85, a 90, a 95, 100 and a 105 gram protein diet! Example:

Diet containing 65 grams of protein

Protein }	Equal quantities {	65 grams × 4	=	260
Fat }		65 grams × 9	=	585
Carbohydrate (four or five times more than protein and fat) . . .		260 grams × 4	=	1,040
				1,885

Diet in pregnancy. The British Medical Association suggests an intake of 96 grams of protein during pregnancy. The pregnant woman's diet has to provide for:

1. The needs of the growing fetus.
2. The maintenance of maternal health.
3. Physical strength and vitality during labour.
4. Successful lactation.

The iron content needs to be stepped up to 15 mg daily; calcium to $1\frac{1}{2}$ to 2 grams to provide sufficient for the primary ossification centres (p. 43) without drainage from the mother's bones and teeth.

Diet in lactation. A diet similar to that in pregnancy is suitable in lactation, and many authorities say it must provide 3,000 kilocalories.

In considering nutrition in relation to health, not only is the content of the diet important, but the cleanliness of the food being served, the manner in which it is served and the emotional frame of mind of the consumer. The surroundings in which the food is served, the efficiency of the teeth or dentures to chew and grind the food, a healthy mouth free from soreness, and the presence of an appetite for food, are also important factors.

DIET FOR THE ELDERLY[104]

When people become frail with the ageing process, they can have difficulty with an adequate intake of food. Sometimes dentures fit less well, due to shrinkage of the mouth. Chewing becomes arduous. Considerable imagination is needed to make a 'soft' diet palatable and to prevent monotony. Drinks can be reinforced with protein powder—Casilan. Fresh fruit needs to be cut into small portions and served as fruit salad. Providing sufficient roughage can be difficult, but it is necessary to prevent constipation—to which the elderly are further predisposed because of insufficient energy to partake of adequate exercise.

WHOLESALE STORAGE OF FOOD

The World Health Organization has developed international food standards of hygiene and these can be obtained from WHO offices in Geneva. Currently, 'Food Hygiene in Britain' is being investigated by the Association of Public Health Inspectors. Parliament is also investigating the subject of date-marking food sold to the public.

As food attracts vermin, it is important that storage places are controlled, so that they conform to minimum standards of hygiene. They must be sufficiently large for order to be maintained, and for like foods to be stored together. The walls should be of impervious material that can easily be washed. Dampness causes moulds to grow on many foods. There should not be a skirting board, for vermin can gnaw through wood. The floor should be of washable, hard material and never of wood. Ventilation must be good and all windows should be protected with fine wire mesh to keep out the flies. Natural and artificial lighting should be provided. Wherever possible

232

goods should be stored in metal or enamel containers to prevent spoilage and contamination by beetles, cockroaches, rats, mice, etc.

Adequate cloakroom, lavatory and washing accommodation must be provided, and the staff educated in their proper usage and encouraged in their important contribution in the prevention of disease.

HYGIENE OF FOOD SHOPS[105]

Rules and regulations are laid down, and such places are inspected at regular intervals. The public are encouraged to make their complaints active and voluble. It is thought that local authorities should be empowered to close dirty cafés, especially in seaside resorts. It takes weeks to prosecute, and finally the owners are punished by a fine—which does nothing to prevent the ill health caused in the meantime. Cloakroom, lavatory and washing facilities have to be provided, and a separate sink for the disposal of water used for washing the floors, walls, etc. All surfaces must be washable; wood-topped tables, etc., are condemned. All food must be covered and must not be handled in the serving. Ventilation must be such that the temperature does not rise sufficiently to spoil the food. All windows must be covered with wire mesh. Animals should not be allowed in food shops. Covered, refrigerated shelves are best for fish and meat. In the organization of the work it is preferable if those actually serving food do not handle the money; it is much more hygienic to have a central cash desk. Clean paper bags should be used for wrapping food, and the practice of wrapping in newspaper is to be condemned.

The public need to become much more outspoken about any deviation from 'clean' food that they find in shops or restaurants. Any complaints should be made immediately to the local Public Health Inspector.

QUALITY OF FOOD

For commercial purposes, colouring matter, anti-oxidants, preservatives, stabilizers, maturing agents, etc. are added to food to standardize and improve appearance, or to extend the 'shelf life'. No two countries can agree on which coal tar dyes are safe and the Food Standards Committee recommend that those at present permitted in Britain should be banned within the next few years. The World Health Organization is working towards an internationally accepted list of food additives.

Adulteration of food, e.g. sand in sugar, potato flour in bread, turnips in jam and starch to thicken cream have become so certain of discovery that their use is no longer worth risking.

Antibiotics and hormones cause spectacular weight increases in farm animals and their use is now widespread throughout the world. However, there is increasing disquiet about these practices.[106,107]

Isotopes (particularly Strontium ^{90} and Iodine ^{131}) in food from nuclear fall-out has caused concern. The Medical Research Council publish recommendations which provide a yardstick by which to assess the significance of figures on the amount of radioactivity in food (particularly milk) published by the Agricultural Research Council.

Pesticides and herbicides are being used increasingly in horticulture. The herbicide —245T, was banned after reports of deformed babies in Vietnam. The Weights and Measures Inspectors take part in on-going surveys of pesticide residues in foodstuffs. The Food and Agriculture Organization in conjunction with the World Health Organization are vigilant about the quality of food throughout the world and offer advice to the individual nations.

Cyclamates, saccharin and sorbitol can all be used as sweetening agents, and are particularly useful for diabetics. However, many Governments have now banned the use of cyclamates. In Britain monosodium glutamate is banned from baby foods.

Increasing pollution of the sea and inland waters has caused an increased mercury content of fish—particularly tuna. Some countries have banned the sale of tuna, others are keeping the situation under surveillance.

The latest concern is about cytoscan cyclophosphanide which when sprayed on sheep, causes the wool to drop off. The question, 'Could this cause baldness, or have other side effects in man who eats the mutton?' was asked in Parliament. It was decided to prohibit this in Britain until more is known about long-term results.

Summary of Food Groups

PROTEINS AND NUCLEIC ACIDS

These form the only source of the body's nitrogen supply, and they build and repair tissue. They are a constituent of RNA, DNA and enzymes, essential for metabolism in each cell. The blood protein must be maintained at 6 to 8 per cent. Proteins are composed of amino acids, of which there are 21, 10 of these being 'essential,' since they cannot be elaborated within the body. Animal proteins contain the essential amino acids, vegetable proteins contain a selection of amino acids. One gram of protein yields 4 kcal and 75 to 105 grams should be taken daily (30 grams equals 1 oz).

In the body, proteins are broken down into their constituent amino acids, which are absorbed into the blood, and taken via the portal circulation to the liver and some is rebuilt into protein. Other amino acids are synthesized by the body cells into the actual proteins required. Of those remaining some are converted into a substance resembling fat, for energy; and some into urea which is excreted via the kidney.

FATS AND FATTY ACIDS

These are known as the heat producers, those of animal origin

234

contain vitamins A and D (calciferol), those of vegetable origin contain vitamin E (tocopherol). Vegetable fat has a predominance of unsaturated fatty acids, deficiency of which may produce defective blood vessel walls that encourages thrombosis. Some authorities advocate that the daily intake should contain two-thirds vegetable and one-third animal fat. One gram of fat yields 9 kcal and 75 to 105 grams should be taken daily.

In the body fats are broken down into fatty acids and glycerine which are absorbed into the lymphatic system, via which they reach the blood stream. They are metabolized to produce heat; any needed by the tissues are laid down; any remaining goes to the fat depots. The complete metabolism of fats into the harmless products of carbon dioxide and water requires a lot of oxygen. If this oxygen is not available poisonous substances are formed and these lead to acidosis. The functions of fat are to act as a food store, to protect delicate organs such as the eye and the kidney and help to keep them in position, to produce heat, to insulate white neurones in the central nervous system, and to form an insulating layer around the body.

CARBOHYDRATES

These are known as the energy producers and are built from a basic unit called a saccharide into monosaccharides, disaccharides and polysaccharides. Monosaccharides are the simple sugars such as glucose, disaccharides are the more complex sugars and polysaccharides are the starches. Two monosaccharides, ribose and deoxyribose are essential components of the nucleic acids RNA and DNA. One gram of carbohydrate yields 4 kcal and 300 to 420 grams should be taken daily.

In the body carbohydrates are broken down into monosaccharides, absorbed into the blood, taken to the liver, metabolized to produce required energy. Glucose is the only source of energy that can be used by the brain cells, which can be permanently damaged by a low blood sugar. That not needed immediately is converted into insoluble glycogen and stored in liver and muscle; any remaining is converted into fat for storage. Excessive carbohydrate can predispose to dental caries, coronary atheroma, diabetes, acne and seborrhoea. Normal blood sugar is 0·08 to 0·12 per cent, and it is maintained by insulin converting glucose into insoluble glycogen, and falling blood sugar causing conversion of glycogen back into soluble glucose.

WATER

Water is essential to life. Of man's 75 per cent water, it is divided between intracellular fluid, extracellular fluid and the transporting liquids, mainly blood and lymph. In the body it is necessary for the digestion, absorption and distribution of foodstuffs; it helps to maintain regular body temperature; it is necessary for elimination of waste products. It is taken into the body as fluids and in food, and is produced within the body from metabolism. One and three-quarter litres (3 pints) of water should be taken daily and an increased amount is needed when there is deficient elimination from the bowel, in fevers and all toxic conditions, to dilute and aid in the elimination of toxins, in rheumatism and gout because of excess uric acid, after excessive loss of fluid as in diarrhoea, sweating, vomiting and bleeding.

VITAMINS

Vitamins are essential accessory food factors, and their absence in the diet leads to deficiency diseases.

Vitamin A. This is fat-soluble and anti-infective. It is necessary for the health of the skin and mucous membranes, and to prevent night blindness. It can be elaborated in the body from the pigment carotene contained in yellow vegetables and fruit. Liquid paraffin interferes with absorption of this vitamin.

Vitamin B. This is water-soluble and is divided into several factors.

Aneurine. Aneurine is the antineuritic factor; in some countries it is called the antiberiberi factor. Severe deficiency causes polyneuritis (which can occur in alcoholics), and beriberi. Previously called vitamin B_1.

Riboflavine. Deficiency of this factor causes cracks at the corners of the mouth. It is destroyed by strong light.

Nicotinamide. This is the antipellagra factor. Pellagra is a disease of the three 'D's'—dermatitis, diarrhoea and dementia. The intestinal bacterial flora synthesize a little nicotinamide.

Folic acid. Folic acid is necessary for blood formation and metabolism of nucleoproteins. Rapidly dividing cells need a lot, therefore pregnancy can give rise to deficiency with consequent anaemia. Cancer cells need a lot, so drugs that interfere with folic acid metabolism can be used as anti-cancer agents.

Pyridoxine. Previously called vitamin B_6 pyridoxine is sometimes given for nausea and vomiting particularly that associated with deep X-ray treatment. Antitubercular drugs can inactivate pyridoxine leading to deficiency. Convulsions are a warning sign.

Cyanocobalamin. Previously called vitamin B_{12} it is the antipernicious anaemia factor. It is the extrinsic factor of Castle, and it needs an acid medium and the intrinsic factor of Castle from the gastric mucosa for its absorption.

There are several other identified factors, but as yet so little is known about them that they need not be remembered by name.

Vitamin C *(Ascorbic Acid).* This is water-soluble and antiscorbutic, i.e. it prevents scurvy. It is present in all citrus fruits and is given to babies; it is obtainable from clinics. It is necessary for healing tissues.

Vitamin D *(Calciferol).* Fat-soluble and antirachitic, it prevents rickets by facilitating the absorption of calcium from the intestine. It is elaborated from ergosterol in the skin when irradiated by ultraviolet rays. Hypervitaminosis D leads to hypercalcaemia.

Vitamin E *(Tocopherol)*. Fat-soluble and thought to be anti-abortive. It is present in wheat germ. Modern thought is that a deficiency of vitamin E and the chemical selium might be the cause of cot deaths in babies.

Vitamin K *(Menadione)*. This is fat-soluble and antihaemorrhagic, i.e. it prevents bleeding. Only absorbed in the presence of bile.

Vitamin P *(Hesperidin)*. Water-soluble and antipurpuric, i.e. it prevents purpura which is the oozing of blood into the skin and mucous membranes.

ELEMENTS

Essential factors for the maintenance of health.

Calcium. This hardens bones, nails and teeth. It is needed for conduction of an impulse along a nerve fibre, for contraction of muscle, and clotting of blood; 9 to 11 mg normally present in each 100 ml of blood. Vitamin D (calciferol) is necessary for its absorption.

Phosphorus. This is needed with calcium for the hardening of bones, nails and teeth; 2·6 to 4 mg normally present in 100 ml of blood.

Sodium. Common salt is sodium chloride, and in the body it regulates the acid-base balance. It is lost in sweat and any excess is excreted by the kidneys. There is one teaspoonful in each pint of blood, or 0·9 per cent.

Potassium. The blood needs to be maintained at 14 to 21 mg of potassium per 100 ml of blood, but most of the body's potassium is within the cells (intracellular fluid). Under conditions of stress, e.g. accident or operation, potassium leaves the cells and increases the amount in the blood, which can cause sudden heart failure. When there is a lowered blood potassium there is nausea, drowsiness, muscle weakness; the blood-pressure is low with a warm dry skin, and heart failure can supervene.

Iron. This element is essential for the haemoglobin of the red blood cells. Deficiency causes anaemia, and this can happen in babies fed on cows' milk because it does not contain iron. When red blood cells are destroyed within the body the iron is saved, when they are lost from the body as in bleeding the iron is lost.

Iodine. Iodine is necessary for the formation of thyroxine by the thyroid gland. Thyroxine controls the rate of metabolism. A simple goitre can arise in response to insufficient iodine, and where there is likely to be an insufficiency, iodine can be added to table salt.

Fluorine. An element present in some drinking waters. It hardens teeth and for this reason can be added to drinking water in deficient areas.

Magnesium. Necessary for normal functioning of neuromuscular tissue. Lack causes muscle spasm, irritability and eventually unconsciousness. Large amounts can be lost in diarrhoea.

ROUGHAGE

Indigestible cellulose that gives bulk to the intestinal content and prevents constipation.

Give the advantages of breast milk.
How much fluid should a baby have in 24 hours?

Man, his Health and Environment

How many kcal should a baby be given in 24 hours?
How many kcal are there in 30 ml (1 oz) of milk?
How much sugar does it take to yield 20 kilocalories?
If you are told the protein content of a diet, how do you work out the rest of the diet and the kcal yielded by same?
What must the diet in pregnancy provide for?

FOOD POISONING

This is a condition of acute inflammation of the lining of the stomach, and/or the bowel as a result of eating or drinking contaminated food. The time of onset of symptoms after ingestion varies according to the type of infection, and the symptoms vary from mild abdominal disturbance and diarrhoea to severe abdominal colic with profuse diarrhoea and vomiting, leading to dehydration terminating in collapse of the afflicted person.[108]

Many types of bacteria (germs) can cause food poisoning. Bacteria are minute unicellular organisms that require food, moisture and warmth for their multiplication, which is accomplished by mitosis every 20 to 30 minutes (p. 8). They are spherical (cocci) or rod-shaped (bacilli) and are 1/1,000th mm in diameter or length, much too small to be seen by the naked eye; indeed a million growing together on a culture plate are about the size of a pin head, and such a cluster is referred to as a 'colony'. It is little wonder that man finds it difficult to appreciate their widespread presence.

Shortly after birth man's skin is 'colonized' by various types of cocci (streptococci and staphylococci), and thereafter can never be sterilized, but throughout life effort has to be made to keep this germ population of the skin within manageable proportions (p. 125). At least twice weekly changing of underclothes and a weekly bath, with a much more frequent washing of the hands, will accomplish this.

The rod-shaped bacteria that can cause food poisoning belong to four families, viz. *Bacterium, Bacillus, Clostridium* and *Salmonella*. Many of these normally inhabit the bowel of human beings and animals, so that contamination usually comes from human or animal excreta. It is only when food contains a large dose of these organisms that disease results. As with cocci, the human body can deal with small numbers, but succumbs to disease if the rod-shaped bacteria are allowed to multiply into millions in the food before it is consumed.

238

SALMONELLA FOOD POISONING

This is by far the most common type, and it is caused by ingestion of large numbers of organisms belonging to the Salmonella group (of which there are 500 different types) and their subsequent invasion of the tissues. An outbreak of *Salmonella typhimurium* poisoning in Glasgow, in 1969, caused 12 deaths. It was traced to infected pig meat. What infected the pigs? Almost certainly bone meal, added to the grain fed to the animals. As previously mentioned the organisms come from human or animal excreta. The onset of symptoms occurs after the germs have had time to multiply within the body (12 to 24 hours), and they are fever, headache, aching of the limbs, vomiting and diarrhoea. These symptoms are less acute than in staphylococcal food poisoning, but the duration of illness is longer and may be from one to eight days.

STAPHYLOCOCCAL FOOD POISONING

This is not due to ingestion of large numbers of staphylococci, but to the *toxin* which they produce in the food in which they have been deposited. This ready-formed toxin gives a rapid onset of symptoms (two to six hours), and they are acute vomiting followed by diarrhoea. There may be abdominal pain and it can result in prostration and collapse. Recovery is usually rapid.

These are the two main types of food poisoning, but many other germs, if present in sufficiently large numbers, can give rise to similar symptoms, and in statistical tables of causes of food poisoning they come under the heading '*Clostridium welchii* and other Organisms'. The incubation period for this miscellaneous group is 8 to 24 hours and the duration of symptoms is 12 to 24 hours. Organisms included in this group are streptococci, paracolon bacilli, coliform bacilli, proteus and others normally present in the intestine.

BOTULISM

This is a rare type of food poisoning caused by the toxin of *Clostridium botulinum*, which is lethal in small doses. *Clostridium botulinum* is an anaerobic, spore forming organism that lives as a saprophyte in soil. It can be recovered from vegetables, fish and marine products. It can grow, multiply and produce toxin at low temperatures, provided there is no oxygen. The spores are heat-resistant. The toxin produced by this organism is probably the most toxic substance known. Botulism is not usually associated with industrially processed foods, but there is considerable danger in home bottling of vegetables. The incubation period is 24 to 72 hours, the symptoms being lassitude, headache and dizziness. Diarrhoea may be present at first, but later constipation is troublesome, and the central nervous system becomes involved giving rise to disturbance of vision, difficulty in speech (dysphasia), difficulty in swallowing (dysphagia), and paralysis of the respiratory centre. If the patient survives, convalescence may take six to eight months.

CHEMICAL FOOD POISONING

This can arise from cooking utensils made of copper or zinc, organic substances used in horticulture, etc., and symptoms arise within an hour or two of ingestion. The tragic poisoning in Morocco belonged to this group and was due to the mixing of 'paraffin' and cooking oil. The World Health Organization's safety limit for mercury is 0·5 parts per million. A committee of experts was set up in Britain in 1971 to analyse for mercury content—flour, potatoes, beef, chicken, eggs, milk, fish, vegetables, fruit and sugar.

A few cases of food poisoning arise each year from the ingestion of poisonous berries and fungi.

Food-borne Infection

As well as the food poisoning that can be caused by many different organisms, there are some specific infectious diseases that can be spread by food. Unlike food poisoning, where the food has to be heavily infected with organisms, these infections arise when only small numbers of germs are present in or on the food, so that prophylactic measures for the former are likely to be adequate for eradication of the latter.

BACILLARY DYSENTERY

Bacillus of Flexner, Shiga and Sonne can all cause this disease, Sonne bacillus being the most usual infecting agent in this country. The incubation period is two to four days and the acute diarrhoea, with blood amd mucus in the stools, may persist for several days.

AMOEBIC DYSENTERY

This disease is caused by one of the pathogenic amoebae (p. 6), *Entamoeba histolytica*. Infection is rare in this country, but a few people are symptomless excretors of amoebae. In contrast to bacillary dysentery, amoebic dysentery has an insidious onset and runs a chronic course. As well as widespread permanent damage to the bowel there is a danger of liver abscess formation.

ENTERIC FEVER

Included in this group are typhoid and paratyphoid fever.
Typhoid fever. This disease is caused by *Salmonella typhi* and the incubation period may be up to 21 days, when there is an insidious onset of general malaise and fever. Intestinal symptoms are not usually severe until the second or third week of the illness.

Paratyphoid fever. This infection is caused by *Salmonella paratyphi A and B*, and usually has a summer prevalence. The incubation period is 7 to 10 days, when there is fever, headache, aching of the limbs, vomiting and diarrhoea. The disease is only distinguishable from Salmonella food poisoning by the longer incubation period and identification of the causative organism from the faeces.

There are a few other infections spread by milk and these are discussed on page 255.

Food Poisoning

CHOLERA

Cholera, which is mainly spread by water, can be spread by contaminated fruit and vegetables, and therefore can be included in a list of food-borne diseases.

VIRUS HEPATITIS[108]

Capable of faecal-oral spread. Contaminated water supplies can cause epidemics of this disease, and of recent years shellfish have been responsible for several outbreaks.

Reservoirs of Organisms causing Food Poisoning and Food-borne Infections

Twentieth-century knowledge compels the realization that some organisms use human beings and animals as a host, and though they do not cause disease in that host, they are capable of doing so in others, when they have been transferred to food which gives them the warmth, moisture and nutrients necessary for their multiplication.

HUMAN RESERVOIRS

Nose. Of normal people 50 to 60 per cent carry staphylococci in the nose. Since all people sneeze occasionally, blow the nose several times daily, and touch the nose (often unconsciously) at an even more frequent interval, there is ample opportunity for nasal staphylococci to arrive in food. This 'carrier' state is not necessarily permanent, but what havoc can be wrought whilst such a person has a 'cold'.

Hands.[109,110] Again many normal people 'carry' staphylococci on their hands, and these germs can penetrate along the pores and hair follicles where they live and multiply. Many nasal carriers are hand carriers.

Skin. Sepsis anywhere on the skin, e.g. pimples, boils, carbuncles, varicose ulcers, cuts and abrasions can result in these germs getting into food. The sufferers' clothing probably becomes heavily infected, and if the germs are not shed directly from the clothing into the food the hands probably act as an intermediary vehicle.

Bowel. Since in the conditions about which we are thinking the bowel exhibits symptoms, it is feasible that it can act as a reservoir

241

of infection. In an outbreak of this type of infection there are four possibilities in the exposed population. A proportion will succumb to the disease, a proportion will have such a mild infection that it is virtually ignored (these are called ambulant cases), some will continue to excrete the organism after recovery from symptoms (these are called convalescent carriers), and some will start to excrete the organism without even the slightest symptom arising (these are called temporary carriers or symptomless excretors).

Fig. 123   Human reservoirs of infection.

Mouth and throat. There are many bacteria forming the natural flora of this region, and occasionally those that can cause food poisoning are to be found here. They can be transferred to food directly by coughing and sneezing, or indirectly by a handkerchief.

ANIMAL RESERVOIRS

The meat of animals can be infected with Salmonella, and as it does not cause spoilage it is extremely difficult to detect. Salmonella are also excreted in the faeces of mice, rats, dogs, cats, turkeys, geese, ducks, pigs, chickens, cattle, horses and sheep. Flies which hatch in a manure heap will therefore be capable of carrying infection to food, and cockroaches have also been incriminated. Many local authorities are failing to carry out the Ministry of Agriculture's suggestion that they should inspect poultry, which is often diseased and is the only meat not required to be inspected by law. A poultry industry spokes-

242

man says that surely housewives know that storing uncooked chicken next to cooked meats is a health risk.

Salmonella have never been found within a hen's egg, but the shell is frequently contaminated by faeces, and it is from this contamination that the germ gets into food. The danger is greatest in dried eggs and these should be used immediately after reconstitution. Frozen liquid egg should not be left out in a warm kitchen longer than the stated defreezing time, otherwise the Salmonella will start multiplying to dangerous numbers.

ANIMAL	GERM	SOURCE OF INFECTION	MODE OF TRANSFERENCE TO FOOD		
Mice	Salmonella	Faeces	Furs and paws smeared with germs ► direct contact →		FOOD
Rats	"	"			
Dogs	"	"	Furs and paws smeared with germs ► human hands →		"
Cats	"	"			
Turkeys	"	"			
Geese	"	"	Flies ————————————→		"
			Flesh infected during evisceration ——→		"
Ducks	"	"			
		Eggs ——————————————→			"
Pigs	"	Faeces	Flies ————————————→		"
		Meat			
Chickens	"	Faeces	Flies ————————————→		"
		Egg shells	Can 'break into' egg ————→		"
		Faeces	Flies ————————————→		"
Cows	"	Meat			"
		Raw milk	Faeces on udders or hairs in milk ——→		"
Horses	"	Faeces	Flies ————————————→		"
		Meat			"
Sheep	"	Faeces	Flies ————————————→		"
		Meat			"

Fig. 124   Animal reservoirs of infection.

From all these observations of germs and how they get into food it will be realized that it is impossible to kill them all, and still leave food palatable for human consumption. The body is well equipped to deal with a reasonable number of germs and only succumbs to disease when the onslaught is unreasonable. We know that germs require moisture, food and warmth in order to multiply, and set out in Figure 125 are the effects of temperature on the growth of bacteria.

Temperature in Relation to the Cooking, Storing and Serving of Food
The availability of refrigeration is different throughout the world. In some countries the refrigerators are oil (paraffin) operated.

From Figure 125 it is evident that if food is cooked and eaten hot no harm is done. If it is cooked, cooled rapidly, refrigerated, carved and served within an hour without handling, no harm results. However, if it is cooked, cooled rapidly, refrigerated, carved on to plates for serving and left in a warm kitchen for several hours, then toxins are given a chance to form and food poisoning can result. Also if food is cooked and allowed to cool at atmospheric temperature toxins are given a chance to form and food poisoning can result. Even if this meat is heated before serving, food poisoning can still result, because staphylococcal toxin can resist boiling for half an hour. In conjunction with the above statements it must be remembered that there is a maximum size for a joint of meat, to ensure that during cooking the centre reaches the required temperature to kill bacteria; smaller joints also ensure quicker cooling.

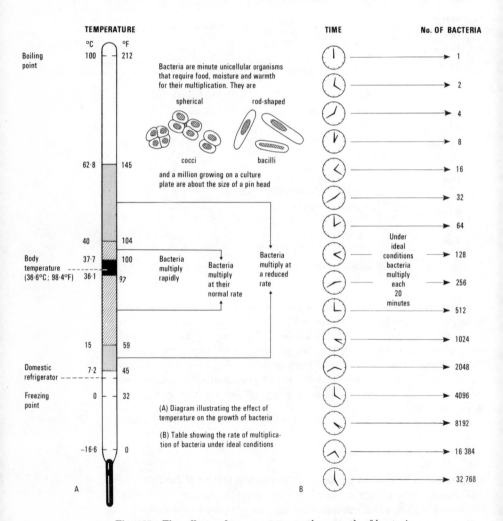

Fig. 125   The effects of temperature on the growth of bacteria.

The savoury foods requiring the greatest thought along these lines are pies, rissoles, sausage rolls, hot dogs, hamburgers and any processed meats.

Sweet dishes incorporating custard, cream (including synthetic varieties) and ice cream must obey the temperature rules, i.e. they must never be left out in a warm room; they must be brought straight to table from the larder or preferably the refrigerator.

Clean food[111,112,113,114,115] and cleanliness in its preparation is such a vast subject that I have summarized the main points in Figures 126 to 129. The Central Council for Health Education's pamphlet—10-*point Code for Housewives*—is obtainable from many local authority clinics.

Summary of Food Poisoning

Food poisoning is manifested as vomiting and/or diarrhoea resulting from eating food contaminated with chemical poison, pre-formed bacterial toxin, or live bacteria.

CHEMICAL POISONING

This is avoided by keeping cooking utensils in good repair, washing fruit and vegetables to rid them of horticultural chemicals, and the vigilance of public health inspectors in analysing food to discover any harmful diluent.

TOXIN TYPE OF POISONING

The worst toxin producer is the staphylococcus, and it can be 'carried' in the nose, mouth, throat and bowel, and on the skin and hands of human beings. Diligence in personal hygiene will cut down the numbers of staphylococci arriving in food. Domestic refrigeration at below 7°C (45°F) will inactivate these germs, and if the food is not being cooked it should be eaten immediately it is removed from the refrigerator. Botulism is rare, but the toxin is lethal in small doses. *Clostridium botulinum* is anaerboic, lives as a saprophyte in soil and its spores are heat-resistant, so that there is danger in home bottling of vegetables.

INFECTION TYPE OF POISONING

Bacterial poisoning is most frequently caused by Salmonellae from human or animal excreta. Diligent personal hygiene will prevent contamination from human excreta, and animals including rats, mice and cockroaches must not be allowed access to food. Animal meat and eggs must be protected from contamination at all stages by a rigid code of hygiene, and all who cook and prepare these foods must understand about the multiplication of bacteria in relation to temperature, and possess integrity in application of this knowledge.

PROVISION OF AN ADEQUATE SUPPLY OF CLEAN MILK FOR THE MAINTENANCE OF HEALTH

The availability of *liquid milk* in each locality of each country depends on many factors, including suitability of land for grazing herds, availability of cattle-food for the winter season, and the country's organization for the production and distribution of the milk. It is difficult for us in Britain to appreciate that there are people who have never tasted fresh milk. Wonderful though it is that the cow processes this excellent nutritive liquid from grass, man in the commercial and nutritional fields is faced with the problem of coping with its perishability. We take for granted a daily delivery of bottled milk but this service is proving to be very expensive to the industry. Glass milk bottles are costly and many of them are broken or simply not returned to the dairies. Some firms are experimenting with plastic bottles of the same size and shape, which can be used with the machinery already installed at the milk bottling plants and still fit into the standard domestic refrigerator. Other firms have turned to plastic-coated paper cartons which can be squeezed without bursting (perhaps in the future our milk will be delivered through the letter box). With more women going out to work, milk is often left standing all day on the doorstep. This leads to a reduction in the nutritive value of the milk as sunlight inactivates the riboflavine it contains. A 'long-life' milk is now on the market and it is thought that this might make possible the introduction of a weekly delivery service, the milk being treated as a 'grocery article'.

The liquid milk market in Britain is about 70 per cent of our total milk production. One hundred gallons of bottled milk weighs one ton and a ton of powdered milk is sufficient to make 1 800 gallons of liquid milk.

Drying of milk. Milk can be rendered to a powder by two processes, that of spray drying whereby it is sprayed into a hot chamber and collected from the base as a fine powder, or that of roller drying whereby a thin film of milk is applied to heated rollers from which it emerges as a fine powder. It is reconstituted at a strength of 1 drachm of powder to 30 ml (1 oz) of water. After reconstitution it must be as carefully looked after as fluid milk. Dried milk is useful wherever there is difficulty in obtaining fresh fluid milk; when transport is of prime importance, e.g. getting enough milk to a scene of disaster; storing milk from a time of abundance for a time when there is less fluid milk available; for feeding babies in any area where the fluid milk supply has become suspect from fall-out from atomic activity. Large stocks are held by the Government for this purpose as strontium^{90} is harmful to growing bones.

Evaporation of milk. Milk is rendered to one-third of its former volume by evaporation of its water content, and it is then hermetically sealed in tins under aseptic precautions. If taken undiluted it is a good source of animal protein in less bulk. Once the tin is opened it must be as carefully looked after as fluid milk. Of recent years it has gained popularity for infant feeding and the formula for babies of varying weight is printed on the tin. It is also more widely used in the home, canteens, aeroplanes, etc.

DO'S AND DONT'S IN A CLEAN FOOD CAMPAIGN TO PREVENT FOOD POISONING

Educate all
people concerned

Cleanliness
of body

Cleanliness
of clothing

Short clean nails
with good cuticles

Wear a neat and
effective head covering

Wash hands after
visit to lavatory
but not in kitchen sink

Wash hands after
using handkerchief
preferably a disposable one

Avoid the unprotected
sneeze

REINFECTION AFTER WASHING MUST BE AVOIDED   Use should be made of

Continuous
roller towel

Hot air drier

Individual
paper towels

Wash hands after
combing hair

Soap can
harbour bacteria

Use liquid soap in
reversible container

All cuts covered
with waterproof dressing

A first aid book
is essential

NO SMOKING

whilst
SELLING
COOKING
or SERVING
FOOD

Saliva contaminated
cigarette

Avoid fingering
the nose whilst
preparing food

Handle cooked food
as little as possible

Receive meat
from a cut-
ting machine
on to a strip
of plastic or
washable material

Do not moisten
finger to pick up
paper in which
to wrap food
Do not use news-
paper for wrapping

Fig. 126

DO'S AND DONT'S IN A CLEAN FOOD CAMPAIGN TO PREVENT FOOD POISONING

Where possible a refrigerated showcase should be installed

ALL FOOD offered for sale should be covered

Perspex cover for sandwiches

Pets not allowed in food shops

No licking of fingers when preparing food

Spoons used for tasting should not be put back in mixture

Sugar and milk kept covered on restaurant tables

Waiters and waitresses not allowed a serving towel with which they are tempted to do odd jobs

Helpings carried on tray to avoid soup and gravy washing over thumb

Always use the handle of cutlery when setting the table

'Thumbs out' is the rule when handling vessels

KITCHEN—walls and floor washable
No flaking of ceiling
Good lighting and ventilation
All working surfaces washable and impermeable

Crockery and utensils should not be stored exposed

Dirty crockery removed when meal finished

Plates scraped, rinsed and stacked tidily. Cutlery stood in vessels of hot soapy water

Two sink unit with sterilizing sink (water—82·2°C ;180°F) from which crockery can be 'drip-dried'

'Drip-dried' crockery stored in dust proof cupboard near sterilizing sink to minimize labour

Dish washing machines are an advantage

Tea towels should be boiled daily

MOPS AND DISHCLOTHS should be boiled daily or should stand in a harmless bactericidal solution. With mops the hands are not unduly exposed to cleaning agents

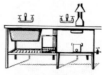

Fig. 127

248

DO'S AND DONT'S IN A CLEAN FOOD CAMPAIGN TO PREVENT FOOD POISONING

WOODEN topped kitchen tables, sinks and draining boards to be discarded

NO HIGH kitchen shelves to collect dust

Stacking of cups and saucers to be AVOIDED— contamination on saucer will be transferred to drinking rim or cup

Fruit and vegetables stacked to avoid bruising

Working surfaces in abattoirs, shops, kitchens etc., wiped down frequently with cloth wrung out of bactericidal lotion

Water used for scrubbing kitchen floors should not be tipped down kitchen sinks Floors should be hosed

Condemn blowing into paper bags to open them before depositing food in them

Dustbins to be kept in good repair and away from kitchens Wash hands after using same

The lid must always be kept on the food refuse bin It must be washed thoroughly daily

Welfare department where medical examination and swab-taking etc., can take place

AVOID VERMIN— Use cement mortar, sheet metal, or mesh to seal openings and protect door edges, meeting points of walls and floors etc. Remove coverage and keep food in metal bins

All manure heaps and breeding grounds for flies eliminated

Fly proofing of kitchen Self closing doors; windows, and ventilation openings, covered with fine gauze

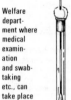

Do not use rat poison containing Salmonella organisms

Careful inspection of meat at abattoirs

Transport in clean vans which can be washed out daily. Van attendants must wear clean clothing

No hair combing at table

Fruit and vegetables thoroughly washed, especially if eaten raw

Mushrooms, watercress, oysters, etc., bought from a reliable source

Children taught to avoid eating wild berries

Fig. 128

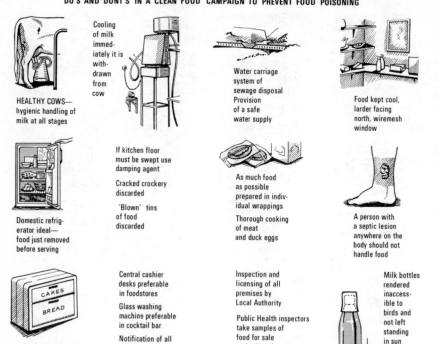

DO'S AND DONT'S IN A CLEAN FOOD CAMPAIGN TO PREVENT FOOD POISONING

HEALTHY COWS—
hygienic handling of
milk at all stages

Cooling of milk immediately it is withdrawn from cow

Water carriage system of sewage disposal Provision of a safe water supply

Food kept cool, larder facing north, wiremesh window

Domestic refrigerator ideal—
food just removed before serving

If kitchen floor must be swept use damping agent

Cracked crockery discarded

'Blown' tins of food discarded

As much food as possible prepared in individual wrappings

Thorough cooking of meat and duck eggs

A person with a septic lesion anywhere on the body should not handle food

Food to be kept in tins to avoid rats, mice, cockroaches, etc.

Central cashier desks preferable in foodstores

Glass washing machine preferable in cocktail bar

Notification of all cases to Medical Officer of Health

Inspection and licensing of all premises by Local Authority

Public Health inspectors take samples of food for sale and submit for public analysis

Milk bottles rendered inaccessible to birds and not left standing in sun

Fig. 129

Condensed milk. This is another process via which milk is reduced to one-third its original volume. In many instances sugar is added. Once the tin is opened it must be as carefully looked after as fluid milk.

Milk from plants. Scientists at the Vegetarian Nutritional Research Centre have produced milk from outer cabbage leaves, pea pods and other green stuffs. It is suitable for babies who cannot take cow's milk and for those with an allergy to cow's milk. It is an invaluable aid to the world's underprivileged children for whom there is not enough cow's milk to go round.

New Zealand has developed a *whole-milk biscuit* to make milk protein available in areas where liquid milk cannot be distributed. It is a practical way of making dairy surplus available to the protein-short countries in the developing regions of the world. The biscuit is fortified with minerals and vitamins and has a 24 per cent protein content. Each biscuit contains the equivalent of nearly 6 ounces of milk. Two biscuits can provide about 36 per cent of the recommended daily protein intake for young children, 50 per

250

cent of the daily need of the main vitamins (except ascorbic acid), and some 12 per cent of the kilocalorie intake. It is tasty and filling. In its special packaging the biscuit can last at least six months without refrigeration. The World Food Programme carried out trials in 12 countries and the children showed increased height and weight. Skin infections that afflicted almost half of them at the start of the trial, decreased to 18 per cent by the end.

Milk is considered a good food for the very young, and throughout childhood at least 568 ml (1 pint) should be consumed daily as it is one of the best sources of calcium and phosphorus, so necessary for the formation of a good strong skeleton. It is for this reason that 190 ml (one-third of a pint) of milk daily is provided free for each school child under seven years. Several countries follow the pattern of providing free milk for school children. During pregnancy and lactation, at least 568 ml (1 pint of milk) which can be obtained at a reduced price should be included in the daily diet. Milk is also useful for invalids and old people because of its liquid state and the ease with which it is digested and absorbed. Having little flavour of its own it can be given in many ways.

Milk is a white emulsion containing the six cardinal varieties of foodstuffs: protein, fat, carbohydrate, elements, vitamins and water.

Besides differing in percentage compositions, human and cow's milk have other differences which need to be considered.

Protein. This is a good place for revision of the different types of protein found on page 220. The curd of human milk is more digestible as it contains more lactalbumin, the more easily digested of the two milk proteins. In the alimentary tract, cow's milk containing more caseinogen may form putty-like masses which give rise to indigestion (dyspepsia), diarrhoea, etc. Some curd may pass unaltered per rectum. When milk is boiled the lactalbumin coagulates to form a skin; this should not be discarded.

Table IV.   Table of comparison of human and cow's milk.

Constituent	Human	Cow's
Protein	1 to 2 per cent	3 to 4 per cent
Caseinogen	$\frac{1}{3}$	$\frac{3}{4}$
Lactalbumin	$\frac{2}{3}$	$\frac{1}{4}$
Fat	3 to 4 per cent	4 per cent
Sugar (Lactose)	6 to 7 per cent	4 to 5 per cent
Water	87 to 90 per cent	87 to 90 per cent
Elements	0·3 per cent	0·7 per cent
Vitamins	A, B, C, D	A, B, C, D
Specific gravity	1,030	1,030
Reaction	Alkaline	Acid
Temperature	Body temperature	Variable
Bacteria	Sterile	+ + +

Fat. The fat in human milk is in a finer state of emulsion, with a greatly increased surface area of the globules exposed to the digestive enzymes. Cream of cow's milk also differs in composition and is liable to set up gastrointestinal irritation if given in excess.

Carbohydrate. This is present in both milks in the form of lactose, but is relatively deficient in cow's milk, leading to constipation if the deficiency is not remedied. Infants tolerate sugar well.

Elements. There is less calcium, magnesium and phosphorus in human milk than in cow's milk, but iron is present provided the mother is taking adequate supplies in her diet. There is a danger of the bottle-fed baby developing an iron-deficiency anaemia if weaning is delayed.

Vitamins. The amount of vitamin C is probably less in cow's milk and therefore a mainly milk diet needs to have addition of this vitamin. Vitamin A varies according to the feeding of the cows and is greatest when they are pasture fed, because of the carotene in the grass. Vitamin D varies according to the season, there being more in the summer when irradiated ergosterol produces this vitamin in the cow. Babies require a supplement of both these vitamins, especially in the winter months.

The Quality of Milk[116]

The law in England states the solids that milk shall contain. The vendor is prosecuted if the standard is below this level.

Many people assess the quality of milk by its cream line. In untreated and pasteurized milk this is determined by the size of the fat globule, and size varies with the different breeds of cattle. Channel Island milk has the largest globule which more readily forms a cream line than milk with high butter fats from other cows. The Food Manufacturing Trade prefers a milk without a cream line. The food trade is the greatest user of sterilized milk because it is homogenized.

Antibiotics in milk. Penicillin is the most effective treatment for mastitis in cattle, penicillin being present in the milk of a treated animal for 48 hours after the last dose. Farmers are advised to withhold such milk from sale because of the danger of: (1) sensitizing some people so that if a large dose of penicillin were needed for treatment, a severe reaction would follow; (2) producing an allergic reaction such as urticaria, dermatitis or other skin rash in people already sensitive to penicillin; (3) transmitting penicillin-resistant staphylococci from cow to man. Broad-spectrum antibiotics have been used for bovine infections—and have been found in the milk of treated cattle. Continued ingestion of small amounts of chloramphenicol by human beings may cause aplastic anaemia. Professor L. P. Garrod recommends that the use of broad-spectrum antibiotics in milk production be prohibited. Dairies now carry out tests on incoming milk for antibiotic content. The Milk Marketing Board run a penalty scheme of price reduction for milk containing antibiotic.

Radioactive substances in milk. The Agricultural Research

Council publish reports on isotopes in milk. The level of isotopes particularly strontium^{90}, iodine^{131} and caesium^{137} has declined since the cessation of large scale testing of nuclear weapons in the atmosphere.

THE CLEANLINESS OF MILK   **Milk**

The attention paid to the breeding and feeding of cattle determines the quality of the milk. The conditions under which the milk is extracted from the cow and treated thereafter determines its 'cleanliness'. This is estimated from the number of coliform bacteria present. The ruling is no coliform bacteria in $\frac{1}{100}$ of a millilitre of milk.

Milk vending machines[117] are creating a problem with regard to the sale of 'clean' milk. Because of their convenience and popularity it is to be hoped that a satisfactory solution to the problem will be found.

DESIGNATIONS OF MILK

The existing designations for milk in England and Wales approved by regulations in 1963, the amending regulations in 1965, and the order in 1967 are as follows:

Untreated milk. When the milk is bottled or cartoned at the farm where it is produced the words 'Farm Bottled' or 'Farm Cartoned' may be added to differentiate it from 'Untreated' milk, which is bottled from churns at a distributor's premises. The words 'Channel Island' may be added if the 'untreated' milk is produced from Channel Island cattle.

Heat-treated milk. Milk which is heat-treated is called pasteurized, ultra-heat treated, or sterilized. The words 'Channel Island' may be added to the designation.

The three heat-treated grades are free from brucellosis. Under the Food and Drugs Act, and subsequent milk and dairies regulations, milk from cows known to be diseased cannot be sold or offered for sale for human consumption. Contagious abortion is not specifically mentioned but could be considered under the designation 'any septic condition of the uterus'.

In Scotland the special designations are as follows:

Untreated milk. This is called premium or standard.

Heat-treated milk. This is called pasteurized, sterilized, or ultra-heat treated.

The conditions governing the licences for their production, processing, and distribution are set out in an order of 1965 and an amendment order of 1966. Milk may also be sold under descriptions specified in regulations of 1967. The descriptions Channel Island, Jersey, Guernsey or South Devon may be used, but it must be sold under the appropriate special designation, that is, premium, standard, pasteurized, etc., as the case may be.

The requirements for premium milk, which relate to compositional standard and hygienic quality, also include a prohibition on the vaccination with live Brucella abortus vaccine of any animal in the dairy herd, but, since the Ministry of Agriculture, Fisheries and Food only authorizes the use of live vaccine on calves under the age of 6 months, this condition has become of academic interest.

Thus in Scotland, as in England and Wales *untreated milk,* if it is the produce of a herd registered and accredited under the *Brucellosis (Accredited) Herds Scheme,* is unlikely to contain the organism, but until the scheme is further advanced the *risk remains* that infection may arise in a herd between periodic testing of the cows. The *heat-treated grades are safe* from the risk of milk-borne *brucellosis* because the process kills any organisms which may be present in the raw milk.

CARE OF MILK IN THE HOME

Milk should be collected from the doorstep as soon as possible after delivery to prevent destruction of riboflavine, birds picking off the silver-foil top, contamination by domestic animals or rodents, and multiplication of germs at atmospheric temperature, this naturally being more rapid in warmer weather. Ideally the outside of the bottles should be wiped and they should then be placed in a refrigerator. Failing this they must be put in a cool, dark place. They may need to stand in a bowl of water with a piece of muslin draped over the top and dipping into the water, which will allow continuous evaporation with consequent cooling of the bottle. The milk should only be poured from the bottle as required, and jugs of milk should never be left in the warm room atmosphere for longer than necessary. When empty, all utensils which have contained milk should be rinsed in cold water to remove the albumin which coagulates on heating, washed in hot soapy water (or detergent) to remove grease and then rinsed in very hot water and drip dried or dried on a *clean* cloth. Bottles should be put out daily for the milkman and not hoarded by the householder.

CARE OF MILK IN THE HOSPITAL

Many hospitals have the milk delivered in large churns and this is satisfactory for kitchen use where a nearby cold room is available, but it is not good practice to pour this milk into ward cans which are too big for the ward refrigerators. Some hospitals have their ward supply delivered in bottles and their care is as for that of milk in the home. Some hospitals adopt the routine of saving their milk-bottle tops for a particular charity.

If milk vending machines are installed in out patients' department, staff residential quarters and canteens, the manufacturer's instructions for hygienic maintenance must be observed.

254

Of the diseases from which cows can suffer only three are important in this context. Tuberculosis from the cow can cause this infection in the bones, glands and joints of human beings. There used to be 4 000 new cases annually, and at least 1 000 deaths. Over the years this has been reduced and all herds are now tuberculin-tested. Streptococcal infection of the udder (mastitis) can cause strepto- coccal sore throat, scarlet fever, rheumatic fever and purpura in human beings. (Antibiotics used for mastitis have appeared in milk and have been blamed for urticaria and dermatitis in human beings.) Abortus fever causes abortion in the cow and undulant, Malta and Mediterranean fever in man. These three are synonymous terms and the term brucellosis is to be preferred, since the infecting organism belongs to the genus *Brucella*. Vaccination of cattle is still on a voluntary, though now free, basis, and has succeeded chiefly in reducing abortions rather than infections in herds. Brucellosis in man causes recurrent attacks of fever and mental depression and it can last for months. It is estimated that 1 000 people per annum are afflicted. As well as being transmitted in milk, it is also contracted by farmers and veterinary surgeons who handle infected cattle.

Water

If any of the personnel handling milk at any stage is a carrier of staphylococci in the nose, throat or skin, the milk can cause food poisoning. Similarly a carrier of the organisms causing typhoid, paratyphoid fever and dysentery can transfer these to milk and cause the respective diseases in man.

There is a remote possibility that any of the diseases carried by water can be transferred to milk if any of the utensils are washed in polluted water. In this country it would only occur from an emergency such as fracture of a water main. Even when this happened during World War II from bombing, the water was kept safe by extra chlorination.

THE PROVISION OF AN ADEQUATE SUPPLY OF CLEAN WATER FOR THE MAINTENANCE OF HEALTH

Water is one of the essentials without which life would languish and become extinct in a comparatively short time. Water enters into every cell of the body and constitutes about 75 per cent of the whole body weight. Being such an important fluid it is necessary for a nurse to have some knowledge of it.

Water is a chemical compound from the union of two parts of hydrogen with one part of oxygen. It expands with heat and contracts with cold until it reaches 4°C, and between 4° and 0°C (freezing point) it expands. This pre-freezing expansion allows ice to form on the surface where the sun's rays can melt it. Had water continued to contract and turn into ice at the bottom, then fish and underwater plant life could never have existed. It is the expansion before ice formation that bursts pipes and cracks roads but the damage is not evident until the thaw comes.

The specific gravity is the weight of a given quantity of a substance compared with that of an equal quantity of water. The weight or specific gravity of water is called 1 000 as a standard of comparison.

When wholesome water is transparent, colourless, tasteless and odourless, it does not contain disease-producing germs. It is a great solvent. Hygienically pure water may contain some dissolved substances, but they must not have harmful properties.

THE USES OF WATER IN THE BODY

1. It makes up two-thirds of the body's weight.

2. Via diffusion tissue fluid gives each cell food and oxygen and receives its waste products.

3. As a solvent it helps to dissolve food particles.

4. It is necessary for many of the chemical changes taking place in the body.

5. It is the chief constituent of blood, thereby allowing circulation and transfer of nourishment to the tissues.

6. It replaces fluid loss caused by excretion from the skin, lungs and kidneys.

7. It aids the regulation of body temperature.

8. It dilutes poisons (toxins) and assists in their elimination.

9. It provides a variety of elements.

10. It prevents constipation.

11. When taken at the end of a meal it washes food particles from between the teeth.

FURTHER USES OF WATER TO A COMMUNITY IN THE MAINTENANCE OF HEALTH

It is essential that we consider the part played by water in maintaining the health of a community. It is a vast subject and after thinking about it you will be able to add items to the following list:

1. Each member of the community needs to partake of 1·75 litres (3 pints) of fluid daily.

2. Food can be washed and cooked, thus rendering any germs, or ova of roundworms, and threadworms, or larvae of tapeworms innocuous.

256

3. Crockery, cutlery and cooking utensils can be thoroughly washed, helping to cut down the incidence of food poisoning.

4. The homes can be kept clean, giving a sense of pride to the occupants.

5. Each person can have a daily bath to prevent body odour. Self-cleanliness is an important factor in self-pride, an essential component of mental health.

6. Each person can wash his hands after visiting the lavatory, and before preparing or eating food, so preventing the spread of enteric diseases and food poisoning.

7. Each person's hair can be washed at least weekly, thus raising individual morale.

8. Clothes can be kept clean and free from smell, thus giving a feeling of self-confidence to the wearer and preventing rotting of clothes from stale perspiration.

9. Food refuse pails can be washed daily, preventing odour and flies.

10. Ashbins can be hosed inside and out, weekly, preventing odour, flies, rats, mice, etc.

11. Streets can be washed daily, preventing dust dissemination with all its attendant evils.

12. The water-carriage system can be used for sewage disposal; drains and sewers can be flushed to prevent odours, flies and vermin. Lavatories can be flushed after each usage.

13. Industry can continue giving economic security to the workers.

14. Workers in dirty industries can wash thoroughly, or bath, before leaving work, thus cutting down the possibility of contact diseases.

15. An adequate fire service can be rendered in emergency.

16. Clinics, nursing homes, surgeries and hospitals can continue to play their part in the prevention of disease and the resumption of health.

17. Windows of factories, offices, shops and houses can be kept clean, admitting maximum light and providing a pleasant environment in which people can live and work.

18. Public wash-houses can continue to provide facilities for the weekly wash, for those people with less suitable home conditions.

19. Trains, buses, trams, cars etc., can be washed inside and out thus preventing infection, providing a pleasant means of transport and preventing deterioration of the said vehicles.

20. Public baths can continue to provide bathing facilities for those who do not have a bathroom at home.

21. Swimming-baths can continue to provide a safe, pleasant relaxation to those who enjoy this hobby.

22. Gardeners can attend to their vegetables, fruit and flowers on a commercial scale earning their living, or on a domestic scale giving pleasure and satisfaction, which are so essential to health.

23. Cattle can be fed, watered and groomed, ensuring a safe supply of milk and meat.

Let us now consider what might happen if there were a shortage of water.

1. Its standards of personal cleanliness would fall.

2. Its standards of environmental cleanliness would fall.

3. Bad habits would creep in, such as lack of hand washing after visiting the lavatory.

Man, his Health and Environment

4. The atmosphere would not benefit by having suspended particles washed down in rain. Not only are these particles irritant to the respiratory system, but their presence robs the air of its freshness and its invigorating power.

5. Industries may have to close down leading to economic insecurity, one of the potent producers of stress, tension and anxiety.

6. Lavatories would only be flushed infrequently encouraging multiplication of germs.

7. Flies would multiply and as they can carry infection to food, the enteric diseases and food poisoning would increase.

8. Mice, rats and vermin of all descriptions would be encouraged by the uncleanliness and the ensuing odours.

9. Germs and other parasites would multiply; infection and infestation would become rife.

10. Each household would draw its ration from a central pipe daily. This could easily become infected in the home from the increased dust, etc. Water is heavy and many flat-dwellers would have to carry it up stairs. There would be some people living alone and incapable of carrying a bucket of water.

11. Scarcity of fresh vegetables, fruit and crops might lead to dietary deficiencies. Prices of such commodities would rise, affecting the lower income groups first, where there tend to be more children.

12. The milk supply may become suspect, from lack of cleanliness of cows, utensils and people handling the milk.

WORLD WATER SUPPLY

We have established that, in terms of world need, water is an *essential* commodity. It is estimated that in many developing countries only about one-third of the urban population has any drinking water supply in the home or on the premises. In some parts of the world, water jars are still carried on the head from a well. Calcutta, one of the world's largest cities, is short of water by 50 million gallons daily for its 5 million inhabitants. Only 1 in 10 rural dwellers can obtain even the smallest amounts of safe water. In the process of industrialization, in which the developing countries are involved, more and more water is needed. As the standard of living rises, tubs and earth closets are replaced by baths and water closets which use much more water. Then come washing machines, dishwashers, family cars and gardens, all needing more and more water. The average daily consumption of water per person is 1 600 gallons in the USA, 90 gallons in Scotland, while England and Wales together use a total of 3 000 million gallons daily. The local authority levies a water rate, which is a flat rate for houses, and a rate calculated by a meter for industrial concerns, hospitals etc. But man has to rely on the natural water cycle for his water supply.

Water cycle. The cycle is achieved by a constant process of *distillation* by the sun's rays. Warm air blowing over the moist surfaces of the globe, i.e. seas, rivers and lakes, evaporates moisture. Added to this is the water vapour given off by plants and mammals. Warm air is light. On rising into higher altitudes this air is cooled and is therefore unable to retain its absorbed moisture, so it is precipitated in the form of dew, mist, rain, snow or hail. At first rain water is pure, but as it washes down suspended matter it can become grossly contaminated, especially in the cities. This covers the *three forms* in which water can exist: *liquid, gas* and *solid.* Man interferes with natural cycles at his long-term peril, but to help Indian agriculture, a harmless chemical is sprayed on water to prevent evaporation.

World distribution of water. Distribution of water on the earth's surface is very uneven. Scientists calculated that 97 per cent of the earth's water is in the oceans, where salt content is too high for human consumption, or for industry or agriculture. *Desalination* is one of the major break-throughs in this century. There are many desalination plants throughout the world using millions of gallons of sea water for local supplies. Students are now able to graduate in Desalination Technology. About 2 per cent of the earth's water is frozen solid, mostly in the great ice sheets that cover Antartica. The remaining 1 per cent is in the lakes, rivers and underground reservoirs. Three-fifths of land on earth is arid. Two American scientists have costed transporting icebergs to these areas and they estimate that it is one-tenth of the cost of the same amount of desalinated water. For areas that have no water to desalt and cannot import icebergs, there is the application of new technology to an old method of *water 'harvesting'.* It is being used in Australia, Israel, USA, and some of the arid regions. It consists of collecting and storing water from sloping land rendered impervious. The catchments can be any size, there are several measuring 10 000 sq. ft. They can be camouflaged to merge with the surroundings. The rest of the world has to rely on *surface water* i.e. all inland waters that are in contact with the atmosphere, river water, and underground water that has percolated through the earth's strata and has collected on an impermeable layer, giving rise to *springs* and *wells.* The Bank of England, the Old Lady of Threadneedle Street, still draws its own water from a well!

Water standards. The World Health Organization sets the maximum allowable limits of all substances in water. Many of these are naturally occurring and depend on the composition of the earth through which the water has percolated. For example in some places water naturally contains *fluorine*, in some instances more than the maximum allowable, when it not only protects from dental caries but it mottles the teeth (fluorosis); in other places the water is deficient in fluorine. Many and heated are the arguments about the advisability of adding fluoride to water[118] to stamp out dental caries.

259

World Health Organization policy is that it is a practicable, safe and efficient public health measure. One hundred years ago similar arguments raged about chlorination of water! With advance in agricultural programmes in an attempt to feed the world's hungry millions, *nitrate levels* in water in some areas (Britain included) exceed the safety level recommended by the World Health Organization. Thus we encourage farmers to grow more crops by using fertilizer, and in so doing we raise the level of nitrate in soil to above acceptable levels! This is one instance which demonstrates the need for specialists to get together so that the situation may be viewed as an interdependent whole and not as a series of separate problems. Clean water cannot be considered apart from refuse disposal, farm waste, trade waste, the world policy for dealing with radioactive substances etc. We have mentioned two substances, fluorine and nitrate, which can be present in water and affect health. There are many others:

1. Disease-causing micro-organisms from animal or vegetable refuse and sewage pollution, the most likely organisms being those causing typhoid and paratyphoid fever, dysentery, virus hepatitis, and cholera. In the first six months of 1971, 27 countries reported to the World Health Organization that they had cholera in their territories, 65 000 cases with 3 000 deaths. In 1970 there were two cases in Cardiff after the people concerned had returned from Tunisia. In July 1971 there were cholera outbreaks in Spain and south-west France, countries which have been free from the disease for many years. This caused a scare in Britain as several Britons had been on holiday in the area of Spain concerned. Two cases of infection were reported in Yorkshire. All media were used to get in touch with contacts.

2. Suspended irritating matter causing diarrhoea.

3. Ova of thread worms, round worms and tapeworms. Piped water supplies in the developing countries has greatly reduced the incidence of diarrhoea and worm infestation in young children.

4. Elements, the most important being magnesium and calcium, which cause hardness of water. They are of current interest because research in Britain, USA and Japan in 1971 showed that residents in towns which have consistently softened their water supplies are more likely to die of heart attacks than those who drink hard water. Water Boards contemplating a new softening process are being advised against it until more is known about the problem. Disadvantages of hard water are that it is wasteful of soap, it impairs the texture and colour of materials, it cannot be used in the dyeing industry, deposition of salts inside vessels 'furs' them leading to use of increased fuel and increased time necessary for heating. This can be dangerous and lead to explosion in large boilers; they are inspected regularly to avoid this. Hard water is an inefficient cleansing agent for the skin and hair, it hardens the outside of meat and vegetables, resulting in less efficient cooking so that the food may be tough and indigestible, micro-organisms and ova of worms may not be killed; the scum formed with soap can block the tap outlets to

baths, sinks, etc. and it calls for extra cleaning with increased use of labour and materials.

5. Increased hydrogen ions causing acidity, so that acid water coming into contact with lead or zinc will dissolve them and can lead to lead poisoning. Peat areas have trouble when gales stir up the peat sediment in the lakes. For this reason, lead pipes and zinc storage tanks are being discarded in favour of copper, steel and synthetic pipes and tanks. In 1970 in Liverpool, out of 47 samples analysed for lead content, 22 had either the international standard maximum allowable limit or slightly more, while three had more than twice the allowable limit. This means that two million people were being regularly exposed to the risk of lead poisoning.

6. There is now a little strontium[90] and caesium[137] in drinking water, there being least in well water. The level of radioactivity is constantly surveyed.

7. There are hormone trace elements in all water, and recently there have been investigations into the oestrogen content, but fears that the increased number of women taking the pill might have caused an increase in the water content, are unfounded.

8. Pollution by mercury from industrial effluent, especially that from wood-pulp plants, has increased vigilance in the paper-producing countries.

POLLUTION OF WATER

If we want the privilege of drinking safe water, then we need to develop an attitude of *personal involvement* in helping to prevent pollution of water. The *Observer* magazine had a scheme for children in 1970. They had to observe a pond, stream, lake, river or canal and answer 12 questions about it. Then they had to write to their local newspaper, and local Water Board, telling each that they had notified the other. It was hoped that if unsatisfactory conditions were found the Water Board might send a trained biologist to investigate. This sort of involvement is very likely to produce a responsible personal attitude to a matter that closely concerns each of us. It would seem to be a very necessary exercise, for in 1971, anyone falling into the Thames was given an antibiotic and antitetanus injection! Also it is much easier to *develop* an attitude, than to *change* one. If we could bring up a generation to be personally involved in the care of the environment, we would be well on the way to solving some of the problems that beset us.

Everyone pollutes: from their excretions, ablutions and domestic waste. It is not the case that everyone except you pollutes! You pollute. I pollute. The industrialist pollutes twice over—personal pollution and plant pollution. The only way that we are going to be able to get enough water to drink is by accepting water that has been used and purified several times over. What is drunk in places near the mouth of a river, will have passed through several pairs of kidneys and factories in places higher up the river. It would seem that there is a case for one Authority looking after water and sewage! The criterion by which effluent from sewage works and factories is judged is its biochemical oxygen demand (b.o.d.), which is a measure of the amount of oxygen needed for a breakdown of the pollutants and chemicals. A river without oxygen cannot clean itself and cannot sustain fish or vegetation. Modern sewage works can treat to a very high order, returning water with a b.o.d. of only 3 milligrams per litre, which is the oxygen level of a healthy river. But how many of Britain's sewage works are modern? River Authorities have appointed Chief Pollution Officers. Under the 1961 Rivers Prevention of Pollution Act, they can refuse to accept too heavily polluted effluent from factories. The firm must then either instal equipment for treating it, or get a local authority to allow it into main drainage that goes to the sewage works, for which service a fee is charged. There are firms that say that they cannot afford to treat effluent properly—it would put them out of business. This could be looked at as a mandate for perpetual pollution, yet in some hard-pressed industries it may be true. If industries go out of business there is unemployment—you see this *interdependence* raising its head again. Is pollution-elimination an inescapable production cost? At present State Aid is as follows: A firm in a development area gets a 40 per cent investment grant on plant costs, which include anti-pollution equipment, but a firm in a non-development area gets only 20 per cent, and a firm gets *no grant at all* if it wants to combine with a local authority for treatment of effluent—often regarded as the best method. A clean country costs money. Someone has to pay. Are you willing to pay heavier taxes for a cleaner country?

The greatest pollution problems have been caused over the last 10 years by increasing use and accidental spillage of oil, pesticides, and other synthetic organic chemicals. They are difficult to remove by treatment and many of them are non-biodegradable. We do not know about their long-term effects if ingested over many years, even in infinitesimal amounts. New methods of analysis are being introduced to identify these substances even in very weak dilution. With this knowledge long-term toxicity tests can be undertaken to assess the significance to health.

A recent report[119] recommends close control of industrial effluents discharged into rivers which are used as a source of drinking water, particularly those effluents likely to contain synthetic organic chemicals, because these materials, sometimes of unknown composition, tend to be stable and biologically potent at very low concentrations. Discharge of crude sewage from boats and other floating craft, into fresh water used for recreation and as sources of domestic water supply, should be prohibited. The law should be amended to require better safety precautions against accidental pollution of water by oil or toxic substances. When a local authority provides a new scheme in fringe and rural areas where cesspools and septic tanks are the means of disposal of domestic sewage, the premises near the line of the sewer should be connected to it at the local authority's expense. Meanwhile the local authorities should provide a free and sufficiently frequent service for the clearance of cesspools and septic tanks to avoid local pollution of ground water and to avoid nuisance. The Report recommended that crude sewage should be discharged to the sea only after screening, comminution, and passage through diffusers on long, carefully sited outfalls.

The difficult job of maintaining the quality of river water is hampered by outdated legislation, which was mainly concerned with faecal contamination and bowel organisms. Now pollution from heavy metals and man-made chemicals is the greater threat. In other words our methods have been overtaken by industrial expansion. But man is not to be out-done for long. The Plessey Organization have tested a miniature computer that automatically analyses samples of water taken at regular intervals from a river. Signs of serious changes are transmitted by radio link or telephone cable to an alarm unit at the headquarters of the river authority. The machine discards samples that pass the tests but retains those that fail to do so, enabling its possible use as a policing system, which it is hoped will act as a deterrent.

Methods of purifying water on a small scale. Rarely are we called upon to purify water for domestic use, but all campers, travellers and climbers need to be conversant with the first two methods.

Boiling. Water which is boiled for five minutes is safe for drinking. The dead germs are still present in the water which must not be used for injection as they can produce a fever reaction, and are therefore termed pyrogens. Should one be confronted with a pail of muddy water it is expedient to filter it before boiling.

Addition of Chemicals. Proprietary tablets of chlorine are available with instructions regarding the quantity of water each tablet will sterilize, and the time it will take. There are also detasting tablets, as many people find the taste of chlorine offensive. Countries offer these tablets to less fortunate countries in times of distress when infected water would be disastrous.

Potassium permanganate can be added until the water is pale pink, and this dilution will sterilize in half an hour. In areas where cholera is rife this dilution is recommended for washing all vegetables.

Domestic Filters. These are used where the only water supply in a rural area is a shallow well. The Pasteur-Chamberland and the Berkefeld types are now out of date, and have been succeeded by the Metafilter and the Stellafilter. These are metal cylinders which are filled with a special powder, and then screwed on to the tap. They have to be emptied, washed, boiled and filled with fresh powder at intervals. There is now available an Ogden water purifier and there is an 'Aquapac'[120] model for travellers. Information about these can be obtained from Safari Water Treatments, Ltd, 299–301 Ballards Lane, London, N12 8NP

Distillation. The water is boiled and the steam passed through a tube, in which it cools and condenses, being collected as fluid at the other end. Distilled water is free from dead bacteria and does not contain any mineral matter. For these reasons it is prepared com-

263

mercially under aseptic precautions for dilution of drugs which are to be given by injection.[121] The proprietary name is Apyrogen.

CONSERVATION OF WATER

Here again the plea is for personal involvement. Careful use of this precious commodity begins with the individual. We simply have to train ourselves to use less water. One water board is using the slogan 'Use water wisely'. Do you keep the tap running while you wash your hands? Do you leave the tap running while you clean your teeth? Do you leave the tap running until the water runs cold for a drink? Ice cubes are more economical—provided you have a refrigerator. Leaking taps should be repaired immediately. Baths and sinks should never be filled too full. Showers use much less water than baths.

It used to be the custom for houses to have the rain water pipe directed into a barrel, and this water was used for washing the skin. Being soft it was much kinder to the skin than hard water. It was used for washing hair and woollen garments. The idea is worthy of revival. Why send all this rain water to the sewage works to be treated and sent to the sea where it is salinated, and then some authority has to pay for removing the salt! Could there be miniature household apparatus so that this water could be used for water closets, baths, car washing, garden spraying etc.?

A similar idea has been suggested, but on a much grander scale. That is a two-pipe system to each house and establishment. One with 'purified' water to be used for anything to do with food, including the washing up, which includes dishwashing machines. One with water which would be used for all other purposes. It seems hardly feasible in a country with an established one-pipe system but it may be that it could be considered in the developing countries.

Having discussed the nutriments needed by the body, the organs that deal with them will now be described.

The Digestive or Alimentary System

The organs comprising this system form a tube about 8 metres long, commencing at the mouth and ending at the anus. Some substances, notably cellulose or roughage, can pass through this tube without being absorbed and therefore never become part of the body structure.

The Mouth (also called the buccal or oral cavity)
Boundaries
Anteriorly. The lips.

Posteriorly. The fleshy projection known as the uvula, and behind this the posterior pharyngeal wall.

Laterally. The cheek muscles. Lying within them are the superior and the inferior maxillae containing the teeth.

Inferiorly. The tongue supported by the mandible (inferior maxilla) and the hyoid bone.

Superiorly. The hard palate formed by the two superior maxillae and the two palatal bones, and the soft palate, which not only has the uvula hanging from it in the centre, but is drawn down on either side into two strong folds, called the anterior and posterior pillars of the fauces. Housed between these pillars on either side is a palatine tonsil.

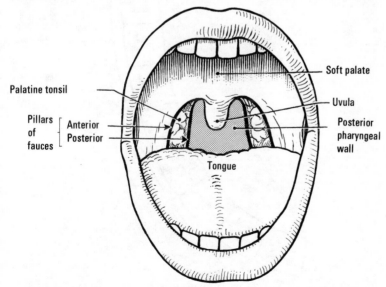

Fig. 130   Diagram of the open mouth.

The mouth is lined throughout by mucous membrane supported by a submucosa of areolar tissue carrying blood and lymphatic vessels, and nerves, to the ever-active secreting cells. The mouth contains:

1. THE TEETH

These are laid down in fetal life, and there are two sets present in the gums at birth. The milk or temporary set start erupting through the gum at about 6 months and should be complete by 6 years. They

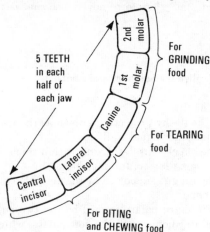

Fig. 131   Diagram of milk teeth.

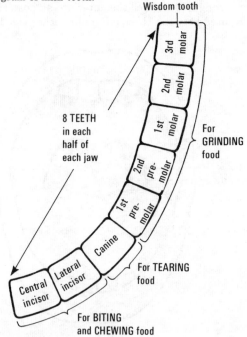

Fig. 132   Diagram of permanent teeth.

266

are then shed gradually as the permanent teeth grow down, and the last of these to erupt are the wisdom teeth which appear in the late teens. The milk teeth are named as in Figure 131.

The permanent teeth are named as in Figure 132.

Structure of a tooth. A glance at Figure 133 shows the crown, covered with enamel protruding above gum level. The fangs, covered with cement, are embedded in the alveolar process of the jaw. The layer below both of these is made of dentine and encloses the pulp cavity, richly supplied with blood and lymphatic vessels and nerves. The area at gum level between the crown and root is referred as the neck. As it is the place where enamel gives way to cement it is a vulnerable area; it is here also that food particles left between the teeth lodge, to be fermented by bacteria in the mouth, with the production of a more acid medium in which calcium is soluble, and thus begins the process of dental caries.

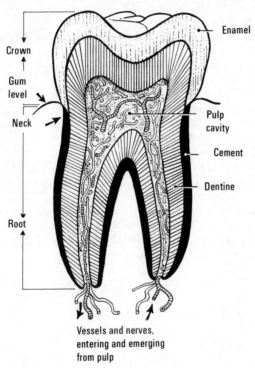

Fig. 133 The structure of a tooth.

2. THE TONSILS

As already stated these are held within the anterior and posterior pillars of the fauces, which are special folds of the soft palate. Strictly speaking they should be called the faucial tonsils. They are collections of lymphoid (reticular) tissue, and together with the palatine tonsil and the adenoids form a protective ring. Can you remember its functions? (p. 18).

The tonsils are often largest in childhood and tend to atrophy in adult life.

3. THE TONGUE

This is a muscular organ lying in the floor of the mouth. Laterally it is attached to the inner surface of the mandible, and posteriorly it is attached to the hyoid bone. Underneath there is a thickened fold known as the frenum. It is covered by mucous membrane (stratified squamous epithelium) with papillae which contain the taste buds. The nerve endings that appreciate texture and temperature are scattered on this surface. Posteriorly there is some lymphoid (reticular) tissue, the lingual tonsil, part of the protective ring already mentioned.

Functions of the tongue

(*a*) The tongue assists in mastication.

(*b*) It assists in swallowing.

(*c*) It assists in articulation.

(*d*) It is the sensory organ for the taste, texture and temperature of food.

4. THE SALIVARY GLANDS

These glands surround the mouth. There are three pairs of them, all secreting 1,000 to 1,500 ml of saliva each 24 hours and pouring it into the mouth. Insufficient secretion is experienced as thirst—a safety mechanism in the body's water balance.

The parotid glands lie in front of the ear, on either side, and discharge their secretion via Stenson's ducts that open into the mouth opposite the second molar teeth.

The submaxillary glands lie at the angle of the jaw, and discharge their contents under the tongue via Wharton's ducts.

The sublingual glands lie under the tongue and discharge their secretion into this area via several small ducts.

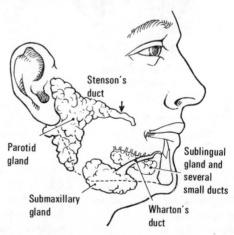

Fig. 134   Diagram of the salivary glands.

Saliva is the secretion from the salivary glands. It is mainly composed of water and is slightly acid (pH 6·8) to allow the contained enzyme ptyalin to act upon cooked starches, converting them into maltose.

Mucus is secreted liberally in the mouth, for purposes of lubrication, and binding of the food into a bolus, in which fashion it is rolled to the back of the tongue, and dropped into the pharynx, from whence it is swallowed.

The Digestive or Alimentary System

We have spoken of some of the functions of the mouth and its contents, and collecting them together we get this list:

FUNCTIONS

1. The mouth assists in the mastication of food and its formation into a bolus.
2. It provides a slightly acid reaction.
3. It provides the first chemical change in the process of digestion, as the ptyalin converts cooked starches into maltose.
4. It takes part in the act of swallowing.
5. The tongue is the sensory organ for the appreciation of taste, texture and temperature of food.
6. It assists in the complex process of speech.
7. It provides a safety airway when the nose is blocked.
8. By its lymphoid (reticular) tissue it protects the body from infection by filtration and the production of lymphocytes.
9. Saliva keeps the mouth clean. (In a 'dry' mouth, sordes form and bacteria flourish.)

BLOOD SUPPLY

Many branches coming from the facial artery, which is a branch from the external carotid.

VENOUS DRAINAGE

Via many branches into the facial, which drains into the jugular vein.

NERVE SUPPLY

The skeletal striped voluntary muscle of the cheeks and tongue, etc., are innervated by white motor neurones belonging to the central nervous system.

Sensation is derived from white sensory neurones, also belonging to the central nervous system.

Secretion is provided by the grey secretory nerves from the autonomic nervous system, partly by reflex action (p. 193).

Before going on to discuss the remaining organs in the digestive or alimentary system, let us pause to consider the care that should be given to the mouth so that it can remain healthy.

Care of the Mouth in the Maintenance of Health[122,123,124,125,126]

As the foundations of the teeth are laid down in the fetus, then the pregnant woman needs to consume a diet adequate in calcium, phosphorus, calciferol (vitamin D) and fluorine[127] for this purpose.

After birth the mouth should be allowed to develop into a normal shape. It is thought by some that a breast-fed baby has a comparatively wider palate and better-shaped arch. Nasal obstruction is to be avoided as the consequent open mouth narrows the palate and heightens the arch (adenoid facies).

The infant's[128] diet should contain adequate calcium, phosphorus, calciferol and fluorine for the health of both sets of teeth. In addition vitamin A is needed for the health of the oral mucous membrane, and riboflavine of the vitamin B complex to avoid cracks at the corners of the mouth.

The baby should be given something hard and clean on which to bite, to improve the circulation and help the eruption of the teeth through the gums.

As soon as possible the child is given his own tooth brush and taught and encouraged to cleanse his teeth from the gum to the free edge. Animals on the handle of the tooth brush and fruit-flavoured tooth pastes are all useful in establishing this habit, which should last a lifetime. An electric tooth brush[129] is said to be slightly more effective than the conventional models. Fluoride-containing toothpastes[130] are available.

When cleansing after a meal is impossible, a drink of water, a mouthwash and the chewing of hard fruit or vegetable, are all useful in removing starch particles from between the teeth. Unilever has recently produced a tablet which, when sucked after a meal, prevents dental caries.[131,132] The provision of tooth picks is favoured in some establishments. Adults should discourage children from eating sweets, 'lollies', fruit drinks, ice-cream, biscuits etc. *between* meals. These things should be part of, or taken at the end of a meal —*before* tooth cleansing. Currently on the market is a new sucrose-free syrup containing Vitamin C, made from acerola cherries. It is claimed that its use will improve dental health since ordinary sugar (sucrose) is a common cause of dental caries. Nothing should be taken by mouth after the last cleansing before sleep.

It is necessary from early childhood to establish the habit of a bi-annual visit to the dentist. Dental treatment in this country is 'free' until the age of 21. When the child goes to school at 5 years, dental inspection will be carried out in the school health programme. Some local authorities issue free of charge, a dental health pack containing tooth brush, toothpaste and 'Happy Smile' cards to children starting primary school. Weekly fluoride mouth rinsing for school children in Norway and Sweden has decreased the work load of dentists in dental clinics. A mouthpiece[133] holding a jelly containing 1·1 per cent sodium fluoride has been held in the mouth of school children for six minutes daily in some schools in New York State, thus decreasing the incidence of dental caries. A dental varnish that hardens into an invisible film in seconds, and prevents food particles gaining contact with teeth, is available. In an attempt to get over to small children the consequence of eating sweets and not cleaning their teeth, a puppet show is visiting day nurseries, nursery schools and infant schools in London.

Natives of a village called Angorma, in New Guinea have perfect teeth until they die. A research worker is conducting tests to see whether trace elements such as cobalt are responsible for the Angorma natives' decay-free teeth.

Dextranase is a preventor of dental decay because it removes the sticky film of sugar and bacteria which clings to teeth. However it has the big disadvantage of being destroyed by heat. The Australian National Health and Medical Research Council have approved the addition to sugar, of a substance that hardens tooth enamel and reduces the acids that cause decay. Work on a similar additive is currently being carried out in Britain at the Royal College of Surgeons. An ultrasonic technique is being used in research work to detect defective patches of enamel, in an attempt to find the factors that cause breakdown of enamel. Research has produced a vaccine that prevents dental caries in monkeys, but it is not yet ready for trial in human beings.

The Annual Reports published by H.M.S.O. show that there has been a marked improvement in the dental health of the adult population since the introduction of the National Health Service in 1948.

Currently, Independent Television is informing the population that dental care is free for pregnant women.

Conditions which can Interfere with the Health of the Mouth

DENTAL CARIES

When starch is left between the teeth it is fermented by the bacteria that normally live in the mouth (natural flora), with the production of an acid and a gas. Calcium is soluble in this more acid medium and so the first stage of dental caries is initiated. The most vulnerable part of the tooth is that nearest the gum, so that infection of the gum can follow. Two-thirds of children under 5 who attend dental clinics need treatment including extractions. The average 5 year old in Britain has seven of his 20 temporary teeth decayed, filled or extracted.[134]

PYORRHOEA

Infection of the gum and dental caries usually precedes this condition.

GUM BOIL

This is abscess formation around the fang of a tooth.

MALALIGNMENT OF TEETH

It is especially important that the permanent teeth should be adequately spaced and in good alignment to give the best chance of the teeth remaining healthy, and to give an adequate contour on which the cheeks can rest. Orthodontology is the science dealing with this, and the service of an orthodontist is employed in some school health programmes.

THRUSH

This condition appears as white patches on an inflamed mucous membrane and is due to the external, vegetable parasite *Candida (Monilia* or *Oidium) albicans.* It is a fungus with yeast-like cells that form filaments. Thrush occurs mainly in bottle-fed babies in poor hygienic conditions. With such a sore mouth the baby feeds badly and can become seriously dehydrated. As well as treatment for the mouth the mother must be taught the hygiene of infant feeding. Local treatment for the mouth is the application of 0·25 to 0·5 per cent aqueous gentian violet; systemic treatment is the administration of nystatin orally.

CHEILOSIS OR PERLECHE

This is an intertrigo at the angles of the mouth with maceration, fissuring, or crust formation. It may result from poorly fitting dentures, bacterial infection, thrush infestation, vitamin or iron deficiency, drooling or thumb-sucking.

ANGULAR STOMATITIS

This comprises fissuring in the corners of the mouth consequent upon riboflavine deficiency. Riboflavine (vitamin B_2) is present in dairy produce, liver and eggs.

APHTHOUS STOMATITIS

This consists of recurring crops of small ulcers in the mouth. It is thought that it has some relationship to the virus infection—cold sore (herpes simplex)—but it is not proven.

The Pharynx

This is the next organ through which the food passes. It is a fibromuscular, funnel-shaped cavity behind the nose and mouth, and it is the common passage for food and air.

POSITION AND RELATIONSHIPS

Anteriorly. The posterior openings from the nasal cavities, sometimes spoken of as the posterior nares. Below these is the palatine arch, with the uvula hanging centrally. Below this is the epiglottis, which is the first part of the larynx or voice box.

Posteriorly. The cervical vertebrae. (Take the opportunity of revising these by turning to page 56.)

Laterally. A tube on either side connects the pharyngeal cavity with the cavity of the middle ear, which is bounded laterally by the tympanic membrane. These tubes are therefore called the pharyngo-tympanic tubes, though many people still use the older name of Eustachian tubes, after the person who first described them.

Inferiorly. The food tube (oesophagus or gullet) leads from the posterior base, and the air tube (larynx) leads from the anterior base.

Superiorly. The sphenoid bone.

STRUCTURE

The pharynx is lined with non-keratinized stratified squamous epithelium which has mucus-secreting glands lying at intervals in the submucosa's areolar tissue. Outside these is a layer of skeletal, striped, voluntary muscle arranged mainly in circles, so that by contraction it can push the food onwards into the gullet (oesophagus). The pharynx continues lubrication of the bolus of food, and moistening and warming of the air passing through it.

The roof of the pharynx contains some lymphoid (reticular) tissue, called the pharyngeal tonsil, enlargement of which gives rise to the condition of adenoids. These can press on the openings into the pharyngotympanic tubes with resultant deafness.

BLOOD SUPPLY

Pharyngeal artery.

VENOUS DRAINAGE

Pharyngeal vein.

NERVE SUPPLY

White motor neurones belonging to the central nervous system to the muscles.

White sensory neurones belonging to the central nervous system.
Grey neurones from the autononic nervous system.

FUNCTIONS

1. The pharynx provides a passageway for food and air.

2. It continues to moisten, warm and filter the air passing through it.

3. It continues lubrication of the food passing through it.

4. During swallowing it allows air to pass along the pharyngotympanic tube to the middle ear to equalize the pressure on either side of the ear drum.

Deglution or Act of Swallowing

The uvula rises on to the posterior pharyngeal wall, thus stopping the regurgitation of food through the nose. Simultaneously the larynx is sheltered under the epiglottis, thus preventing food from entering the air tube (larynx), as illustrated in Figures 135 and 136. Rarely is there interference with this mechanism, but when there is, it has disastrous results. Some infectious diseases can cause paralysis of the soft palate.

The remainder of the food tube has a common basic structure with variations in each area to fit the particular functions.

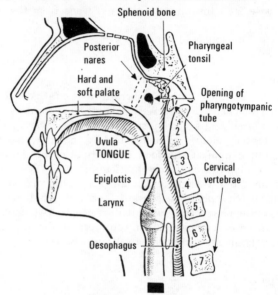

Fig. 135   Diagram of the pharynx.

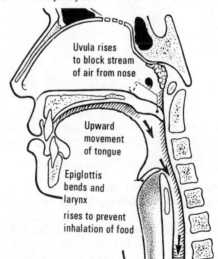

Fig. 136   Pharynx during the act of swallowing.

Basic Layers of Tract

It is *lined* throughout *by mucous membrane*, supported on a *sub-mucosa* of areolar tissue to carry the nerves, blood and lymphatic vessels. Lying outside these there is *internal, plain, involuntary muscle* arranged in an inner circular and an outer longitudinal layer. It is supplied with grey motor neurones from the sympathetic and parasympathetic portions of the autonomic nervous system. The sympathetic portion inhibits peristalsis, whilst the parasympathetic encourages peristalsis. For revision of the action of this plain muscle layer see page 105.

The Oesophagus

This is commonly called the gullet. It commences in the neck and passes through the thorax.

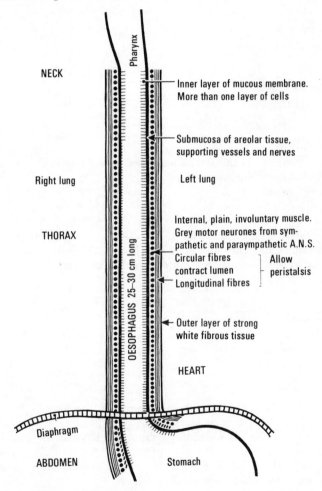

Fig. 137   Diagram of the oesophagus.

POSITION AND RELATIONSHIPS

Anteriorly. The larynx and trachea (windpipe) in the upper portion, and the heart in the lower portion.

Posteriorly. The dorsal or thoracic vertebrae.

Laterally. The lungs.

Inferiorly. The oesophagus passes through the diaphragm and is continuous with the stomach.

Superiorly. The oesophagus is continuous with the pharynx.

VARIATIONS IN STRUCTURE

Throughout its length the mucous membrane lining has more than one layer of cells, as the fibres in the food are still somewhat coarse. It has an outer layer of strong white fibrous tissue, which blends with that of the other tissues of the neck and thorax.

BLOOD SUPPLY

Oesophageal arteries.

VENOUS DRAINAGE

Oesophageal veins.

NERVE SUPPLY

Grey neurones from the sympathetic and parasympathetic portions of the autonomic nervous system.

Sensory neurones.

FUNCTIONS

The oesophagus transmits food by peristaltic action from the pharynx to the stomach, continuing to lubricate same.

The Stomach

This organ is of variable shape according to its contents, and lies in the upper abdominal cavity, more to the right than the left.

RELATIONSHIPS

Anteriorly. The muscles of the abdominal wall (p. 98).

Posteriorly. The pancreas and spleen.

Laterally. On the right side the liver; on the left side the pancreas and spleen.

Inferiorly. The stomach is continuous with the duodenum, and the transverse colon passes across at this level.

Superiorly. The stomach is continuous with the oesophagus. The

276

right lateral portion of the liver overlies the uppermost part of the stomach, and the diaphragm is in close relationship. The heart and left lung are above the diaphragm.

VARIATIONS IN STRUCTURE

The mucous membrane lies in impermanent folds called rugae; these disappear when the stomach is distended. The glands secrete (*a*) mucin, (*b*) gastric juice which contains pepsinogen to convert protein into peptones and rennin to convert soluble caseinogen into insoluble casein, and (*c*) hydrochloric acid to increase the acidity of the food, disinfect same and control the opening of the pylorus. Pepsinogen in the presence of hydrochloric acid becomes pepsin. The glands also secrete hormones (gastrin, gastrozymin) that keep up the flow of gastric juice until the meal is digested. The lining makes the anti-anaemic factor, previously called the intrinsic factor of Castle. It is essential for the absorption of an extrinsic factor taken in food, and now known to be vitamin B^{12} (cyanocobalamin) which prevents pernicious anaemia.

The Digestive or Alimentary System

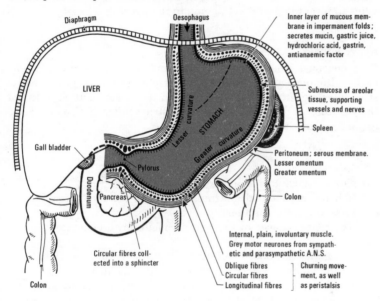

Fig. 138   Diagram of the stomach.

In the *muscle layer* there are extra oblique fibres to produce churning. The circular fibres are collected into a weak sphincter at the cardiac orifice, i.e. where the oesophagus enters the stomach, and a much stronger sphincter at the pyloric orifice, i.e. where the stomach is continuous with the duodenum.

The *outermost layer* is composed of *serous membrane*, the peritoneum. It is formed in two layers, that closely investing the organ being called the visceral layer, and that reflected to line the containing cavity being called the parietal layer. The function of the peritoneum is to allow movement without friction.

There are two special portions of the peritoneum:

1. The lesser omentum passing from the lesser border of the stomach to the liver.

2. The greater omentum, hanging down like an apron from the greater border of the stomach, and traversing upwards again as a double fold, to be inserted into the transverse colon.

The omentum is sometimes called the 'policeman' of the abdomen, for it can wall off an area of infection, e.g. when an appendix perforates, and thus it can prevent a widespread peritonitis.

BLOOD SUPPLY

Gastric artery from coeliac artery from aorta.

VENOUS DRAINAGE

Gastric vein, which joins splenic, superior and inferior mesenteric veins to form the portal vein going to the liver.

NERVE SUPPLY

Grey neurones from the sympathetic and parasympathetic portions of the autonomic nervous system. The vagus nerve (parasympathetic) stimulates motility and acid production in the stomach. Severing of the vagal nerve fibres will minimize gastric acidity, but can give rise to dilation of the stomach.

FUNCTIONS

1. The stomach acts as a reservoir for food.

2. It has a mechanical churning action on the food, breaking up the fibres and softening them.

3. The heat produced by this activity melts any fat contained in the food.

4. It acidifies the food, disinfecting same and terminating the action of ptyalin, which can only act in a slightly acid medium.

5. By the action of rennin it converts soluble caseinogen into insoluble casein, which can then be acted on by pepsin.

6. By the action of pepsin, it converts proteins into peptones.

7. It secretes the hormones (gastrin, gastrozymin), which maintain the flow of gastric juice until the meal has been digested. The first flow is secreted as a reflex action (p. 193).

8. It secretes the intrinsic factor which is essential for the absorption of the extrinsic factor taken in food. The latter is cyanocobalamin (vitamin B_{12}) and it is essential for the prevention of pernicious anaemia.

The Small Intestine

This long tube, 6 metres or more in length, lies coiled up within the abdominal cavity. For purposes of description it is divided into three parts, though these are continuous.

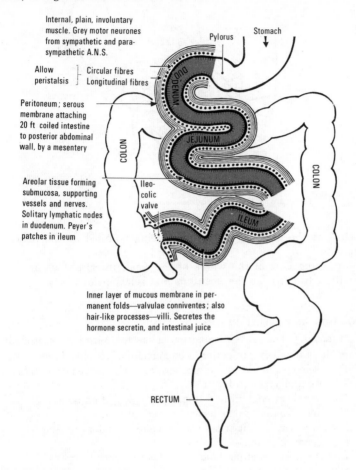

Internal, plain, involuntary muscle. Grey motor neurones from sympathetic and parasympathetic A.N.S.

Allow peristalsis ⎱ Circular fibres / Longitudinal fibres

Peritoneum; serous membrane attaching 20 ft coiled intestine to posterior abdominal wall, by a mesentery

Areolar tissue forming submucosa, supporting vessels and nerves. Solitary lymphatic nodes in duodenum. Peyer's patches in ileum

Pylorus   Stomach

DUODENUM

JEJUNUM

COLON

COLON

Ileo-colic valve

ILEUM

Inner layer of mucous membrane in permanent folds—valvulae conniventes; also hair-like processes—villi. Secretes the hormone secretin, and intestinal juice

RECTUM

Fig. 139   Diagram of small intestine.

The duodenum. The duodenum comprises the first 30 cm and is like a horseshoe in shape. It leaves the pylorus and encircles the head of the pancreas to which it is firmly attached. It lies mainly to the right of the midline, and is continuous with the jejunum distally. On its medial surface it receives the bile duct from the liver and the pancreatic duct from the pancreas at a common opening called the ampulla of the bile duct (Fig. 140).

The jejunum. The jejunum comprises the next 2·5 metres of this tube and is continuous with the duodenum above and ileum below; this structure lies in coils.

The ileum. This comprises the last 3·5 metres of the tube. It is continuous with the jejunum above and enters the large bowel

279

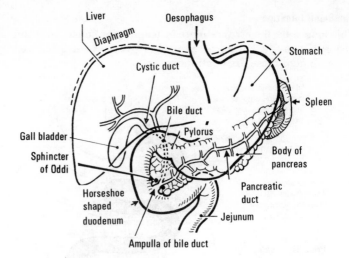

Fig. 140   Diagram of the duodenum.

below at the caecum, the opening being guarded by the ileocolic valve.

Speaking now of the whole of the small intestine, we are in a position to consider the positions and relationships.

POSITION AND RELATIONSHIPS

Anteriorly. The greater omentum, a special 'apron' of peritoneum, capable of localizing infection. The muscles of the abdominal wall.

Posteriorly. The kidneys, ureters and the adrenal glands immediately above the kidneys.

The vertebral column and upper portion of iliac bones.

Muscles forming the posterior abdominal wall.

Laterally. On the right side the ascending colon, on the left side the descending colon.

Inferiorly. The rectum, bladder and reproductive organs.

Superiorly. The transverse colon, pancreas, spleen, liver and gallbladder.

VARIATIONS IN STRUCTURE AND FUNCTION

Mucous membrane. This lies in permanent folds called *valvulae conniventes,* this arrangement increasing the surface area for digestion and absorption.

Superimposed on these permanent folds to further increase the surface area are millions of hair-like structures called *villi,* each containing blood vessels and a special lymphatic vessel called a lacteal for the absorption of the digested food. The villi are covered with columnar epithelium that has a striated border to increase absorption of food. In some diseases, notably sprue and coeliac

280

disease, these villi are damaged and the surface becomes smooth. As a result there is lack of absorption, leading to malnutrition. A greater amount of food passes into the large bowel causing diarrhoea.

This lining secretes hormones (secretin, enterogastrone, cholecystokinin, pancreozymin, villikinin, duocrinin), that keep up the flow of pancreatic and intestinal juice until the meal has been digested. The mucous membrane also secretes *intestinal juice* (succus

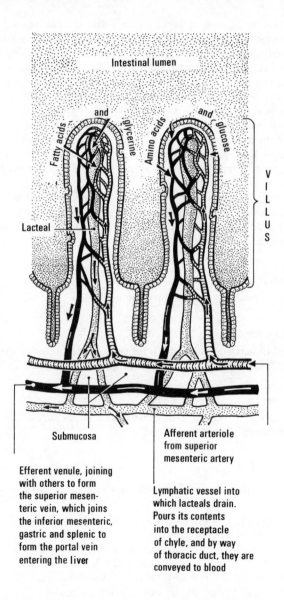

Fig. 141 Diagram illustrating the absorption of digested food into the intestinal villi.

entericus), which is alkaline in reaction and contains the following enzymes:

Maltase which converts maltose into glucose.

Sucrase which converts sucrose into glucose and fructose.

Lactase which converts lactose into glucose and galactose.

Erepsin which converts peptones into amino acids.

Enterokinase which activates the trypsin contained in pancreatic juice.

If lactase is absent at birth, undigested lactose passes into the large intestine and causes diarrhoea, with subsequent malnutrition. It can only be overcome by omitting lactose from the diet.

Submucosa. This contains *solitary nodes* of lymphatic (reticular) tissue in the duodenum and the upper portion of the jejunum, for further protection against infection, by filtration and the production of leucocytes (lymphocytes).

The reticular tissue is collected into *Peyer's patches*, particularly in the ileum, and these are the seat of infection in typhoid fever. You will realize that you are never going to be allowed to forget this subject, as you will require to recall it again and again in senior lessons.

Peritoneum—outermost layer. To allow the freedom of movement necessary for intestinal function, this 6-metre tube is suspended from the posterior abdominal wall in a double fold of peritoneum known as the *mesentery* (Fig. 33).

BLOOD SUPPLY

Superior mesenteric artery from the aorta.

VENOUS DRAINAGE

Superior mesenteric vein which unites with the inferior mesenteric, gastric and splenic veins to form the portal vein going to the liver.

NERVE SUPPLY

Grey neurones from the sympathetic and parasympathetic portions of the autonomic nervous system.

FUNCTIONS

These are mainly concerned with the digestion and absorption of food:

1. Bile, secreted in the liver, is poured in from the bile duct to the duodenum. With the detergent action of its alkaline salts, it emulsifies and saponifies fats, breaking them into tiny droplets, thus preparing them for the action of pancreatic lipase.

2. Pancreatic juice is also poured into the duodenum. It is alkaline in reaction and brings about the following changes:

Amylase, converts sugars and starches into monosaccharide, mainly glucose.

Trypsin, when activated by enterokinase from the intestinal juice, converts peptones into amino acids.

Lipase, converts the already prepared fats into fatty acids and glycerine.

3. Intestinal juice is alkaline in reaction and brings about the following changes:

Maltase, converts maltose into glucose.

Sucrase, converts sucrose into glucose.

Lactase, converts lactose into glucose.

Erepsin, converts peptones into amino acids.

Enterokinase, activates trypsin from the pancreatic juice.

4. The permanently puckered mucous membrane is capable of tremendous absorption.

Glucose and *amino acids* are absorbed into the *blood capillaries* in the villi; these join ever-increasing vessels until they become the superior mesenteric vein, which joins the inferior mesenteric, the gastric and the splenic, to form the portal vein going to the liver (Figs. 150 and 152).

Fatty acids and *glycerine* are absorbed into the *lacteals* of the villi, which pour their contents into the receptacle of chyle on the posterior abdominal wall; from here they are conveyed via the thoracic duct and poured into the blood stream at the union of the left subclavian with the left jugular vein. (See Figs 108, 141 and 151).

The Large Intestine

This tube is in fact about 180 cm long, but it only covers an area of about 90 cm in the abdominal cavity. This is accomplished by some of the longitudinal muscle fibres being arranged in three tape-like bands, on to which the 180 cm of bowel are gathered, so forming sacculations, where the gas produced by the intestinal bacteria can collect, until a suitable time for evacuation. The varying portions of the bowel are designated as in Figure 142. It forms a 'frame' with the small intestine packed into the middle as the 'picture'.

VARIATIONS IN STRUCTURE

Mucous membrane. The goblet cells are more numerous to produce the mucus needed to lubricate passage of the drying faecal matter.

Submucosa. Here the areolar tissue contains extra lymphoid (reticular) tissue in the region of the vermiform appendix.

Muscle layer. The longitudinal fibres are collected into three tape-like bands forming sacculations.

In the anal canal the circular fibres are collected into the internal and external anal sphincter muscles. When passing a tube per rectum it is important to remember that the external sphincter is voluntary muscle and with encouragement a patient will be able to relax it. But the internal sphincter is of involuntary muscle, over which the

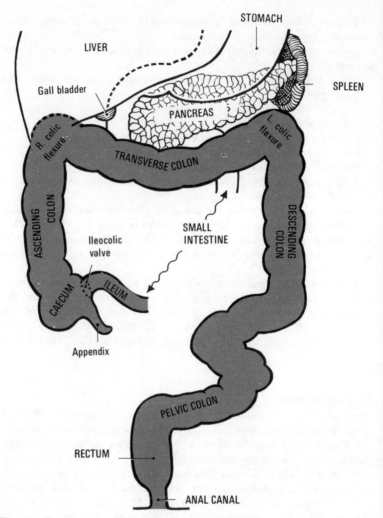

Fig. 142   Designation of the parts of the large intestine.

patient has no control. To appreciate the various degrees of piles, one has to remember that the external sphincter is covered by skin and the internal sphincter is covered by mucous membrane.

Peritoneum—outermost layer. The large intestine is suspended from the posterior abdominal wall by a double fold of peritoneum called the mesentery. (A good description of the latter can be found in *Applied Anatomy for Nurses,* by Bocock and Haines, 4th edition. Edinburgh and London: Churchill Livingstone).

BLOOD SUPPLY

The superior mesenteric artery supplies the first two-thirds: the inferior mesenteric artery supplies the lower third.

The superior and inferior mesenteric veins which join the gastric and splenic to form the portal vein going to the liver.

NERVE SUPPLY

Grey neurones from the sympathetic and parasympathetic portions of the autonomic nervous system.

FUNCTIONS

1. To convey indigestible residue from the ileocolic valve to the rectum for excretion.

2. The lower portion is essential for the act of defaecation.

3. The excretion is a thickened fluid at the ileocolic valve; absorption of fluid takes place along the tract as faeces is a soft solid.

4. The natural bacterial flora synthesizes several vitamins, notably B complex and K.

From research using radio-opaque markers mixed with food, we now know that food does not pass along the colon in orderly sequence. It mixes, one meal with another, one day's intake with the following. Thus from a batch of markers given to a normal person at one meal, the first appeared in the stool three days later, but

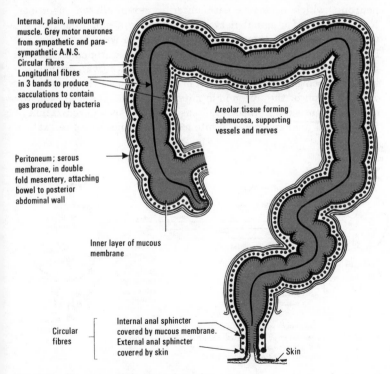

Internal, plain, involuntary muscle. Grey motor neurones from sympathetic and parasympathetic A.N.S.
Circular fibres
Longitudinal fibres in 3 bands to produce sacculations to contain gas produced by bacteria

Areolar tissue forming submucosa, supporting vessels and nerves

Peritoneum; serous membrane, in double fold mesentery, attaching bowel to posterior abdominal wall

Inner layer of mucous membrane

Circular fibres

Internal anal sphincter covered by mucous membrane.
External anal sphincter covered by skin

Skin

Fig. 143   Structure of the large intestine.

80 per cent did not appear until five to seven days after intake. In a constipated person, the first markers of a batch appeared from eight to nine days later and 80 per cent up to 15 days after intake.

Composition of Faeces

Faeces is mainly made up of water, indigestible residue from food and millions of bacteria both living and dead. The normal intestinal bacteria do good work, as long as they only inhabit the intestine. They can cause inflammation elsewhere in the body, e.g. in the bladder. Their presence is used as a test for faecal contamination of water. Faeces is coloured by bile pigment (stercobilin). It contains some bile salts, intestinal secretions, mucus, leucocytes, shed epithelial cells and inorganic matter, chiefly calcium and phosphates.

The Act of Defaecation

In the baby this is a reflex action. The dropping of the faeces from the sigmoid flexure into the rectum sets up a stimulus of distension which is conveyed, via a posterior sensory afferent neurone, to the subsidiary defaecation centre in the spinal cord. Here the stimulus is conveyed from the posterior to the anterior root by a connector neurone, which arborizes with anterior motor efferent neurones, forming part of a spinal nerve. These end in a ganglion near the vertebral column if they are going to arborize with grey neurones belonging to the sympathetic portion of the autonomic nervous system; they end in a ganglion within the bowel wall if they are going to arborize with grey neurones belonging to the parasympathetic portion of the autonomic nervous system. These grey motor neurones cause simultaneous increased peristalsis and relaxation of the internal anal sphincter. White motor neurones cause relaxation

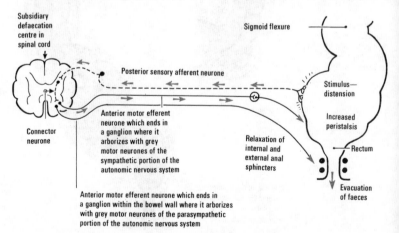

Fig. 144   Diagrammatic illustration of defaecation.

of the external anal sphincter, so that faeces is evacuated. As maturation of the neuromuscular mechanism involved in the above action is reached, the conscious brain can inhibit the desire to defaecate until the lavatory is reached.

ACCESSORY DIGESTIVE ORGANS
The Liver

This is the largest gland in the body. It lies immediately underneath the diaphragm on the right side, a portion stretching over to lie on top of the upper portion of the stomach. Its under surface is not normally palpable below the right costal margin. With a little thought and the help of Figure 145, you will be able to decide what lies anteriorly, posteriorly, at the right side, at the left side, inferiorly and superiorly to the liver.

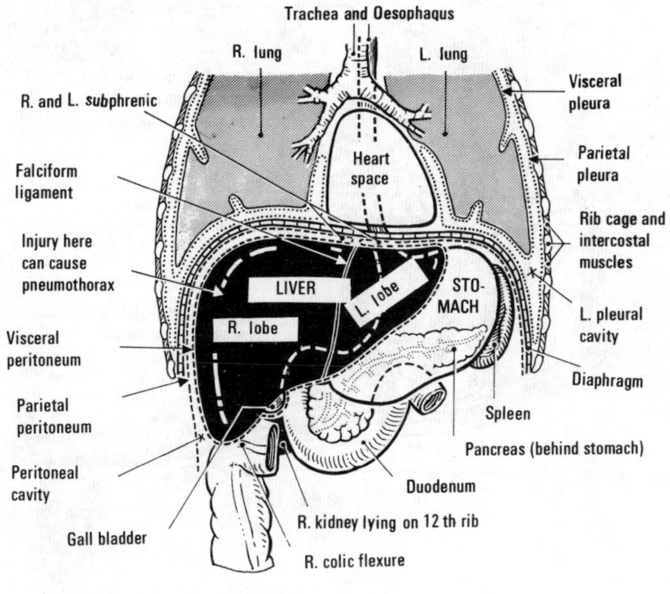

Fig. 145   Relationships of the liver.

STRUCTURE

The liver is divided into *two lobes* by the falciform ligament stretching from the diaphragm. On the under surface of the right lobe is the pear-shaped gall-bladder, with the cystic duct leading from it, and joining the hepatic duct from the liver, thus forming the bile duct conveying bile to the duodenum. The left lobe stretches over the midline and overlies the upper portion of the stomach.

Peritoneum covers the major portion of the *external surface*, and is reflected at the falciform ligament to form the right and left subphrenic spaces. You will learn later about a subphrenic abscess, and you will need to recall the position of the subphrenic spaces.

287

Inside the peritoneal covering the liver tissue is bounded by a *capsule of white fibrous tissue.* The two lobes are further divided into *hexagonal lobules.* The efferent blood supply, i.e. the *portal vein* carrying deoxygenated (venous) blood rich in foodstuffs, and the *hepatic artery* carrying freshly oxygenated blood, break up into smaller vessels that travel between the lobules and thus earn the name of *interlobular arteries* and *veins.* These vessels then send branches into the centre of the lobule, and in this way the blood comes into close contact with millions of liver cells, so that these liver cells may perform their complex functions. Reticulo-endothelial cells perform a phagocytic function (Fig. 4).

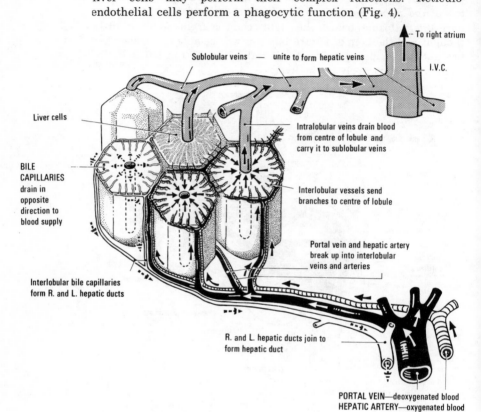

Liver cells

Sublobular veins — unite to form hepatic veins

— To right atrium

I.V.C.

Intralobular veins drain blood from centre of lobule and carry it to sublobular veins

BILE CAPILLARIES drain in opposite direction to blood supply

Interlobular vessels send branches to centre of lobule

Portal vein and hepatic artery break up into interlobular veins and arteries

Interlobular bile capillaries form R. and L. hepatic ducts

R. and L. hepatic ducts join to form hepatic duct

PORTAL VEIN—deoxygenated blood
HEPATIC ARTERY—oxygenated blood

Fig. 146   The liver lobules.

The vessels draining this blood are placed centrally within the lobule and thus earn the name of intralobular veins; several intra-lobular veins unite to form sublobular veins, which eventually unite to form the *hepatic veins* carrying blood to the *inferior vena cava,* which actually traverses the posterior, medial surface of the liver.

The fine *bile capillaries* start within the substance of the lobule and traverse in the opposite direction to the blood supply. They pour their contents into the *interlobular bile vessels,* which eventually unite and leave the liver as *right* and *left hepatic ducts,* soon joining to become the *hepatic duct,* which joins the *cystic duct* from the gall-bladder to become the *bile duct* via which bile is conveyed to the duodenum.

In the interests of developing good methods of description, and for revision purposes, we will include the following headings.

BLOOD SUPPLY

Deoxygenated (venous) blood, rich in foodstuffs via the portal vein.

Freshly oxygenated blood via the hepatic artery, a branch of the coeliac artery from the aorta.

VENOUS DRAINAGE

Via hepatic veins into the inferior vena cava.

NERVE SUPPLY

Grey neurones belonging to the sympathetic and parasympathetic portions of the autonomic nervous system.

Sensory neurones.

FUNCTIONS

1. The liver stores the anti-anaemic factor, cyanocobalamin or vitamin B_{12}. It is absorbed from food when the intrinsic factor is secreted by the gastric mucosa. The liver liberates the anti-anaemic factor to the red bone marrow as needed, for the conversion of a pro-erythroblast into a normoblast in the production of red blood cells (Fig. 82).

2. It stores vitamins A and D.

3. It helps in the destruction of worn-out red blood cells; stores the iron from this process and liberates it to the red bone marrow as needed, for the conversion of a normoblast into a non-nucleated red blood cell (Fig. 82); it stores some of the pigment from this process, and excretes the remainder as colouring matter in bile.

4. By the action of insulin the liver is capable of converting soluble glucose into insoluble glycogen for storage. A lowering blood sugar causes conversion of insoluble glycogen into soluble glucose, by glucagon, so that, by these actions and those of the growth hormone (p. 392), the blood sugar* is maintained within the normal limits of 0·08 to 0·12 per cent.

5. It converts glucose into fat.

6. It secretes bile.

7. It deals with the amino acids brought via the portal vein, synthesizing some of them into actual proteins required by the body, e.g. serum albumin. Others are converted into non-essential amino acids. Some are converted into a fat-like substance for production of energy. The remainder are deaminated to form urea.

---

* Fructose and galactose are converted by enzymes into glucose in special cells along the villi. Either of these enzyme systems may be deficient. Accumulation of galactose in the blood damages the brain, and is one cause of mental subnormality. The brain relies on *glucose* for its source of energy.

289

8. It synthesizes nucleic acids.

9. It desaturates fats so that the body can utilize them more easily for the production of heat or for storage. Synthesizes fatty acids and phospholipids—fat molecules in the cell wall (p. 5).

10. It produces anticoagulant substances, notably heparin.

11. It produces fibrinogen and prothrombin. Vitamin K is necessary for production of the latter.

12. It produces antibodies for protection of the body against infection.

13. It inactivates many hormones and drugs. (Special attention to drug dosage is necessary for patients with liver damage.)

14. It produces heat as a result of this tremendous chemical activity.

The Gall-bladder

This is a small pear-shaped muscular bag lying on the under surface of the liver, its 'tip' in contact with the anterior abdominal wall in the region of the ninth costal cartilage on the right side.

In structure it is similar to other hollow organs, i.e. on its external, anterior surface it has the serous membrane—peritoneum; within this there are two layers of internal, plain involuntary muscle supplied with grey motor neurones from the autonomic nervous system. These muscle fibres relax to accommodate bile from the liver, and on contraction they expel the bile via the bile ducts into the duodenum. Within the muscle layer lies a submucosa of areolar tissue supporting blood and lymphatic vessels and nerves. Adherent to the submucosa is a mucous membrane lining capable of tre-

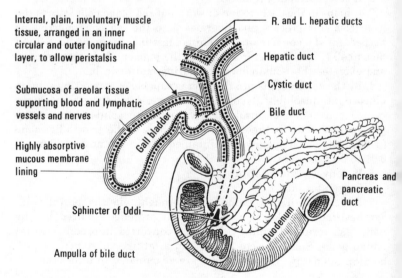

Fig. 147 The gall-bladder and accompanying ducts (extrahepatic biliary apparatus).

mendous absorption of fluid, so that the bile in the gall-bladder actually becomes concentrated.

The cystic duct leaves the neck of the pear-shaped gall-bladder and joins the hepatic duct coming from the liver, to form the bile duct that passes behind the duodenum, and actually through the head of the pancreas, to join the pancreatic duct to form the ampulla of the bile duct controlled by circular muscle fibres collected together to form a sphincter, named after Oddi, who first described it. The presence of food in the duodenum, notably fat, causes relaxation of this sphincter and a pouring of bile on to the duodenal contents.

THE COMPOSITION OF BILE

Like most body fluids it is mainly composed of water, in which alkaline salts and waste pigment from natural haemoglobin breakdown are dissolved.

FUNCTIONS OF BILE

1. The alkaline salts emulsify and saponify the fats preparing them for the action of pancreatic lipase.

2. The pigments provide colouring matter for faeces and urine.

3. It is necessary for the absorption of menadione (vitamin K) from the intestine.

4. It has deodorant properties.

5. It acts as a natural aperient.

N.B. Because ampicillin is excreted in bile, it is often used for infections of the biliary tract.

The Pancreas

Here we are going to consider the pancreas as an accessory digestive organ, since it pours out pancreatic juice to assist in the process of digestion, but it can equally well be considered a ductless gland, i.e. part of the endocrine system, for it makes an internal secretion called insulin that passes directly into the blood passing through the organ (p. 391).

POSITION AND RELATIONSHIPS

The head of the pancreas is encircled by the duodenum, these structures lying in the epigastric region (Fig. 148).

The 'body' of the pancreas passes below the stomach in a posterior direction, so that the 'tail' of the pancreas touches the spleen.

STRUCTURE

The pancreas is sometimes described as a tongue-shaped organ, comprising a head, body and tail. It is made of glandular or secretory tissue and is highly vascular. The secretory cells concerned with the production of pancreatic juice are placed around ducts communicating with one main duct, the pancreatic duct via which the pancreatic juice is poured into the duodenum, there to play its part in the process of digestion.

The secretory cells concerned with the production of insulin are found scattered throughout the gland in tiny patches called the islets of Langerhans.

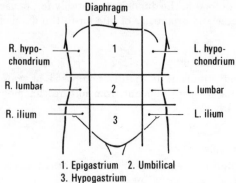

1. Epigastrium 2. Umbilical
3. Hypogastrium

Fig. 148 Anatomical regions of the abdomen.

BLOOD SUPPLY

Pancreatic branches from the splenic artery coming from the coeliac artery, which comes from the aorta.

VENOUS DRAINAGE

Via pancreatic branches into the splenic vein, which joins the gastric, and the superior and inferior mesenteric veins to form the portal vein, carrying deoxygenated (venous) blood, rich in nourishment to the liver.

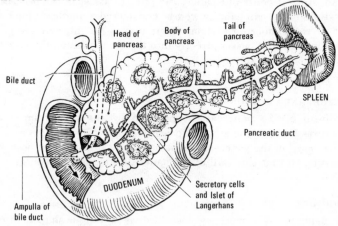

Fig. 149 The structure of the pancreas.

292

Grey neurones from the sympathetic and parasympathetic portions of the autonomic nervous system. A few sensory neurones.

FUNCTIONS

1. The pancreas secretes pancreatic juice. (Can you remember the reaction of this juice, the names of the enzymes contained therein, the food groups upon which these enzymes act and the changes brought about by their action? (p. 282).) A 'sweat' test, measuring the chloride content of sweat, can detect cystic fibrosis in the first week after birth. Such children are deficient in pancreatic enzymes and have to take them orally.

2. It secretes insulin[135] and glucagon (p. 391) which is absorbed directly into the blood passing through the gland.

The Digestive or Alimentary System

Care of the Digestive System in the Maintenance of Health

We have already discussed the important psychological and sociological implications of food for each individual (p. 217). Children from an early age need to learn the necessary rituals associated with 'clean' food and the serious implications of 'dirty' food (p. 238) Hygiene of the kitchen sink is discussed on page 327. It is important to buy food at 'clean' shops and restaurants and not to patronize 'dirty' ones. Wash fruit and vegetables, especially before eating them raw. Cook food well, especially bacon and pork. As instructed previously in this book, pay special attention to keeping the mouth and teeth in a healthy state. Wash hands before preparing food, after visiting the lavatory and before eating food. Arrive at the table anticipating enjoyment of the meal Never sit down to eat in 'anger'. Chew food properly, taking sufficient time over a meal. Do not imbibe excessive alcohol as it can cause cirrhosis of the liver. Have typhoid injections when travelling abroad, except to North America, Scandinavia and the Low Countries.

For most people in the western world, it is customary to think of defaecation as taking place when sitting on a water-flush lavatory. Two-thirds of the world population use privies and latrines. The latter can be merely a hole in the ground, so that defaecation takes place with a person in the squatting position. Many people think that this gives maximum use of the muscles and the weight of the abdominal organs to assist in expulsion.

Theoretically there is a more liberal attitude to 'toilet training' in the west, but mothers still experience social pressure to have their babies 'clean' as soon as possible. Two articles[136,137] show how two mothers coped with this, and also the variable age at which normal children gain control. Some mothers 'pot' their babies from an early age, and are proud that they have not had a dirty napkin since. They have been lucky and have caught the faeces being expelled as the result of the gastrocolic reflex, a mechanism whereby there is increased peristalsis after food enters the stomach. It is important to be consistent towards the child about defaecation. He can only be bewildered if there is sometimes pleasure (when he uses his pot) and sometimes displeasure (when he uses his napkin). There has to be a strong association in the child's mind between the pot and defaecation, and the ability to 'tell' by gesture if not words, before he can be 'clean'. If there is anxiety when the child does not defaecate daily, he is able to effect the behaviour of others, and this produces the first glimmering of independence. Many traits in an adult are attributed to attitudes acquired during establishment of voluntary control over defaecation. The adult objective is to have an individual routine of evacuating the bowel and not to consistently ignore the desire to evacuate, e.g. because of pressure of time. Much harm has been done by the belief that every person must evacuate the bowel every day. We now know that no harm comes from a longer interval of evacuation—the important thing is that it is *regular for that person.*

Conditions which can Interfere with the Health of the Digestive System

Those conditions interfering with the mouth have already been mentioned (p. 271).

Food poisoning has been dealt with (p. 238).

CONSTIPATION

This is an implied chronic condition of infrequent and often difficult evacuation of faeces, due to insufficient food or fluid intake, or to sluggish or disordered action of the bowel musculature or nerve supply, or to habitual failure to empty the rectum. The following regime helps to prevent constipation:

1. Increase the amount of roughage in the diet (p. 230)

2. Increase the fluid intake to more than 1·75 litres (3 pints) daily.

3. Increase the amount of vitamin B complex to ensure health of the supplying nerves (p. 225).

4. Increase the amount of daily exercise to ensure good muscle tone.

5. Rebuild the daily habit of evacuation at a time that fits in with the day's routine, and preferably after a meal to take advantage of the gastrocolic reflex, i.e. increased peristaltic waves in the colon on entry of food into the stomach.

If the condition does not respond to these measures then medical opinion must be sought.

PILES (HAEMORRHOIDS)

This is a condition of varicose veins in the anal canal, and is frequently caused by constipation. If there is bleeding from a pile it is imperative that medical opinion is sought.

Threadworms *(Oxyuris vermicularis).* To the naked eye these look like small white threads, the male being 0·4 cm and the female 1·25 cm long. They inhabit the large intestine, caecum and appendix and can be seen wriggling about in newly passed motions. Nocturnal migration of these parasites to the skin around the anus, where the gravid female deposits her ova, causes intense irritation with scratching, restlessness, bed wetting and sometimes convulsions. The eggs adherent to the fingers can easily be swallowed the next morning resulting in reinfestation. Each egg is small, oval and contains a coiled-up embryo; after ingestion these eggs hatch out in the duodenum, and the larvae after inhabiting the small intestine migrate to the caecum, where the majority of adult worms live.

The Digestive or Alimentary System

Diagnosis and cure can be confirmed by placing a piece of Sellotape over the anus at night, removing it the next morning, sticking it on to a glass slide and sending this to the laboratory.

It is only in the last decade that drugs dealing adequately with this condition have been available. Prior to that prolonged bowel treatment and an elaborate hygiene aimed at the prevention of reinfestation were necessary. With the new drugs (Antepar, Pripsen, etc., given under medical supervision) the hygienic measures recommended are thorough washing of the hands and scrubbing of the nails on rising in the morning; the whole family must be treated at the same time. Viprynium embonate is given in a single dose in some countries for disinfestation. Unlike Antepar and Pripsen, viprynium embonate is ineffective against round worm.

Roundworm *(Ascaris lumbricoides).* The common roundworm, with its curious life cycle through the lungs, has a general resemblance to the earthworm, and is 15 to 20 cm long. It is not infrequently associated with nervous disorders. Apart from passing the worm(s), the main evidence of active infestation is the presence of ova in the faeces.

The female produces many minute, oval, brownish eggs which are discharged with the faeces of the host. These eggs are taken into the body with contaminated food (this often results from manuring gardens with human excrement) and hatch out in the upper part of the small intestine. The free embryo promptly pierces the mucous mebrane, passing to the liver via the portal circulation, and is subsequently carried in the blood stream to the lungs. Here it pierces a lung capillary to gain the lumen of an air sac, causing bleeding in transit. Finally it passes up the bronchi and trachea to be swallowed. On reaching the intestine for a second time (having developed in its migration), it gradually attains adult form. The adult worm can be passed per rectum, or it can crawl through the pylorus and be vomited. The life span of an adult round worm is thought to be less than one year. Larvae can, however, survive in soil for four to five years.

The above life history recalls the heroic experiment of the Japanese investigator, Koino, who voluntarily swallowed 2,000 ripe ascaris eggs. Six days later, pneumonia accompanied by a temperature of 40° C (104° F) set in and lasted seven days. Sputum was profuse from the eleventh to the sixteenth days and contained ascaris larvae of which 202 were counted. Fifty days after the infestation, a total of nearly 700 ascaris varying in length from 3 to 8 cm were voided.

Washing of the hands after visiting the lavatory, a safe water supply, attention to the disposal of sewage and thorough washing of vegetables, especially if they are going to be eaten raw, help to prevent this condition. Infestation requires medical treatment, the most popular drugs in Britain being Antepar and Pripsen.

Tapeworms (*Cestodes*) (*Taenia solium*, pork; *Taenia saginata*, beef). These are composed of numerous segments united to form a tape-like chain. The head or scolex should now be called the holdfast on both etymological and functional grounds. It bears four sucking discs and a part adapted to pierce called a proboscis, on which are 10 to 20 hooklets. The terminal segments or proglottides contain eggs and are mature. They separate from the body of the worm at the rate of about six daily. They are either evacuated in human faeces or squirm through the anus and fall to the ground, their passage causing anything from slight discomfort to acute embarrassment, and are eventually digested in the stomach of an animal. Here the embryos are set free and pass onwards through the intestinal wall to get into the musculature of cattle, especially the muscles of the jaw, the diaphragm and the heart, where they turn into bladder-like cysts 1 to 1·5 cm ($\frac{1}{2}$ in) in diameter. Local authority meat inspectors routinely incise the jaw, diaphragm and heart of freshly slaughtered cattle. When cysts are found the carcase is condemned or refrigerated for three weeks, depending upon the extent of the infestation. Public Health Inspectors in one area found 150 infested beef carcases in one year. This represented 2·4 per cent of the total number of cattle inspected. The 'cyst' requires to be eaten in a raw or half-cooked state by man, then in his alimentary tract the holdfast appears, followed by terminal mature segments, and the cycle is complete.

The symptoms produced by such infestation are usually vague and indefinite, the patient seeking treatment because he has seen segments in his stool. There may be abdominal pain, nausea and diarrhoea. Any person reporting segments needs to be advised to see a doctor. Treatment is not successful until the holdfast has been removed.[140] If mepacrine is used to gain expulsion, it dyes *T. saginata* which then fluoresces under Wood's light making identification of the holdfast in faeces easier and more pleasant than sieving the faeces through black muslin. If niclosamide (Yomesan) or dichlorophen (Anthiphen) is used, the tablets are chewed and swallowed in the morning. The patient continues his normal life. The remains of the worm are passed in a semidigested state within a day or two.

Prevention of this condition is by careful inspection of meat at the abattoirs, and thorough cooking of beef and pork, remembering that the inside of a joint never reaches as high a temperature as the outside. Fortunately the pork tapeworm is only seen in this country as an occasional import. Its danger lies in the fact that man as well as the pig can become infected with the larval stage and the cystic stage can be in brain tissue with disastrous results. The laboratory can distinguish the tapeworms from inspection of segments. Meticulous personal hygiene is essential to prevent eggs from an infested person being deposited on food.

Rarely in this country, but more frequently in sheep-rearing countries, the *Taenia echinococcus* uses man as an intermediary host, the encysted larvae producing a hydatid cyst, which requires surgical removal without rupture.

Muscleworm *(Trichina spiralis)*. Man is usually infested with this small worm by eating raw or undercooked pork, but the disease has been traced to rat-infested areas. There is diarrhoea, vomiting and fever when the larvae are in the intestine and migrating through its wall. The larvae become encysted in muscle causing intense pain and swelling accompanied by fever. The infestation is called trichinosis and death may ensue or the disease may abate over a long period, the cysts in muscle eventually becoming calcified.

Prevention is along the lines of thorough inspection of meat, washing and cooking of vegetables and meat, and meticulous personal cleanliness of all concerned in the preparation of food.

Hookworm *(Ankylostoma duodenale)*. Hookworm is still one of the most common and serious infestations in tropical countries. A few infestations are acquired in Britain each year. The hookworm is a tiny worm being only 1 cm long. The larvae hatch out in water and moist earth. They penetrate the skin of a human being at the feet and ankles and gain the lymphatic stream, via which they are poured into the blood stream, and arrive in the right side of the heart to be distributed to the lungs. Here they pierce the capillary wall and the wall of the air sac, and travel up the bronchi, trachea and larynx, to be swallowed. When they reach the duodenum they are fully developed, and they attach themselves to the mucous membrane by means of their hooked teeth and suck blood from their host. They discharge large numbers of eggs in the faeces of their host; these reach moist earth, where they develop into larvae, and so the cycle is repeated.

The first essential in prevention is the proper disposal of sewage—including latrines, which are still the means of disposal of excreta in areas with heavy hookworm infestation. Next is wearing of adequate footwear in areas known to be infested. Next comes a clean water supply, especially for washing vegetables and fruit that has grown near the earth. Personal cleanliness and cleanliness of food is very important in preventing infestation which can cause gross anaemia. A single dose of bephenium (Alcopar) or tetrachlorethylene will so reduce the number of worms that they no longer cause anaemia. Such patients should have an adequate intake of iron-containing foods.

Whipworm[141,142] *(Trichuris trichiura)*. Whipworms are thicker at one end than the other. They inhabit the caecum of sheep, goats, swine and dogs. Infested animals excrete Trichuris ova, and man, especially children in rural areas, can become infested. The whipworms adhere to the rectal mucosa and cause diarrhoea, with or without bleeding. The most recent drug—dithiazanine, must be given under medical supervision. All previously mentioned hygienic measures must be taken to prevent reinfestation and spread of worms to other human beings.

Bilharzia (*Schistosoma, blood flukes*) There are several varieties of Schistosoma, the most common of which is *Schistosoma haematobium*. It is found in Africa and Asia Minor. In the human being, the gravid female Schistosoma migrate to veins of the bladder and deposit eggs in the smallest venules. Eggs secrete a substance which enables them to penetrate the vessel wall and enter the bladder, to be voided in urine. Inside the eggs the ova hatch into miracidia which swim about in water and mud and become attached to snails which are their intermediate hosts. On the skin of the snails the miracidia develop into sporocysts that develop into cercariae. If cercariae come into contact with human skin, they penetrate to the blood vessels and are carried to the liver where they multiply, starting off another cycle. Gravid schistosoma in the bladder cause intense irritation and inflammation causing increased frequency of passing urine, difficulty and pain on passing urine, and the passage of blood in the urine.

The cycle for *Schistosoma mansoni* is exactly the same, but the gravid female favours the large intestine and rectum, and the ova are excreted in the faeces. Infested persons pass blood in their stools, and have pain on defaecation. There is ulceration of the bowel and there can be cirrhosis of the liver.

In the prevention of these conditions, the use of latrines rather than the bush or fields is imperative until such time as a water carriage system of sewage disposal can be installed. Washing of hands after urination and defaecation is essential. Areas where snails are endemic should not be used for planting crops. All people should wear shoes. Immersion of feet and bathing in water that is suspect should be prohibited. Boil all drinking water in suspected areas. Cook food properly. Wash all salads and fruits before eating them. Treatment is antimony in some form, which has to be given by injection. There are two drugs that can be taken orally: lucanthone (Nilodin) and niridazole (Ambilhar).

There are other worms, flukes and schistosoma.[143]

SUMMARY OF THE DIGESTIVE OR ALIMENTARY SYSTEM

The organs form a tube, 8 metres long, beginning at the mouth and ending at the anus.

THE MOUTH

This is lined with mucous membrane; it contains the teeth, the palatine tonsils and the tongue. Three pairs of salivary glands pour saliva into it, the enzyme ptyalin contained therein converting cooked starch into maltose, in a slightly acid medium. The mouth is necessary for mastication, speech and swallowing, and provides a safety airway when the nose is blocked. Its lymphoid (reticular) tissue protects the body from infection.

Care of the mouth

1. Avoid nasal obstruction in the formative years.

2. Consume vitamin D so that calcium and phosphorus in the diet can be absorbed to harden the teeth. Take vitamin A for healthy mucous membrane, and riboflavine to avoid cracks at the corners of the lips.

3. Cleanse at least twice daily, brushing the teeth from the gum to the free edge with a suitable brush and a good dentifrice.

4. When cleansing after a meal is impossible, take a drink of water, use a mouthwash, and chew hard fruit, vegetable, chewing gum or special tablets to remove starch particles from between the teeth. Starch is fermented by the bacteria normally present in the mouth, producing acid in which calcium from the teeth is dissolved, with resultant dental caries.

5. Do not eat or drink (except water) between meals and after the last cleansing before sleep.

6. Visit the dentist bi-annually.

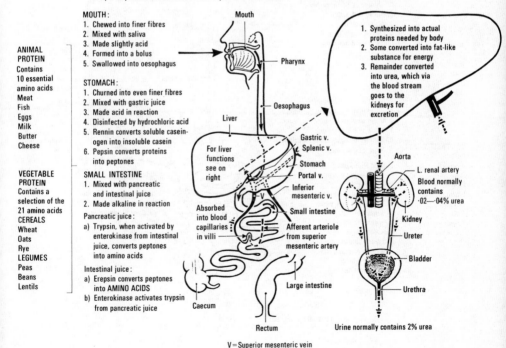

DIGESTION AND ABSORPTION OF PROTEIN

Body's only source of NITROGEN. 1 Gram yields 4 kilocalories. 75–105 Grams daily. 300–420 kilocalories

ANIMAL PROTEIN
Contains 10 essential amino acids
Meat
Fish
Eggs
Milk
Butter
Cheese

VEGETABLE PROTEIN
Contains a selection of the 21 amino acids
CEREALS
Wheat
Oats
Rye
LEGUMES
Peas
Beans
Lentils

MOUTH:
1. Chewed into finer fibres
2. Mixed with saliva
3. Made slightly acid
4. Formed into a bolus
5. Swallowed into oesophagus

STOMACH:
1. Churned into even finer fibres
2. Mixed with gastric juice
3. Made acid in reaction
4. Disinfected by hydrochloric acid
5. Rennin converts soluble casein-ogen into insoluble casein
6. Pepsin converts proteins into peptones

SMALL INTESTINE
1. Mixed with pancreatic and intestinal juice
2. Made alkaline in reaction

Pancreatic juice:
a) Trypsin, when activated by enterokinase from intestinal juice, converts peptones into amino acids

Intestinal juice:
a) Erepsin converts peptones into AMINO ACIDS
b) Enterokinase activates trypsin from pancreatic juice

Mouth
Pharynx
Oesophagus
Liver
For liver functions see on right
Absorbed into blood capillaries in villi
Caecum
Large intestine
Rectum

V = Superior mesenteric vein

1. Synthesized into actual proteins needed by body
2. Some converted into fat-like substance for energy
3. Remainder converted into urea, which via the blood stream goes to the kidneys for excretion

Gastric v.
Splenic v.
Stomach
Portal v.
Inferior mesenteric v.
Small intestine
Afferent arteriole from superior mesenteric artery

Aorta
L. renal artery
Blood normally contains ·02—·04% urea
Kidney
Ureter
Bladder
Urethra
Urine normally contains 2% urea

Fig. 150 Summary of protein.

299

As an exercise to check that adequate learning is taking place, use the following framework to describe the following organs:

Position and relationships.
Structure.
Blood supply.
Venous drainage.
Nerve supply.
Functions.

Pharynx.
Oesophagus.
Stomach.
Small intestine.
Large intestine.
Liver.
Gall-bladder.
Pancreas.

Be sure that you can describe in your own words the acts of swallowing and defaecation, and the composition of bile and faeces.

What can you do to keep your digestive system in a healthy state? Can you discuss the parasites that can infest the intestinal tract?

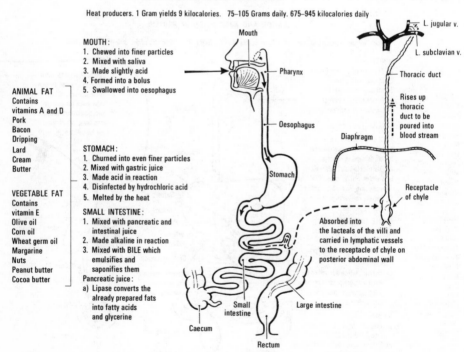

Fig. 151 Summary of fats.

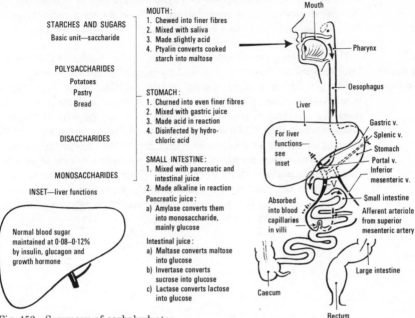

DIGESTION AND ABSORPTION OF CARBOHYDRATES

Energy producers. Assist in the complete oxidation of fats. 1 Gram yields 4 kilocalories. 300–420 Grams daily.

STARCHES AND SUGARS
Basic unit—saccharide

POLYSACCHARIDES
Potatoes
Pastry
Bread

DISACCHARIDES

MONOSACCHARIDES

INSET—liver functions

Normal blood sugar maintained at 0·08–0·12% by insulin, glucagon and growth hormone

MOUTH:
1. Chewed into finer fibres
2. Mixed with saliva
3. Made slightly acid
4. Ptyalin converts cooked starch into maltose

STOMACH:
1. Churned into even finer fibres
2. Mixed with gastric juice
3. Made acid in reaction
4. Disinfected by hydrochloric acid

SMALL INTESTINE:
1. Mixed with pancreatic and intestinal juice
2. Made alkaline in reaction

Pancreatic juice:
a) Amylase converts them into monosaccharide, mainly glucose

Intestinal juice:
a) Maltase converts maltose into glucose
b) Invertase converts sucrose into glucose
c) Lactase converts lactose into glucose

Mouth
Pharynx
Oesophagus
Liver
For liver functions— see inset
Gastric v.
Splenic v.
Stomach
Portal v.
Inferior mesenteric v.
Absorbed into blood capillaries in villi
Small intestine
Afferent arteriole from superior mesenteric artery
Large intestine
Caecum
Rectum

Fig. 152   Summary of carbohydrates.

Table V.   Summary of the chemical digestion of foodstuffs.

ORGAN	JUICE SECRETED	ENZYMES CONTAINED THEREIN	MEDIUM IN WHICH ENZYME WORKS	FOODGROUP ACTED UPON	CONVERTED INTO:—		ABSORBED INTO:—
Mouth	Saliva	Ptyalin	Slightly acid	Cooked starch	Maltose		
Stomach	Gastric	Rennin	Acid	Protein-soluble caseinogen	Insoluble casein		
		Pepsin	Acid	Protein	Peptones		
Pancreas	Pancreatic	Amylase	Alkaline	Carbohydrate	Glucose		
		*Trypsin	Alkaline	Peptones	Amino acids		Blood capillaries in the villi; thence via superior mesenteric and portal veins to the liver
		Lipase	Alkaline	Fats	Fatty acids & glycerine	Lacteals in the villi; thence through receptaculum chyli and thoracic duct to bloodstream	
Liver	Bile	——	Alkaline	Fats	No actual conversion, but preparation so that pancreatic lipase can act		
Small intestine	Intestinal Succus entericus	Maltase	Alkaline	Maltose	Glucose		
		Invertase	Alkaline	Sucrose	Glucose		
		Lactase	Alkaline	Lactose	Glucose		
		Erepsin	Alkaline	Peptones	Amino acids		
		*Enterokinase	Alkaline	Peptones	Amino acids		

* Trypsin activated by Enterokinase

301

CARE OF THE DIGESTIVE SYSTEM IN THE MAINTENANCE OF HEALTH

The offering of food to people is a symbolic gesture of friendship and the psychosociological implications of food as they refer to each culture should be remembered. Attempt to anticipate food with pleasure and do not eat when angry.

The necessary rituals associated with 'clean' food and the serious implications of 'dirty' food should be learned at an early age.

The activity of 'toilet training' in the socialization process should be consistent; the child will be confused if there is pleasure when he uses his pot and displeasure when he uses his napkin. If there is anxiety when the child does not defaecate daily, he can use the withholding of faeces as a behaviour that establishes his independence. We now know that no harm comes from a longer interval of evacuation—the important thing is that it is *regular for that person*.

Several conditions can interfere with the digestive system such as constipation, piles, threadworms, roundworms, tapeworms, muscleworms, hookworms and whipworms.

Sanitation and Disposal of Refuse

Wherever there are people, their personal excretion of *faeces* and *urine* has to be dealt with. There is also *household refuse* consisting minimally of that from cooking and cleaning. All countries pass through a similar pattern of development in dealing with these. The currently developing countries may be able to benefit from mistakes made by the more developed countries. One hundred years ago in Britain, each house or group of houses had an *ashpit* (midden) in which refuse was deposited. In many instances the ashpit was built at the back of an *earth closet* (privy—in some countries called a latrine), the double service being known as a *'privy midden'*. The structure, use, maintenance and emptying of the ashpits, privies and privy middens was left to individual enterprise. Then the local authorities collected the contents of the *privy buckets* and sold it to farmers for fertilization of crops. There are still outlying cottages and crofts that are inhabited by people managing with these minimal facilities. The large developing countries of today are faced with up-grading similar minimal facilities on a large scale, and it is inevitably a long, slow and expensive business. Trying to keep a nation healthy without piped water and adequate sanitation is a formidable task. Many are the stages through which a country passes to adequate sanitation—making laws, deciding how much of its revenue shall be spent on these very necessary social services, and how much shall be spent on industry which will bring the country more money to provide its inhabitants with these necessary amenities. Having attained the amenity of a *water carriage system of sewage disposal,* sludge boats take the end result out to sea. On their way they pass cargo vessels bringing in to the country chemicals which are the artificial substitute for the organic matter being dumped into the sea! Adequate sanitation decreases the incidence of bowel diseases and worm infestations, so that more children attain adult years. There are more people needing water and sanitary services!

Technology of industry has progressed at such a rate that every country is now faced with problems of dealing with 'disposable' articles and plastics *that will not rot.* Currently research is directed towards plastics that are *biodegradable,* i.e. can be eaten up by bacteria—in rubbish dumps and sewage plants. Thriving industry means that more people have more money to spend on things that create more refuse. They no longer make-do-and-mend—they throw away. There is a quicker turnover of bulky articles such as furniture, cookers, refrigerators, washing machines and cars. And householders expect the local authority to deal with the increased refuse, but many people do not expect to pay increased rates and taxes for the increased work.

Most countries report that the quantity of domestic refuse has doubled in the last 20 years. So that sheer *volume* has added to the problem as well as the *indestructibility* of many of the items. Industry produces *trade waste* and it, together with *sewage effluent* causes disastrous *ecological upset*. A recent Working Party on Refuse Disposal recommended licencing of disposal sites and methods of disposal, and licencing at the waste-production source, so that the industrial waste-producer must declare where and how his wastes are to be disposed of. The ecological upset is the result of pollution of the environment—today's in-phrase. For many it means the polluting that other people do. Dr B. R. Brown, scientific adviser to the Greater London Council, says that the general public is unaware of, and in the main indifferent to, waste disposal and the true urgency of the situation has not been appreciated hitherto, except by relatively few people. However, a subject 'the environment and pollution' is being taught in some schools. The Working Party putting forward the suggestion say that one of the main ideas behind the course is that the environment can no longer be left to run itself. It must be managed, and we are trying to teach the managers. The Royal Sanitary Association of Scotland thinks that children should begin to learn about pollution and become familiar with its problems during their primary school years.

HOUSEHOLD (DRY) REFUSE

In different countries and in different stages of a nation's development that which is included under this heading varies. It is impossible to list every article, but in Great Britain in 1976, the following forms the bulk of the refuse collected by the Councils' dustmen: Ashes (decreasing with most methods of central heating), food wrappings (increasing with the increased use of 'instant' foods), peelings from fruits and vegetables, dead flowers, tins, bottles, jars, broken crockery, leaking pans, kettles, etc., dust from sweepers and vacuum cleaners, used paper (handkerchiefs, serviettes, hand and kitchen towels), clean papers and magazines. Some Councils issue paper drawsheets and disposable polythene sheeting for incontinent patients being nursed at home. Non-returnable bottles have greatly increased household refuse in the last few years.

In most areas the householders are requested by the Council to tie papers and magazines into bundles, and to put them out separate from the bin or sack, on the day of collection. Profit from collecting, sorting and baling is small, but valuable material is returned to industry (lessening import) and the cost and pollution from incinerating is avoided. Except in smokefree areas householders are requested to burn as much refuse as possible, for not only is this the most hygienic method of dealing with rubbish, but it cuts down the cost of the service to a minimum. In 1976, with the trend towards lessening atmospheric pollution this would appear to be a questionable request.

As a rule the Councils' dustmen will not accept *garden refuse*, and householders are left with the choice of burning it, or making it into a compost heap. A leaflet, *Give up Smoking Bonfires* giving directions for composting kitchen and garden wastes can be obtained free, by enclosing a stamped addressed envelope to the Director-Secretary, Henry Doubleday Research Association, Bocking, Braintree, Essex. This leaflet suggests that much that goes into our refuse bin, can and should, where possible, go into a compost heap.

Waste disposal units attached to the kitchen sink deal admirably with household waste. They can strain the sewage system in areas where they are popular, because of the amount of water used in flushing them after grinding.

In a few areas the Council arranges to collect *food refuse* (sometimes called *pig swill*) deposited by the householder in a separate pail. It is either used on farms owned by the Council, or sold to farmers, to provide a source of revenue.

When moving to a new area it is wise to find out the local Cleansing Department's rules and regulations about refuse. The following is typical for a large local authority. 'Charges will be made for removing large quantities of "do-it-yourself" materials, old cars, large quantities of garden refuse, and fabrics, fittings, or fixtures of houses and offices. If the owners concerned deliver old cars or garden refuse to cleansing depots or tips, no charge will be made. Among articles for which no charge will be made for removal from the kerbside are furniture, washing machines, refrigerators, stripped wallpaper, car tyres, small quantities of garden refuse, and do-it-yourself materials under three cubic feet.' Because of accidental death of children trapped in disused refrigerators, it is suggested that the door be removed before putting them out for collection.

The Westminster City Council used the vacuum system of refuse removal in a development of flats in central London. The system was developed in Sweden. Suction conveys refuse from individual chutes, along underground pipes to central disposal points, there to be incinerated. There is no odour, even on hot summer days, and in the winter collection is not disrupted by weather emergencies. Refuse can be collected on Saturdays, Sundays and holidays without difficulty.

With the Garchey system of dealing with household refuse, the householder's only responsibility is to dispose of his refuse down a specially constructed sink. The waste water runs off into the drains as usual: the refuse is sucked along pipes into a drying unit, from whence it is conveyed to a furnace. As yet the method has not been developed for the conventional single-storey and double-storey houses but immediate, mechanical removal of household refuse ought to be standard practice in future housing projects. An expert said, 'Refuse disposal is where sewage disposal was 100 years ago—we're still carrying it out in buckets.' And yet the Garchey system does not answer all the problems. As there are no open fires, difficulty has been encountered in disposing of sanitary towels, dressings, incontinence pads, etc. A resolution sent to the Minister of Health by the Royal College of Nursing and National Council of Nurses of the United Kingdom (Rcn), has resulted in several Councils now instructing householders to collect these items in large plastic bags and put them out for the daily (twice weekly or weekly) special collection which is taken to an incinerator. In other areas householders are instructed to put the large plastic bag out with the dustbin at the weekly or twice weekly collection.

Individual collecting articles. That the time is ripe for reconsideration of the article used by householders for short term storage of household refuse is acknowledged by the Institute of Public Cleansing. Manchester Cleansing Committee instituted a large scale pilot scheme in which the householders used paper sacks. An excellent account giving the pros and cons is set out in a film-strip.[144,145] The Medical Officer of Health for the Borough of Hornsey wrote in his annual report, 'It appears that the paper bag system of refuse collection is not quite so popular—at least among flat dwellers—as enthusiasts for this system would have us believe. Hornsey Borough Council received a petition from 91 of the 104 tenants of Council flats in the Stroud Green Area objecting to the continuation of the paper bag system of refuse collection 'on grounds of nuisance and prejudice to amenity' and suggesting that the original system of refuse disposal by way of refuse chutes be reinstated. The Council however thought that disposal by chutes was thoroughly unsatisfactory and that the experiment with the paper bag system should be continued. The paper collection bags are to be placed on the balconies and in selected positions for ground floor flats'. After using sacks for many years, New York city is finding the rodent menace of the 1970s unmanageable and are advocating *metal* garbage bins. The more conventional collecting articles are called *refuse, dust, ash* or *garbage bins* according to custom prevailing in the area. It is thought that within five years the average family's dustbin will have to be twice as big. The dustbin may be the Council's or the householder's property. In favour of the former policy are the following facts:

1. They are all the same size; this may be necessary for the type of collecting van in use.

2. Maintenance in a satisfactory state, and renewal, are not dependent on the householder's (*a*) income, e.g. an old age pensioner may have difficulty in saving the money for a new bin, (*b*) standard of values, e.g. one person may consider a new hat more important than a new dustbin.

3. Full bins can be collected on to the lorry or van, and clean ones left in their place. This avoids emptying the bins into the van on the street with possible 'scatter' in windy weather.

4. The Council can provide facilities for washing, disinfecting and drying the bins.

In the past it was thought that to withstand our climate, refuse bins had to be made of galvanized iron, but those made of heavy plastic material are said to be satisfactory, are lighter to carry and are not as noisy in use. Rubber lids on galvanized bins are less noisy. Refuse bins should have a 5 cm rim at the base to avoid wear and tear. They should be cylindrical in shape, the 'cylinder' being in one piece, so that they are easily washed out with a hose or a mop, and present no corners in which germs and debris can lodge. They should have a handle on each side for easier lifting, and a well-fitting lid to exclude temptation to domestic animals, rats, mice and flies. Loss of lids because of the elements, vandalism and human carelessness have increased the popularity of bins with attached lids. The type in Figure 153c has two handles for contamination in use and is particularly difficult to empty and wash. All material within bins should be wrapped and dry, as moisture encourages the rapid multiplication of germs and hastens the process of putrefaction, with resultant foul

Fig. 153   A, Dustbin with strong 2-in. rim at base.   B, Dustbin mounted on wheels.   C, Dustbin with attached lid.

smell. They should never be overfull as this defeats their object. They *should* stand in a suitable place in the shade, protected from the wind, and away from kitchen, larder or nursery windows, *but* some householders have no alternative to keeping the refuse bin in the kitchen or living room and consequently feel a need to wash the bin after emptying. But where can this be done—in the hand basin, the kitchen sink (in which food is prepared) or the bath? Such people should make their plight known to the Local Cleansing Committee, so that public pressure might result in improved sanitary arrangements.

Bins are collected regularly (once or twice weekly) and the house-holders are usually responsible for putting them out on the road or street to save the dustmen's time, and make the service as economic as possible. Some householders put the bin out the evening before collection morning, when the two strongest members of the family are available to carry it. Where the house door opens directly on to the street, bins are not allowed on the pavement during the night and these areas are 'collected' early—before the morning traffic. Not only does this mean that some people have to get up earlier than is their custom on bin morning, but those afflicted with 'low back pain'

307

find their backs most stiff and painful in the mornings—and least capable of carrying a full bin, especially from a flat above the ground floor, and a second descent and ascent is necessary to retrieve the bin after emptying. (Sacks avoid this second journey.) In a villa-type house where there is difficulty, such as a person living alone, the bin can be mounted on wheels, as illustrated in Figure 153B, or the Boy Scouts encouraged to make carrying the bin their 'good deed,' thus fostering good community living.

Collecting carts (vans). Over the years several patterns have evolved. The features that need consideration are:

1. There must be minimum effort for those emptying the bins or lifting the sacks into the van. The 'height' of the lift must be such as to avoid strain.

2. There should be a separate collecting place for clean papers, magazines, etc.

3. The cubic capacity of the van should be such that a minimum number of journeys have to be made to the final disposal area.

4. Avoidance of 'scatter' as the bins are emptied. Sacks avoid this nuisance.

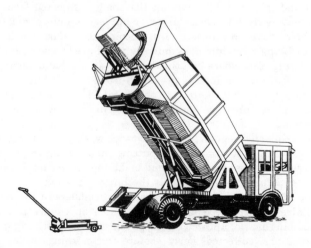

Fig. 154 Household refuse collecting van that minimizes manual lifting of bins and scatter of rubbish.

5. The emptying processes of bins and vans should be as free from hazard to the operators as possible. One is appalled to see the amount of dust that collects on the operators' footwear, clothes and caps. There must be considerable inhalation of dust and one wonders if dustmen are prone to pneumoconiosis. It would seem that sacks would score on lessening contamination of attire and lungs.

5. Cleansing. It is essential that these vans are cleaned daily.

Final disposal of household dry refuse.

1. Incineration, with or without previous salvage disposal. With salvage, the refuse passes along a conveyor belt, where metal objects are removed by magnet. Rags, bones, bottles and jars are hand sorted. The remainder arrives at the furnace, into which it is fed and out of which it emerges as clinker which is sold for road making.

2. Controlled tipping into disused quarries, mines, gravel pits, etc. When full, after time for subsidence, this land is usable.

3. Pulverization, after which the powder can be spread over suitable land without causing a nuisance.

4. Trash trains—American scientists are working out plans for these, to haul refuse from cities, to disposal plants in the country. Such is the magnitude of their household refuse problem.

The following extracts are taken from childrens' essays. 'I wouldn't like to be a dustbin man, collecting smelly rubbish.' 'It must be awful being a dusbin man, because women give you cheek.' It would seem that there is a need for helping children to develop a better image of the dustbin man who performs an *essential service* for us, and *helps to keep the nation healthy*.

OUTDOOR REFUSE

We cannot leave this area without considering the refuse that arises when human beings are out of doors. Throughout the towns, cities and country, the Councils place various types of receptacles at strategic points, and as a nation we are increasingly being urged to 'Keep Britain Tidy'. We are asked not to throw out of the window of a moving car—cigarette ends, cigarette cartons, wrappings from sweets and chocolates, fruit skins, etc. And not to throw away *any* rubbish in the countryside. Plastic bags have been swallowed by animals, then as they breathed, the bag inflated and choked them. Animals have suffered terrible leg wounds from standing in an empty tin and not being able to throw it off. Flesh wounds have also been inflicted by broken glass. Anti-litter campaigns have been organized by many local authorities. Indeed in some areas bye-laws make the throwing of litter on the streets a finable offence. In the countries that have adopted this code nationally, the streets are pleasantly free from litter. We must each do everything we possibly can to prevent our streets and beautiful countryside from becoming litter strewn. We do not want to be ranked as a litter-tolerant nation.

Returning to the types of receptacle provided by the Councils for deposition of outdoor litter, Figure 155 will give you some idea of the

Fig. 155   Four types of receptacle for refuse.

variety. As you travel round the country look out for the different kinds and observe their good and bad points.

It is fashionable to have 'sponsored' walks for charity. Useful abolition of many dangerous eye-sores could be rendered by 'sponsoring' *litter-bagging walks*.

HOSPITAL REFUSE

With the increased use of paper towels, serviettes, handkerchiefs, etc., this presents an ever-increasing problem. That which is of a *household nature* is collected into a dustbin in each ward and department, and is taken daily, on a trolley kept solely for that purpose, for emptying and incineration. It is best if the lay administrator accepts the responsibility of the bins being hosed and dried at a central point, before return to the ward. If the bins are returned unwashed, then the domestic staff must be taught to swill them out with a mop, and leave them to dry in front of an open window.

There are also available metal frames, from which disposable sacks are suspended. The *inside* of the metal rim must be covered by the sack as shown in Figure 156. A foot-operated, metal lid is incorporated. When full the sack can be removed and incinerated without any labour or atmospheric pollution. It would seem that in future ward construction, an incinerator will be a 'must'.

Each ward and department has its *food-refuse* pail, and again these have to be collected daily; otherwise the same rules of cleanliness apply.

The *special refuse* arising in hospitals (and when patients are cared for at home[146,147]) is that from dressings, swabs, bandages, incontinence pads, disposable articles such as waxed sputum cartons, tinfoil gallipots, syringes, needles, glass ampoules, etc., and encasing plasters that have been removed from limbs. Even if a hospital cannot make arrangements for incineration of its total refuse it must incinerate its special refuse. Collection of each bit of refuse into a

310

paper disposable bag, and strong paper linings for the dressing bins, have done much to make this a more sanitary procedure over the last few years.

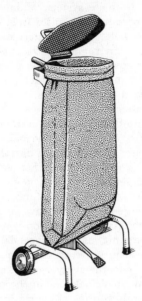

Fig. 156   Metal frame with disposable sack and foot-operated lid.

EXCRETAL REFUSE

As faeces can contain the ova of worms, and the germs causing typhoid and paratyphoid fever, gastroenteritis, dysentery, cholera, infective hepatitis and poliomyelitis, it is important that its safe disposal, together with that of urine, is accomplished, so that the community is not exposed to threats of outbreaks of these diseases. Where a piped water supply is available a *water carriage system* of sewage disposal should be installed. The World Health Organization's recent figures state that 85 per cent of the world's population have to depend on the most primitive methods for the disposal of excreta and refuse.

Water-carriage system of sewage disposal.[148,149] In this system, within the building, there are receiving receptacles which have a water seal at the base before connecting with pipes on the outside wall. The water seal prevents contamination of the air in the room containing the receiving receptacle, by air from the ensuing system of pipes, which join increasingly larger pipes, until finally they connect with the main sewer leading to the sewage works or disposal plant. (The Sewage Works Committee of several Councils is being renamed The Water Pollution Control Committee. This trend is to be welcomed in that it clarifies the *object* of treatment of sewage.) This is the best method of disposal, but unfortunately not only does it deal with excretal refuse, but also with trade wastes,

311

street and roof washings, and waste from sinks and baths. *Trade wastes* are of such variety it would be the best policy if all trades accepted responsibility for rendering waste innocuous before disposal into a public drainage system. Mainly because of mercury pollution of water by industrial waste from wood-pulp plants the United Nations has held several international conferences on pollution of the human environment. *Street and roof washings* are of variable quantity in our inclement weather. After storms, the quantity is increased to proportions which are unmanageable at the sewage (water pollution control) works. After the storm water tank is full of sewage-contaminated storm water, the overflow is often run off, untreated, into the river. The question arises, should street, roof and waste water be drained into the main sewer? It would seem preferable to drain only excretal refuse into the main sewer, and to drain *street, roof* and *waste water* together, as such waste does not need to pass through settling tanks to precipitate organic matter; the grit from street washings would be uncontaminated with faecal matter; domestic detergents and excessive soap would be unable to interfere with the bacterial activity upon which the proper rotting of faecal and other organic material depends. Because Paris has a separate drainage system, tours of 'Les Egouts' (the sewers) start from the Place de la Concorde every second and fourth Thursday in the month at 2, 3, 4 and 5 o'clock, and at the same times on the last Saturday of each month. Stories of people living in sewers during wars, etc. are more credible after such a tour.

Receptacles for excretal refuse. *Water-closet.*[150] As distinct from the earth-closet, this room can be included within the structure of the building, as it is hygienic and does not create a nuisance, but it should not open directly from a room in which food is prepared. One wall must be an outside wall, down which the connecting soil pipe can run. This arrangement also allows adequate window space to assist in ventilation and lighting, and allows the use of a ventilating brick high up in the wall. The window space should be not less than 2 sq ft, and frosted glass is usually inserted; artificial lighting should be provided, and the two safest methods are that which operates as the door opens, and that which operates from a pull-cord just within the door. The floor space should be such that the user's outdoor clothes can be manipulated without contacting the lavatory pan. Pegs should be provided on the back of the door for reception of outdoor coat, handbag, umbrella, gloves, shopping bag, etc. A shelf is useful for some of these articles. The toilet-paper dispenser should be within hand reach of a person sitting on the lavatory pan, and in some countries paper seat-covers are provided for each user with a receptacle for their disposal. A medicated, water-soluble paper lavatory seat guard has recently been marketed in Britain.

Ideally hand-washing facilities should be within the water-closet and should be used immediately after use of the lavatory pan, so that the cistern and door handles remain uncontaminated. In the interests of community health, a handless method of operation for the cistern *and the door* would be ideal. The walls and floor should be of a

312

washable material; in some public and communal water-closets the walls are not continued to floor level, so that the total floor area can be hosed, this being economical of labour. In some of them an attendant wipes the lavatory seat with a cloth wrung out of bactericidal solution after each user. A lavatory brush in a suitable container of disinfectant encourages immediate removal when there has been inadvertent soiling of the pan. Paper bags and a sanibin or incinerator should be provided for used sanitary towels in public and communal lavatories, e.g. in factories, offices, hotels, restaurants, schools, hospitals, etc. Incinerators can cause an unpleasant smell. A new unit disposes of sanitary towels in less than one and a half minutes by pulverizing them in a stream of water which washes away through the drains, leaving no residue. Some firms offer a chemical disposal unit that they service regularly. One unit has a capacity suitable for 40 females for 28 days. These units need no flues, gas or electricity connections. There are usually several water-closets adjoining, and the dividing walls are not continued to the ceiling. This gives an unorthodox but secondary means of entrance should an emergency arise within the bolted closet, such as the user having a heart attack or pulmonary embolism, etc. In some new hospitals the pins of the door hinges are removable from the outside, so that the door can be removed to gain access in emergency. There are minimum regulation numbers of lavatories that have to be provided, according to the number of people using the building. With the complete change in the type and treatment of hospital patients in the last 15 years, many wards built 50 years ago have totally inadequate sanitary arrangements. The American pattern has a bathroom, and lavatory adjoining each patient's room. With the availability of deodorizers and air fresheners, this does not create a nuisance.

The local authorities in Britain are requested to provide public conveniences,[151,152] and some of them redeem some of the money expended, by having a 'penny-in-the-slot' machine on the door. When this fashion spread, it was increasingly realized that it could cause suffering and embarrassment to anyone bereft of 'one pence'. Women's organizations did much to bring this to the notice of the then Ministry of Health, and now a steadily increasing number of Councils are removing 'penny-in-the-slot' machines. Councils are requested to provide facilities for wheelchair customers.

Not all authorities provided hand-washing facilities, and again much work was carried out by women's organizations including the Royal College of Nursing[153,154] (now the Royal College of Nursing and National Council of Nurses of the United Kingdom (Rcn)), and this resulted in the Councils being requested by the Minister to provide adequate hand-washing arrangements and to display encouraging notices regarding their use. At the time of going to press there are still some Councils that have not complied with this request. (One Council provided cold water only, and was promptly accosted by members of the Royal College of Nursing. The Council had refrained from installing hot water, as several cases of scalding

313

of children had been reported by Councils that had installed hot water! Motto: there are two sides to every question, therefore never jump to a hasty conclusion.)

Instead of a 'penny-in-the-slot' on each lavatory door, some Authorities provided this centrally, and the operator then proceeded through a turnstile to an area from which two rows of lavatories opened. This was found to be a multiple hazard.

1. Failing and absent sight, and inability to read (this may be due to language difficulty) may render instructions useless.

2. The operator must be in possession of 'one pence'.

3. The operator must have the required intelligence and strength to operate the turnstile.

4. The mechanics can go wrong and one finds that in spite of following instructions in the town to get to the public convenience, one cannot get in to use it.

Women became outraged at this further indignity and again banded together, the result of this being that all authorities have been requested to abolish turnstiles.

Lavatory Pan.[155] Of whatever material this is constructed, it must have an impervious, smooth surface that will stand up to frequent cleansing with an abrasive such as Vim, without becoming roughened or cracked, which would encourage the harbouring of germs. Vitreous china and glazed stoneware fulfil these requirements.

The pan is made complete with its 'trap,' a U-shaped bend that maintains a 5 cm (2-in) water seal (Fig. 157). The back of the pan is

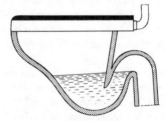

Fig. 157   Section of lavatory pan showing a water-seal.

vertical and dips down into this water seal. The surface area presented by the water seal in the bottom of the pan should be adequate to receive the excreta, without soiling the pan. The jets should be evenly spaced around the flushing rim, to cleanse the whole pan. The most modern fitting is to cantilever the lavatory pan from the wall (Fig. 159B) thus minimising the surface area for cleaning, with no impediment to floor cleaning. Otherwise the lavatory pan has to be supported on a pedestal (Fig. 159A and 159C). With the cantilevered lavatory seat the cistern has to be enclosed within the wall (further minimising cleaning) and is operated by foot-pedal. The amount of water delivered from the cistern at each release should be a minimum of 9 litres (2 gallons). With the pedestal lavatory pan the cistern can be placed high with a chain-pull (Fig. 158A)—this was thought to give greater momentum to the flush—or in accordance

314

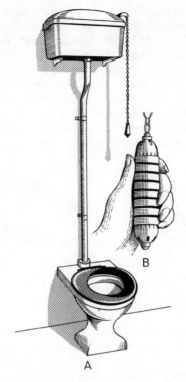

Fig. 158   A. Water lavatory with high cistern and a chain-pull.   B. Hygienic
chain-pull.

with modern thought, it is placed 60 to 90 cm (2 to 3 ft) above the pan.
Figure 159A has a hand lever for operation. A hygienic germ-destroy-
ing handle is available (Fig. 158B). It comprises two plastic caps and a
spindle holding eight chemically impregnated rings. It is like the

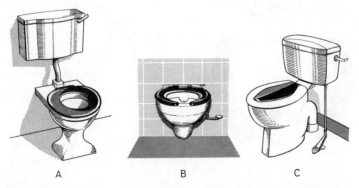

Fig. 159   A. Water lavatory with low cistern and a hand-lever for operation.
B. Water lavatory cantilevered from wall. Foot-pedal for flushing cistern.
No impediment to floor cleaning.   C. Exposed cistern's foot-operated flushing
mechanism.

315

ordinary chain-type handle and has an S-hook for quick attachment.
A model is also available for low-level cisterns. The rings are pleasantly perfumed, keep the handle free of germs, disinfect the hand
and refresh the atmosphere. Some models have been tried with
a push-button release and Figure 159c shows an exposed cistern's
foot-operated flushing mechanism. (The remaining problem to be
solved is prevention of contamination of the lavatory door handle.)
A hinged seat of highly polished wood or bakelite is placed over the
pan when in use. It has been decided that there is less risk of perineal
contamination in women if this seat is horse-shoe shaped (Fig. 160B),

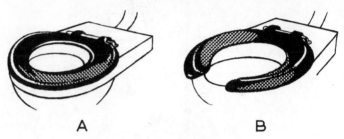

A   B

Fig. 160   Hinged lavatory seats.   A, Complete.   B, Incomplete.

and most hospitals and midwifery units endeavour to supply this
type. A hinged lid over the seat, is provided with some units. Research workers are interested at present in the ill-effects[156] of
infected droplets that are produced by the excretory effort of the user
and when a lavatory pan is flushed. A closed lid will help to offset
the latter hazard.

A Specitest (Fig. 161) is similar to an ordinary lavatory pan, but
has a shelf at the front to take a waxed container, which can provide a specimen of a patient's urine for testing.

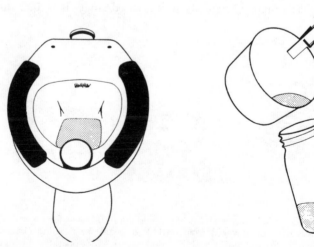

Fig. 161   A specitest.

During the cleansing of a lavatory pan, after the application of an abrasive such as Vim or Harpic, the water seal needs to be disturbed by the lavatory brush, otherwise a permanent water line will mark the pan. In the interests of prevention of accidents, Chloros or Domestos or any other chlorine preparation must not be used with Harpic, as the combination liberates poisonous chlorine gas.

There is a vacuum lavatory, the Bastat Electrolux Vacuum System. It looks like an ordinary lavatory, makes slightly more noise when flushed, but not for as long as a water lavatory. Sewage and waste entering the system are collected in a vacuum tank, and can either be discharged to a gravity sewer; transported directly to a treatment plant; collected by a tanker lorry to be discharged into a treatment works, or treated by a packaged *in situ* purification plant. It is useful in areas where water is limited and where the existing sewage system is overloaded.

Bidet. A low-set trough-like basin in which the perineum can be immersed, whilst the legs are outside and the feet on the floor. This is being tried out in some operating theatre suites to cut down cross infection from perineal staphylococcal carriers. They are becoming increasingly popular in maternity hospitals. Some bidets are fitted with a thermostatic control and a knob to adjust the flow of water and the height of spray. Removal of faeces from napkins and bed linen in the home is more easily accomplished using a bidet, rather than a lavatory pan. If families are to be encouraged to care for their elderly relatives at home, it may well be that bidets will become standard equipment.

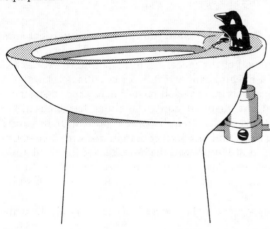

Fig. 162 A bidet.

Sluice. Some people call this a slop sink. It is provided in maids' pantries, nurseries and in hospitals. It is especially important that the material of which it is made should not crack or become roughened, which would encourage a collection of germs. It has a water seal which is placed centrally or to one side, and the U-bend containing the minimum 5 cm (2-in) water seal is made in one piece with the sluice pan.

317

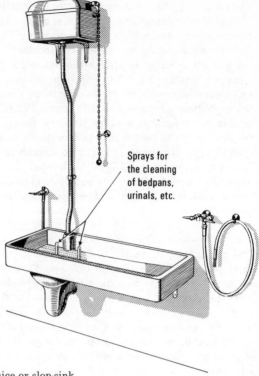

Sprays for
the cleaning
of bedpans,
urinals, etc.

Fig. 163   Sluice or slop-sink.

That in the hospital ward annexe has two vertical sprays incorporated, to cleanse the inside of an upturned bedpan and urinal. This allows removal of the faeces from the pan to the water seal, as in a lavatory; a flushing rim and 9 litres (2 gallons) of water are necessary to remove it into the soil pipe. Where paper bedpan covers have been tried out, disposal down the sluice has blocked same. Most units provide a bin for their disposal at present. In hospitals in which the fouled linen that has been contaminated with faeces, urine, blood, vomit, etc., is still dealt with before it leaves the ward, the sluice has yet another use. The fouling agents are removed with a stiff brush under running water. At the end of the process the sluice must be flushed.

There are many *disadvantages to the open sluice,* the main of which are:

1. The faeces is broken up by a forceful spray and is in contact with the atmosphere for longer than in a lavatory.

2. The above process causes more smell.

3. There is more likelihood of faeces adhering to the sluice pan, and this means manual removal, using a mop soaked in a disinfectant solution.

4. The under-surface of the bedpan rim usually needs to be mopped for complete cleanliness.

5. Sluicing of foul linen can spread pathogenic organisms into the atmosphere, and leave them on the sluice pan.

Bedpan Washers. These are undoubtedly a great improvement on the sluice, though the ideal pattern has not yet evolved. Their *advantages* over the open sluice are:
1. There is less smell.
2. There is no atmospheric spray.
3. Sterilization is effected if steam is provided after the flush.

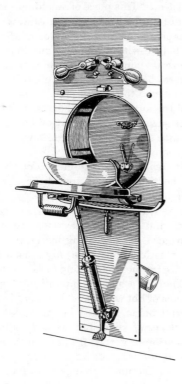

Fig. 164   Wall bedpan washer.

The principle is that, within a closed chamber the contents of the bedpan are removed automatically, the bedpan is thoroughly washed out, then subjected to steam for sterilization. Placed centrally on the floor of this closed container is a water seal which leads to the soil pipe outside. The upright model can be partially incorporated into the wall, thus saving space. The down-dropping entrance is heavy, making closure difficult and often causing it to end in a bang. There is considerable noise from the placing of the stainless steel bedpan within the metal-holding prongs on the stainless steel lid. The box-like model takes up considerable floor space, the lid is foot-operated, but being heavy, there is often a bang as it contacts the base. In both models the powerful flush and the ejection of steam into the closed cabinet create noise.

319

Though paper bedpan covers are sold as 'disposable,' the drains from the bedpan washers have been found inadequate to deal with them, and a separate disposal bin has been provided in most areas.

Summary of the disadvantages of bedpan washers:
1. The lids are very heavy for manipulation.
2. The lids are noisy in the closing process.
3. There is considerable noise from insertion of the bedpan, the powerful flush and the steam.

Disposable Bedpan Units. This piece of machinery is a recent innovation and as yet is only provided in a few hospitals. The disposable bedpan is made of papier mâché in the shape of the stainless-steel slipper bedpan. It is placed within a plastic mould for use, and then removed complete with its contents, and placed in the disposal unit which looks like a metal box with a side opening. It has a water seal at its base and is connected to the soil pipe. Metal 'teeth' grind the bedpan to a pulp and the automatic flush drives it into the soil pipe, and on its way to the sewage (water pollution control) works.

The only *disadvantage* seems to be the noise, which is much worse than that made by a bedpan washer.

The foregoing *five receptacles* for excretal refuse, i.e. the lavatory pan, bidet, sluice, bedpan washer and disposable bedpan unit *connect* up *to the soil pipe.*

Receptacles for waste water. *Wash-basins.* These can be bought in various sizes, shapes and colours. They must be made of an impervious, smooth-surfaced, hard-wearing material, the most recent addition being of the plastic variety. The *plug* is usually attached to a *chain*, and both require frequent cleansing to avoid them becoming a harbour for germs. There is a black composition plug which is to be avoided, as with the use of hot water and detergent it softens and can mark delicate material being washed in the sink. Recessed areas accommodate the *tablets of soap*, but as soap allows adherence, growth and multiplication of germs, it has been found preferable to supply *containers for liquid or powdered soap* over wash-basins where the prevention of cross-infection is an important factor. Hot and cold water are usually supplied via the conventional, chromium-plated *screw tap*, the free-standing delivery spout in Figure 166F being more hygienic than the low-set one in Figure 166A; brass taps should be a thing of the past! In the prevention of scalding accidents, incorporation of a thermostat into hot water taps is being considered. The *levers* that can be operated by the *elbow* are preferable (Fig. 166B), and avoid contamination of a screw tap by germ-laden hands. Where the hands have to be washed under running water, as in the operating theatres and dressing-rooms, a central, thermostatically controlled, elbow-operated tap is essential (Fig. 166c). A disadvantage in some models is too forceful a spray or jet. A modern sink in which the force of the spray is constant is illustrated in Figure 166H. In the newest model (Fig. 166I) when the hand is placed under the faucet, the water is automatically turned on. When the hands are dried in the hot air duct on right of wash-basin, water ceases to flow from the faucet. Where communal wash-basins are provided a push-release tap (Fig.

166E) is sometimes used to avoid flooding and waste of water, as can happen when taps are inadvertently left running. An extra requisite with the communal wash-basins is hand-drying facilities. In the interests of communal health, the communal towel is out. Individual small towels are acceptable, but they are a great tax on the laundry, and small articles tend to get lost. The continuous roller towel, whereby a clean portion is dispensed for each user, is favourable. Paper towels cut out laundry costs, as does the foot-operated hot-air dryer.

The *outlet* from the wash-basin is guarded by (*a*) a *sieve*, and it must be remembered that that which is caught in the sieve needs manual removal, a pair of old forceps being ideal for this purpose; (*b*) a U-shaped pipe leading to the waste water pipe on the outside wall. This ensures the minimum 5 cm (2 in) *water seal*. It has a *removable screw cap* at the base of the U for cleansing purposes. In older wash-basins an overflow pipe is incorporated to avoid flooding, and the sink should be filled so that this is flushed and cleaned out with a bottle brush at least once daily. New models (Fig. 166H and I) do not have an overflow exit.

REINFECTION AFTER WASHING MUST BE AVOIDED

Use should be made of

| Continuous roller towel | Hot air drier | Individual paper towels |

Fig. 165   Various methods for safe hand-drying.

Where pleasing design is important and in the interests of hygiene, the pipes should be hidden in a removable 'pedestal' (Fig. 166G). If they are boxed in (Fig. 166F), then it too must be removable, so that the trap is accessible. Where the sinks are to be used for tooth cleansing, hair washing, etc., they can be incorporated into various designs of splashboards (Fig. 166G), some having a toothbrush rack, shelf for toilet articles and a mirror. Wash-basins can also be fitted with flexible shampoo sprays.

Now that the lavatory and hand washing facilities have been discussed, this seems a suitable place to be reminded that a sanitary survey of schools in 1966 disclosed that in one town, nine primary schools had *outdoor* sanitary blocks, some without artificial lighting or heating. In many schools hand washing facilities were apart from the lavatory accommodation. In some schools there were insufficient basins and hot water supplies, and even insufficient soap. In almost

all schools the only method of hand drying was the completely unsatisfactory roller towel. One wonders how many parents are aware
of these conditions. In such schools any attempt at health education
is nullified.

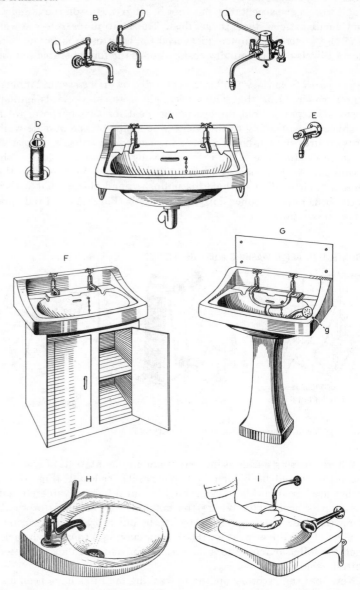

Fig. 166   A to I.   A, Wash-basin with screw taps.   B, Lever-taps that can
be operated by the elbow.   C, Taps with central thermostatic control and
lever.   D, A detachable type of overflow.   E, Water is only obtainable when
button is depressed.   E, Wash-basin with cupboard surrounding U-bend
leading to waste water pipe.   G, Pedestal wash-basin with splashboard and
shampoo spray.   H, Modern sink in which force of spray is constant.
I, Modern tapless sink.

This room is taken so much for granted by those who have one that we seldom think of the 700 000 houses in Britain that still have neither baths nor hot water laid on. In most modern houses the bathroom is small and compact, this being a desirable feature where there is no central heating! The hot-water cistern is often enclosed in a cupboard, and this provides an airing place for clothes, etc., and also warms the atmosphere in the room. In a small bathroom where extra heat is required it can be supplied by fitting a special type of bulb to the light fixture. This is the same type as has been perfected for the poultry farmers and pig breeders. Electrical-heating apparatus is not considered safe, but there are several safe types of heated towel rails, which serve the dual purpose of warming the atmosphere and drying the towels after use—especially useful where there are several members of a family. In larger bathrooms a comfortable warmth can be supplied by the modern gas-convector heater.

The bathroom must be provided with a window in good working order, as it is desirable to open same after bathing to let out the steam, which otherwise deteriorates decoration. Frosted glass is usual in this window. The bathroom must have artificial lighting; electricity is best and the safest method of operation is a pull-cord outside the door.

Nowadays there is a tremendous choice of materials for interior decoration, which are steam resistant and which do not cause condensation. The wall nearest to the bath needs to be impervious to cope with splashes; tiles are very suitable and will stand up to frequent cleansing. Most floors are washable, with non-skid rugs placed at strategic points.

A stool or chair should be provided for use whilst the bather is drying and attending to his feet. Bending down for this purpose should be discouraged, especially in older people who may have an increased blood-pressure. If the bathroom cupboard is to be used as a repository for tablets and medicine, then it must be placed so as to be inaccessible to a child.

Organization within the household is necessary so that each member can readily identify and keep separate his toothbrush, face and body flannels, and face and bath towels; some drying arrangement usually needs to be incorporated for the latter articles.

In communal bathrooms arrangements have to be made for the provision of a separate bath mat for each user, as there is a danger of spreading such conditions as verruca and athlete's foot.

A wash-basin is usually incorporated into a bathroom, and much controversy exists as to the inclusion of the lavatory pan in the family bathroom. There are pros and cons and I leave you to work them out.

Old people should be discouraged from locking the bathroom door, and privacy can be accorded them by putting an 'engaged' notice on the door. This method is also employed in hospital.

The bath. This can be bought in various sizes, shapes and colours. It must be made of a hard-wearing, smooth-surfaced, impermeable

Sanitation and Disposal of Refuse

323

material. The plug is attached to a chain which can be fitted to the chromium-plated guard at the entrance to the overflow pipe. This may or may not connect with the waste-water pipe. An overflow exit is omitted in modern baths. (The plumbing system can be complex, and a nurse is not expected to know the details of drains and pipes, but she is expected to know about the hygiene of the room in which the receptacle for waste water is placed, and she must be able to give advice about the types of receptacle for waste water. She must be able to teach people how to use them to best advantage, and how to keep them clean,[157] so that they do not become a hazard to health.) The outlet is guarded by a sieve and U-shaped pipe leading to the waste-water pipe on an outside wall. Hot and cold water are provided in the conventional chromium-plated screw taps, but where the water is likely to be dangerously hot, the addition of a central mixing pipe will avoid scalding accidents. Detachable tap keys are provided in children's hospitals and homes, and where the mentally subnormal are cared for.

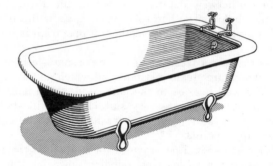

Fig. 167   Diagram of an open bath.

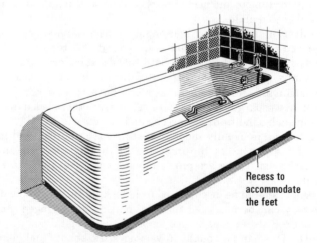

Recess to
accommodate
the feet

Fig. 168   Diagram illustrating a fitted bath.

In most bathrooms the bath is placed along one wall, and in days gone by it was mounted on ornate legs (Fig. 167). Not only did this involve labour to remove dust from between the nooks and crannies of the moulding, but some ingenious method had to be found for removing dust from the narrow space between the bath and floor. The wall behind the bath was difficult to decorate and even if tiled, difficult to keep clean. In this enlightened labour-saving age all these deficiencies have been remedied by the 'fitted' bath (Fig. 168). In many older houses the 'open' bath has been converted into a 'fitted' bath, as a labour-saving device. However, this does away with the rim of the bath, which previously could be used as a leverage device, consequently side handles are now incorporated into the fitted bath. If the fitted case is at right angles to the floor, the cleaner of the bath is at a great mechanical disadvantage. This can be overcome by sloping the side case, or recessing the last few inches to accommodate the feet. This is also useful when helping a person out of the bath. Where people frequently need helping in and out of the bath, as in hospital and eventide homes, it is best to place the bath so that it is accessible from both sides. Another great advantage for older people is the sunken bath, or one designed so that they can be in the sitting position (Fig. 169). Some local authorities provide these in their old age pensioners' bungalows.

Much vexation and frustration can be avoided in a family if each member is taught to accept the responsibility of leaving the bath and

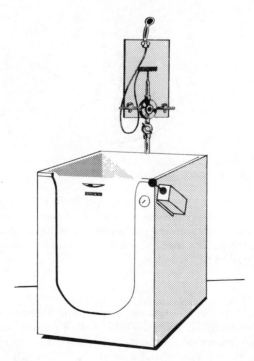

Fig. 169　A medicbath.

washbasin clean and the bathroom tidy. Each member must realize that the bathroom cannot be monopolized by one person for long periods.

In communal bathrooms a mop in a canister of soapy solution encourages the users to mop the bath after use. In hospital a disinfectant is often added to this solution, and 1 per cent formalin is very efficient.

A bath can be used as a shower if it is fitted with a thermostatically controlled spray and a plastic curtain. For the advantages of a shower, see page 126. Currently on the market is a portable shower (Fig. 170) that can be used in the bath, or in a basin with as little as two inches of water, thus making it useful for rheumatic patients, the elderly, campers, caravanners and those living in bed-sitters, etc. The spray is controlled by a foot pump.

Fig. 170   Portable shower bath.

Another portable version is shown in Figure 171. No plumbing is required. Simply fill container with water at required temperature. A few strokes of the pressure pump gives 10 minutes' spray. There is an on/off control. The frame and tub are collapsible for storage and can be erected in two minutes.

THE KITCHEN

This room should be as pleasant, as light and as well ventilated as possible, for many of the housewife's activities take place here. If doors and windows are not sufficient to keep the atmosphere free from steam during cooking and washing, an electrically operated fan is desirable.

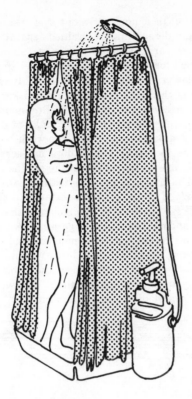

Fig. 171   Portable shower bath.

It is often best to place the sink underneath the window, certainly
not in a corner. If the sequence of dirty dishes, washing process in
the sink, washed dishes on draining board, then crockery cupboard
can be achieved, it is economical of labour. With the modern trend
towards drip-drying crockery, a double sink, one side soapy water,
one side hot rinsing water, and a crockery rack that can be closed in,
is desirable. All surfaces should be of an impermeable material that
is easily kept clean, formica being ideal. Cupboard space should be
adequate and heavy articles should not be placed in high cupboards.
The cooker should be on the same side as the sink to avoid unnecessary lifting of heavy pans and to cut down the risk of scalding
accidents. There should be no exposed shelves to collect dust. The
floor is best made of, or covered with, a washable material. Adequate
artificial lighting must be provided, and adequate and safe electricity
plugs for all the gadgets used therein. The most modern plugs are
switchless, shuttered and recessed into the wall. The interior decoration should be such as to withstand steam and prevent condensation.
The kitchen sink. All that was said about the wash-basin applies
equally well to the kitchen sink, but there are a few further points
worthy of note. Only that which concerns food should be dealt with
in the kitchen sink. Water that has been used for washing walls,
floors, etc. should not be put down the kitchen sink. The lavatory pan

327

is the best solution in a house without a slop-sink. If, and when, excretal refuse and detergents are drained separately, slop-sinks in houses will be necessary. Dishcloths (or mops) should be kept *solely* for equipment used for cooking and eating and should *not* be used for wiping dust-laden surfaces in the kitchen. It is important that the kitchen sink is the right height for the worker, so that it does not impose any unnecessary muscle strain. Ergonomists study people working at sinks, and one can now buy sinks of different heights. If one has to use a sink that is too low, the wash bowl can be raised on a tray, such as in Figure 172. It does not scratch the sink. A draining board on one or both sides of the sink is necessary. These should be made of a smooth, impervious material, which can be easily dried after use. Wood is not a good substance as it roughens with scrubbing and takes time to dry. The moisture that it holds is conducive to growth and multiplication of germs. Nowadays the draining board and kitchen sink are made in one piece, and the spaces below utilized for cupboard, but leaving the U-trap with its removable screw cap accessible. As previously mentioned the double sink unit has much to recommend it, until such time as dish-washing machines become a standard part of kitchen equipment!

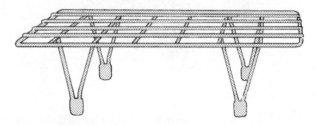

Fig. 172   Tray for sink.

Dish-washing machines. Several patterns are available for domestic use, and in institutions they are indispensable. They have to be included in a list of receptacles for used water and they are fitted to the waste-water pipe.

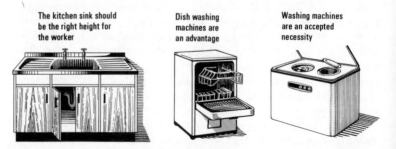

The kitchen sink should be the right height for the worker

Dish washing machines are an advantage

Washing machines are an accepted necessity

Fig. 173   The kitchen unit, dish-washing machine, and washing machine.

Washing machines. These are now accepted as a necessity and not a luxury, and it is only right that they are available to housewives to

328

take the drudgery out of washing. Most of them do not require separate plumbing, but are filled by rubber tubes attached to the kitchen sink taps, and are emptied by tube and pump into the kitchen sink. A few models, however, are connected to their own waste pipe, and are therefore included in a list of receptacles for used water.

Water Pollution Control (Sewage) Works

Excretal refuse from lavatory pans, sluices, bedpan washers and disposable bedpan units goes into the *soil pipe*. Waste water from washbasins, baths, showers, kitchen sinks, dish-washing machines and washing machines goes into the *waste-water pipe*. The heading gives the *raison d'etre* for these works. These pipes connect underground, and receive pipes from street drainage, and, in some instances trade wastes, until they are connected to the disposal works. Here the sewage goes through several stages, so that the effluent is rendered non-putrescible, i.e. it will not take oxygen from the river into which it is piped, otherwise there will be interference with fish and plant life—ecology rearing its head again. Those interested in the health of sewer workers should read the report by HMSO.[158] The serious problems of sewage disposal are dealt with in an article by Gunn.[159]

A recent development is the LYCO system. It is specially designed to provide sewage treatment at an economic cost for factories, schools and communities with populations of up to 10 000. The equipment consists of metal tanks, pumps, and high speed fans, which process the waste into almost pure water before discharging it into sewers or rivers.

During the treatment of sewage, samples are taken at each stage and many times throughout the day. If the Water Pollution Control (Sewage) Works is too small to warrant the expense of its own laboratory, then the local council makes arrangements to use a nearby laboratory. *Money spent on sewage disposal is money well spent,* as it helps to *prevent* the following *diseases:*

Typhoid and paratyphoid fever
Dysentery—amoebic and bacillary
Salmonella food poisoning
Cholera
Virus hepatitis (catarrhal jaundice)
Parasitic infestation of the bowel—all types of worms
Any diseases that can be spread by flies—which probably includes poliomyelitis, as the virus is excreted in faeces
Weil's disease, a type of jaundice caused by a spirochaete excreted by rats. A workman with skin abrasions can be inoculated with this spirochaete from rat-infested sewers.

Returning to the Working Party Report[158] it says that sewage disposal should be considered as part of the whole water cycle, together with water conservation, and the control of quality and quantity of flow in our waterways. It points to inescapable changes needed in financial priorities, administration, enforcement and in *public attitudes*. That includes you and me!

Conservancy System

We must now learn how the people who have no water carriage system, dispose of their excretal refuse.

Earth or pail closet, or privy. This must be at least 2 metres (6 ft) away from the dwelling and 12 metres (40 ft) away from a well or stream. The walls and roof should be in good repair, and adequate lighting and ventilation should be provided. The walls should be lime-washed or treated with some other washable material. The holed wooden seat must have a lid, and a thick coat of paint will discourage germs from adhering to it. The floor should be of impervious material. There should be a properly fitting door with a bolt or lock for privacy. Toilet paper should be provided within convenient distance for the user, as in any other lavatory. A wire-mesh door and window frame will help to keep out flies.

The excreta falls into the earth, or a pail, and should be immediately covered with earth which acts as a deodorizer and assists bacterial decomposition of the excreta. Emptying is usually the householder's responsibility, and disposal is effected by burying where it will not contaminate nearby water.

A refinement can be installed whereby on pulling a lever, earth is delivered to cover the excreta (Fig. 174A).

An amusing account of the construction of a privy is to be found in *The Specialist*.[160]

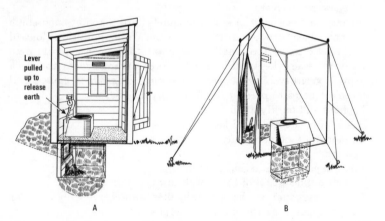

Lever
pulled
up to
release
earth

A   B

Fig. 174   A, Earth closet or privy.   B, Temporary privy or latrine.

Temporary privy or latrine. For camping purposes the conservancy system has to be adapted to fit the situation. A trench is dug in the earth, wooden seats erected over this, privacy gained by canvas and poles, and an overall waterproof covering. The removed

earth is scattered on the excreta after each use of the *temporary
privy* or *latrine*, and the remainder at the end of the camping period,
most campers taking a pride in leaving a field as clean and tidy as
they found it.

Chemical closet. Over the last 30 years *chemical closets* have come
to be the most favoured of the conservancy methods, and provide the
sanitation for aeroplanes, long-distance buses and caravans. They
are available as portable or permanent structures, and the special
fluid in the bucket deodorizes, liquefies and sterilizes the excreta.
Into the bargain it discourages flies. Disposal is therefore no great
hazard to health.

Work is in progesss to produce a self-flushing, chemical action,
portable loo, that is light enough to carry on the roof of a car or other
vehicle. The collapsible frame and walls of the cabinet are of fibre-
glass. It is being produced in this country to meet the needs of
agricultural workers—at shows and for extra labourers during fruit
picking and harvest time. It will of course be useful in other countries
for people who now have privies and latrines. (The International
Sanitary-ware Association is trying to introduce the word Loo for
lavatory throughout the world.)

Fig. 175   One type of chemical closet.

Hand-washing facilities are usually limited where the conservancy system is the only method of sewage disposal, and great
vigilance and perseverance is necessary to see that children acquire
the habit of hand washing after visiting the closet.

The disposal of sanitary towels must be accomplished separately,
by burning. A closed bin for their collection is as acceptable as in a
water-closet.

Railway closet. Excreta as it is disposed of from railway lavatories
is flushed from the receiving receptacle by water, and conveyed to
the earth between the lines. Passengers are requested not to flush
the lavatory when the train is stationary so that insanitary conditions do not arise where there is habitation. Currently, British
Rail are experimenting with anti-pollution equipment for train
lavatories. Hand-washing facilities are usually adequate.

Before leaving this subject I must draw your attention to a currently developing hazard—excretal contamination of fields near lay-bys. Dr B. H. Chantrill wrote to the British Medical Journal, 'On the A1 road for instance, now that the towns and villages have been by-passed, one travels south from Darlington well over 161 km (100 miles) before reaching the next public lavatory at Newark on the A1 or at Nottingam via the A614. The stinking middens of yester-year referred to in your article have indeed already returned—try hopping over the hedge bordering some lay-bys, but look before you leap! The British Medical Association's Public Health Committee has asked the Minister of Health to consider urgently the question of increased provision of sanitary accommodation on main roads.

Air or Atmosphere

Just as we discussed the composition of food before we discussed the digestive system, here we are going to discuss the air we breathe into our lungs 12 to 20 times per minute, before we explore the respiratory system. Not only do we need to know the composition of the atmosphere, but we need to understand the factors that characterize 'fresh' air, 'stale' air, a 'relaxing' atmosphere and a 'bracing' atmosphere. We need to know why some atmospheres which are good for the majority of people are bad for a few people. We need to know about the factors that cause pollution of the atmosphere, and what can be done to prevent *pollution*, so that understanding the 'breath of life' we can preach the gospel of health to all with whom we come into contact. Why is the 'breath' of life so essential? First and foremost because it supplies our bodies with oxygen without which the cells cannot live. (If brain cells are deprived of oxygen for longer than three minutes, irreparable damage is done.) Breathing is the means by which the body gets rid of some of its heat, excess carbon dioxide and water, the latter two being end-products of metabolism.

HUMIDITY

Saturation point can be defined as the point at which water vapour in the air begins to condense. It is also known as 100 per cent *relative*

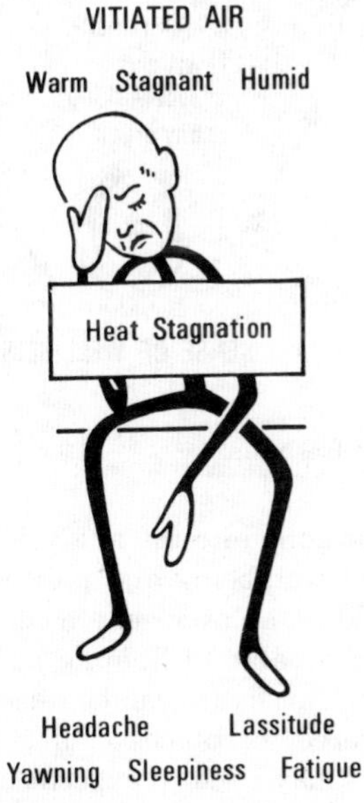

VITIATED AIR

Warm Stagnant Humid

Heat Stagnation

Headache   Lassitude
Yawning Sleepiness   Fatigue

Fig. 176   The effects of vitiated air.

humidity. When air has no water vapour, its relative humidity is 0 per cent. Human beings are more comfortable in an atmosphere of *low* relative humidity than in one of *high* relative humidity. *Too dry* an atmosphere is irritant to the respiratory passages, which fact you will remember as you learn to moisten oxygen before administering it to your patients. *Too moist* an atmosphere is unable to absorb the 1 to 1½ litres of fluid given out via the lungs and skin of each person each day. The perspiration remains on the skin, the skin is robbed of its function of cooling the body by evaporation of its perspiration, and so heat stagnation ensues and it can lead to death. Headache, faintness, nausea and lassitude are warning signs. If moisture condenses on a bedroom window during the night, it is important to open it during the day to allow the room to dry out. Similarly steam on kitchen windows indicates inadequate ventilation, and if it cannot be prevented by the intelligent use of windows, a suction fan is needed.

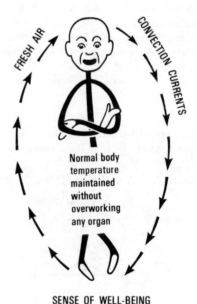

SENSE OF WELL-BEING

Fig. 177   The effects of fresh air.

Much research has gone into the analysis of an atmosphere which will give a feeling of well-being and prevent lassitude and fatigue. With regard to humidity, an atmosphere containing 65 to 75 per cent of the moisture it is capable of holding before becoming saturated approaches the ideal. This figure can be raised for places of sedentary occupation and lowered for places of heavy physical work. Occupational health units may need to experiment with a hygrometer to determine the most comfortable humidity for their workers.

The heat of the atmosphere is gained indirectly by radiation from the sun, the rays of which do not heat the air as they pass through, but they do heat the land and water covering the earth's surface. The air passing near to this warmed land and water is in turn warmed, becomes lighter and rises, cooler air flowing in to take its place. This provides a continuous circulation of air by convection currents. According to geographical location, some countries enjoy a warmer and some a colder climate[161]—and inhabitants have to learn to maintain health, and patients have to be cared for in these, as well as in the temperate climate of our country.

Air or Atmosphere

CONVECTION CURRENTS

These form the basic principle in all forms of ventilation of buildings. Inlets for air are placed low; heating units are low so that the incoming air will become warmed and rise; outlets for vitiated air are placed high, thus ensuring a continuous circulation. On a bigger scale, air in natural motion can be a breeze, wind, gale or hurricane. Again, people living where these are common have to learn to cope with them, in the interests of health.

COMPOSITION OF THE ATMOSPHERE

Table VI shows this:

Table VI   Composition of the atmosphere.

Constituent	Inspired Air	Expired Air
Oxygen	20 per cent	16 per cent
Nitrogen	78 per cent	78 per cent
Carbon dioxide	0·04 per cent	4·04 per cent
Argon and other rare gases	1 per cent	1 per cent
Water vapour	Variable	Saturated (It is for this reason that the surgeon warms the mirror before inspection of the pharynx and larynx)
Temperature	Rarely exceeds 21° C (70° F) in Britain	Body temperature, 37° C (98·4° F)
Bacteria	Variable	Increased in respiratory infection

Oxygen. This is a clear, colourless gas essential for all forms of combustion. The air at the top of high mountains contains less oxygen, hence the use of oxygen from cylinders in high altitude climbing and flying. People who live continually in a rarefied atmosphere compensate by producing more red blood cells. The blood absorbs about 400 litres of oxygen a day in the lungs, as muscles need a large supply during contraction. Vegetation releases oxygen during the day, but absorbs oxygen at night. It was for the latter reason that plants and flowers were removed from the sick-room at night, but it is now known that the amount they absorb is very small. However, the custom prevails in many hospitals as the perfume may become 'heavy' in the still night air, and because of centralizing the task of daily vase washing and disposal of dead flowers. In nursing you will learn to recognize the signs of shortage of oxygen (not because of shortage of atmospheric oxygen but because of the body's inability to use the oxygen, there being many reasons for this) in a human being and your responsibilities when administering oxygen.

Nitrogen. This is found as a chemical element everywhere and is necessary for the existence of life, but that in the atmosphere cannot be utilized by the human body, and it has therefore to rely on protein for its nitrogen supply. Nitrogen dilutes the oxygen in the atmosphere thus checking the rapidity of combustion.

Carbon dioxide. This is a heavy gas produced whenever the two elements carbon and oxygen combine, as in respiration, combustion and decomposition. The greatest concentration of carbon dioxide is that found in submarines, but the men easily withstand an atmosphere of 3 parts per 100. Carbon dioxide is not poisonous until present to the extent of 5 parts per 100.

Plants breathe it in, retaining the carbon and setting free the oxygen during daylight (photosynthesis); the process is reversed at night. The respiratory centre in the medulla oblongata receives impulses from cells that are affected by the amount of carbon dioxide in the blood. Inhalation of carbon dioxide is a useful respiratory stimulant.

Bacteria. Mountain and sea air contain very few bacteria; the air over open fields contains more, the air of towns still more, and the air of crowded rooms, especially crowded dusty rooms, contains most of all. Prevention of the scattering of dust is therefore an important measure in maintaining the hygiene of the atmosphere, and this principle needs to be remembered and put into practice during many domestic and catering activities.

The force of gravity acts on bacteria so that floor dust is a menace, and sweeping without 'laying' the dust by using a dampening agent is to be frowned upon; indeed the only safe way of removing dust is by suction.

Season and temperature affect the number of bacteria in the atmosphere, there being less on cold, wintry days and more on warm, sunny days.

FRESH AIR

Fresh air is that which allows the body to maintain its normal temperature without overworking any organ, and which is sufficiently stimulating to produce a sense of well-being. The aim in the ventilation of dwellings is to supply fresh air and remove used air, so that the inside air is kept as near as possible to the freshness of the outside atmosphere.

To secure efficient ventilation it is necessary to have gently moving air to prevent stagnation, to prevent excessive moistness or dryness, to maintain an even temperature [in Britain 18° C (65° F) is acceptable; in America 21° C (70° F) to 24° C (75° F)] and to keep dust down to a minimum. Each person requires 3,000 ft³ of air hourly. Air can be changed three times hourly without causing a draught, so that 1,000 ft³ of air space is allowed for each occupant. This is usually reckoned as 100 ft² of floor space.

STALE AIR

This term is synonymous with stagnant air, and not only does it contain more bacteria and water vapour (if there are people present), but in it smells are more pronounced—that of dust in an empty room, that of stale tobacco smoke and decomposing perspiration in an occupied room.

RELAXING ATMOSPHERE

Some low-lying sheltered areas are less exposed to the stimulating effects of the winds, and are said to be 'relaxing'. People convalescing from chest and heart diseases are well advised to do so in a place which has a less exhilarating atmosphere. People who have earned their living by doing hard physical work throughout the year, and those of declining years who are less capable of physical activity, will do well to holiday in a relaxing atmosphere.

BRACING ATMOSPHERE

Included in this term are the fells, the hills and the glens with the wind whistling through the heather, and the seaside with its exhilarating winds and tides, which benefit all healthy people, who, released from their work for a while, seek stimulus to their body that they might return freshened in mind.

Pollution of the Atmosphere[162,163]

This has now become a universal problem concerning every nation. It is a large part of the price being paid for *rapid technological advance*. The World Health Organization has set up an international network of monitoring and studying air pollution. The two international centres are in London and in Washington. There are three regional centres in Moscow, Nagpur and Tokyo and 20 laboratories in various parts of the world. The laboratories feed information to the centres. The World Health Organization says that despite general opinion, air pollution is not the 'privilege' of highly developed countries, but is a problem in the large cities of Calcutta, Sao Paulo and Taipeh. The World Health Organization is still trying to reach

337

agreement about the level of pollution that constitutes a health risk. Any radioactive pollution of the air will be monitored. The organic dust content, chemical and gaseous composition, and heavy poisonous metals in the form of air borne particles of lead, cadmium, nickel, zinc, copper, cobalt and other substances are detected. Measurements are made of the atmosphere, and by examination of soil, vegetation, wild life and some farm animals. Contour maps are made of the distribution of pollutants. These maps already show how agents can be spread for 20 to 30 miles from their source, even in relatively still air. Increased oxidants, irritants created by the action of sunlight on nitrogen oxides discharged through the use of coal, gas and oil, were held to be responsible for irritation of the eyes, nose and throat, prevalent in recent smog formation in America and the Far East.

Each time an airliner crosses the Atlantic its jet engines burn up 35 tons of fuel thus using oxygen and producing thermal pollution. The amount of carbon dioxide in the world's atmosphere is increasing. This could raise the mean global temperature by 3° C by the year 2000. American scientists have shown that the aerosol question will interfere with temperature more seriously than the carbon dioxide content. Aerosols are collections of liquid and solid particles dispersed in the air. They are so small that they float for a very long time. The most commonly experienced aerosol is industrial smog. It reflects the sun's heat. The earth robbed of this heat is cooler. The heat radiated from the earth as infra-red rays can pass through the aerosol layer. This upsets the balance of heat. If no steps are taken to curb the spread of aerosols in the atmosphere, a cooling of the earth by $3\frac{1}{2}$° C in 50 years seems inevitable. This could start another ice age. The already growing ice caps at each pole would probably continue to develop even if the aerosol layer were destroyed. The greatest contribution to the prevention of this would be replacement of fossil fuels by nuclear power. The dangers of manipulating water resources are also being considered from the point of view of change of climate.

Each country makes its own laws about pollution of the atmosphere. Condensation around the solid particles in smoke produces fog. When the severe fog caused 4,000 deaths in Greater London in 1952, a newly coined term came into being—smog. It serves to remind all that fog is proportional to the amount of smoke. This smog epidemic was a prelude to the Clean Air Act of 1956. Another Clean Air Act of 1968 has given local authorities much wider powers than they had, and many companies are now accepting the responsibility of helping in the fight against air pollution.

In the Acts, *smoke* heads the list of pollutants. It arises from the combustion of coal for domestic, commercial and industrial purposes; from the burning of wood—used in some industries, for example the 'smoking' of food; from the combustion of diesel and other oils used in heavy road vehicles, trains and in central heating systems; from tobacco, which pollutes confined air in any building. It is estimated that over the country as a whole three-quarters of all

338

the smoke produced at the present time comes from domestic sources, and this smoke is almost always discharged at low level. The Minister of Housing gave 82 local authorities in the so-called 'black areas' until 1966 to produce plans for smoke control. London was one of the black areas. Its atmosphere is now less polluted and it registers increased direct sunshine even in the winter. In 1976, smoke control still remains a permissive and not a statutory duty of local authorities. It is estimated that it will be the late 1980s before the whole of Britain is a 'smoke-free' area.

The soot in smoke blackens the interior of the lungs, and the solid particles have an abrasive action on the respiratory mucous membrane so that germs can enter more easily, there to set up inflammation and disease. Repeated attacks cause scarring and fibrosis, thus reducing the lung's efficiency and rendering it more susceptible to further infection—a vicious circle. Chronic bronchitis eventually gives rise to chronic heart failure, so that it is right that any measures which can be taken to cut down smoke should be taken, in the interests of health as well as economy. Chronic bronchitis causes 25,000 to 30,000 deaths and a national bill of £72 million annually. These are hard facts and do not take account of the suffering and emotional upheaval in the families. Work goes on constantly at the Medical Research Council's Air Pollution Research Unit at St Bartholomew's Hospital, London. Research into production of portable apparatus for recirculating cleaned air is going on at the Paddington and Kensington Chest Clinic, London. It is hoped that the machines will remove both particulate and gaseous irritants from

Fig. 178   Air conditioner that removes over 90 per cent sulphur dioxide.

the air. The project is sponsored jointly by the North West Metropolitan Regional Hospital Board and Messrs. Vokes of Guildford, Surrey. Another firm have marketed a room air conditioner (Fig. 178). It is claimed that it is capable of washing and humidifying the air in a self-adjusting manner and is intended for small rooms where silent operation is essential. It removes over 90 per cent of the sulphur dioxide from the air. Another air cleaning unit is the Vacuair V41. Apart from washing the polyurethane filter and replacement of an activated carbon cartridge, maintenance is not required. Five minute's attention, therefore, per year, should be sufficient. The unit is portable.

Smoke incurs a greater expenditure on the washing and cleaning of personal clothing, household furnishings, interior and exterior decorating, and maintenance of buildings because of erosion. During fog there is also the use of extra artificial light, the disruption of outdoor work, transport by land, sea and air, bringing in its train the increased risk of accident. Smoke is an effective barrier to the sun's rays so that not only do we lose their gladdening effect upon our lives, but also their bacteriostatic and bacteriocidal effects, and their ability to irradiate the ergosterol in the skin with production of vitamin D.

Vehicular pollution. The exhaust given off by vehicles contains carbon monoxide. In the body this combines with haemoglobin to form a stable compound, so that the haemoglobin is not available for carrying oxygen round the body. Each country has different rules about exhaust pollution. In Britain, the 1976 models of petrol-driven cars and light vans must not emit more than 23 grams of carbon monoxide and 2·2 grams of hydrocarbons per mile of driving. This compares favourably with the 1968 to 1969 standards of 34 for the former and 3·3 for the latter. Lead has been added to petrol to improve its performance. The skeleton of modern man contains up to one-third of the level of lead considered by The World Health Organization to be toxic. Lead is absorbed more readily by inhalation than by ingestion. Many people believe that there is need to control lead additives to petrol.

Other pollution from industrial processes. Particles of stone, metal, linen, flax, hemp, wool, arise in many industries, but the problem usually remains confined to the factory, and every measure possible should be taken to prevent silicosis in the workers. Silicosis has crippled men since the days of the Pyramid builders. Egyptian mummies have shown traces of silicosis. Coal miners, stone masons, potters and steel-grinders all know the disease well. In the 19th century there was a move to pass a law to allow only criminals to grind knives. More than half the knife grinders died before the age of 30. The most recently incriminated substance with regard to atmospheric pollution is *asbestos*. A *Code of Practice for Asbestos Workers* has been established and it extends to builders and plumbers using asbestos for lagging pipes. All waste must be immediately collected into sealed containers. People living near asbestos factories were at risk before the dust precautions were adhered to.

It was thought that workers' families were at risk from dust taken home on the workers' clothing. Over the years various dust filter machines have been developed but the greatest breakthrough came from Reading University, a photo-electric dust monitor. Knowing the danger areas in factories and mines, dust filter machines can be fitted. Workers in these industries are encouraged to have regular chest X-rays.

There are some *organic* dusts that pollute the atmosphere and thereby put workers at a health risk. *Dust of hay* can produce farmers' lung; the fungus *spores of mushrooms* can produce a similar condition in the lungs. Maple-strippers' lung is caused by inhalation of *sawdust*. Bird-fanciers' lung is caused by *feather dust*. These all produce an allergic immunological response in the respiratory membrane. Another organic pollutant is the *Bacillus subtilis* used in the enzyme or biological washing powders. Though these can cause skin reactions in sensitive people, their greatest danger is from inhalation—and again it is an allergic-like reaction.

In the country there is the added dust from pollen to which some people are sensitive, and it sets up such conditions as dermatitis, hay fever and asthma. There are no wide-scale measures for the prevention of pollen pollution of the atmosphere, and each individual has to solve his problem, which may be removing the particular flower from his own garden, or residing in the town during the hay-making period and so on. Evening papers and the radio announce the pollen count in some areas. The count is sponsored by the Asthma Research Council and is carried out by the Wright-Fleming Institute at St Mary's Hospital, Paddington. Severe symptoms can be expected when the count reaches 50 (calculated as pollen grains per cubic metre of air averaged over a 24-hour period) and sufferers are advised to remain indoors or avoid country areas where pollen is likely to be more intense. Many people find the forecasts a valuable aid in planning their daily routine and have been saved from days of sneezing misery.

The advent of toxic chemicals used for spraying in horticulture and agriculture has brought a hazard of atmospheric pollution to factory and field workers. Strict regulations are enforced in such areas.

Questions have been asked in Parliament about local pollution of atmosphere in hairdressing establishments. Samples of air from hairdressing establishments have been collected to discover the concentration of spray. Over 300 hairdressers divided into three groups, those using shellac-based hairsprays; those using PVP (polyvinyl pyrrolidone) and those using both types. Hair sprays have not been completely exonerated but current thought states that lung disease in hairdressers is not a major occupational hazard.

Some factories emit a foul smell which in itself may not be harmful, but it may interfere with the appetite of residents and on this score effort should be made to overcome the nuisance.

At the atomic centres there is a slight risk of radiation polluting the atmosphere. The 'fall out' is on to vegetation and so into man. In the case of milk the cows have eaten the polluted grass. Strontium^{90}

341

causes the greatest worry, as it simulates calcium and is deposited in human bone, there to interfere with the blood-forming marrow. The latest report of the Agricultural Research Council's Radiobiological Laboratory states that the concentration of strontium^{90} in Britain's milk supplies has fallen to the level of 1961 before the resumption of nuclear tests. The concentration reached a peak in 1964 and has been falling since.

Respiratory pollution. Warm, saturated expired air is projected 150 to 180 cm (5 to 6 ft) in ordinary quiet breathing, the distance being increased in talking, shouting, coughing, sneezing and singing. If an agar plate is exposed in a room where people are chattering, and it is then incubated, it is proof of the millions of germs present in the atmosphere. Another danger is that these germs can fall to the ground, be dried by the air and can then be blown around as dust.

The *oxygen and carbon dioxide content* of the atmosphere *is never changed to a dangerous level,* but the temperature and humidity rise, and the body cannot lose sufficient heat and moisture, so there is headache, faintness, nausea and lassitude. The respiratory mucous membrane is lax and atonic and therefore more susceptible to the millions of germs floating around; had it been turgid and tonic with fresh air it would have been able to resist the germs. Other gaseous constituents of the atmosphere corrode stonework, the damage being there for all to see, while the damage to the human respiratory tract remains unobserved until it produces disease.

Decaying organic matter. This may give rise to unpleasant smells. When decayed to the point of pulverization it may be blown about as dust. Probably the greatest danger of decaying organic matter is its attraction for flies and rodents, discussed on page 479.

PREVENTION OF ATMOSPHERIC POLLUTION.

The use of smokeless fuel minimizes the production of smoke, therefore attempts should be made to awaken the public's conscience by education, especially that of the building trade regarding the installation of apparatus suitable for burning smokeless fuel. Sufficient supplies of smokeless fuel should be made available at a comparable price to coal, to encourage householders to change their apparatus. A Government grant for those on a small fixed income would encourage them to change over. The centralizing of heating for rooms and hot water can be achieved in blocks of flats, and this principle may prove possible for other housing projects of the future. Meantime the use of gas and electricity for central heating can be considered. It may be that in the near future clean heat and power from atomic energy will be available on a large scale for domestic purposes.

Great strides have been made in the extraction of grit from the smoke of commercial and industrial furnaces, the use of mechanical stokers, and training of boiler-house staffs. The emission of dark smoke is now an offence.

Electrification of railways will make a large contribution to a clean atmosphere; cleanliness of the engines and keeping them in good working condition does lessen the blackness of the smoke meantime.

With the advent in the last 20 years of lighter and more easily laundered fabrics for furnishing and personal wear, and the production of a great variety of soaps, soapflakes, soap-powders and detergents it has become much easier for people to keep themselves, their clothes and their environment free from dirt and dust. Modern buildings no longer have ledges and cornices that serve as dust traps; built-in cupboards and baths further reduce dust collection and are a great blessing to the housewife, especially at spring-cleaning time.

In the towns, as previously mentioned, the street-sweeping carts are fitted with perforated pipes to spray water and lay the dust as they go along on their cleanliness mission.

Prevention of pollution of the atmosphere costs money. Money for research to find out how to render waste innocuous. Money to provide the facilities advocated by research to lessen the problem, to bring about a change of attitudes in the workers to adopt the advocated routines—be they wearing masks, setting filter apparatus in motion, donning special clothing, bathing before leaving work, or whatever. Public opinion may well prove to be the strongest tool in bringing about air fit to breathe. At the beginning of 1971, a picket line of angry housewives blockaded the entrance to a factory they accused of polluting the atmosphere they had to breathe. For the first time a working class neighbourhood showed that it does not accept industrial pollution as part and parcel of its everyday life. And something was done about it! Public opinion could well demand less pollution from petrol and diesel engines.

PREVENTION OF POLLUTION OF HOSPITAL ATMOSPHERE

This brings us to the very important consideration of the prevention of dust contamination of the atmosphere in hospital. This problem has been recognized for many years, and damp sweeping and dusting taught; also it was known that dressings could not safely be exposed for at least one hour after the cessation of domestic activities in the ward. Then came greater privacy for patients by the erection of cubicle curtains and it was realized that unless they were changed weekly (daily around a patient with an infection) they greatly increased the risk of cross infection by increased dust dissemination. Simultaneously it was found that blankets returned from the laundry could still contain staphylococci and the incriminating evidence of dust from blankets is under each bed after the bed has been made. How that fluff flies about on a windy day, and defies all attempts to get it safely deposited in the dustbin where it can do no further harm! So the problem had to be tackled at source, and bactericidal detergents were added to the rinsing water for blankets; then it was discovered that if oil were added to the rinsing water it considerably reduced the dust whilst handling the blankets. Eventually the firms that make blankets became interested (for hospitals must be amongst their best customers) and they manufactured cotton cellular blankets, which are gradually replacing the woollen ones in most hospitals today. Cotton blankets disseminate some dust, but as they can be boiled without deteriorating or shrinking, the dust is less likely to contain bacteria.

Floor dust also came under scrutiny, and research has found several more satisfactory methods of removal that can be used until such time as money can be spared to supply an adequate number of suction cleaners and washers, which constitute the safest means of removal of floor dust. Oiling and waxing of the floors are inter-- mediate measures and they act because the dust does not rise so easily from them, and when it is swept from such a surface it forms little balls and does not fly about. The two-mop method must be used wherever floors are mopped, so that the dust and bacteria lifted on the first mop are immediately placed in a pail of bactericidal solution, whilst the area is dried with a second mop, which is not exposed to as heavy a contamination as the first. At the end of the proceedings both mops should be thoroughly washed and dried in the open air. If they are put away wet, the few germs on them will have multiplied into dangerous millions, and will then be applied to the ward floor in an emulsion, in the belief that the floor is being cleaned. I have been deliberately facetious to show that a willingness to work is not sufficient; each member of the team must understand their work so that the patient is not endangered from lack of knowledge. The provision of a dust-free atmosphere is such a complex business that it is by far the best policy to remove the patient requiring a wound dressing to a treatment room where the atmosphere can be controlled.

As to the prevention of droplet contamination of the atmosphere, it is the rationale behind the wearing of masks.

Pollution of the Atmosphere as it affects Health[164]

1. It can cause headache, lassitude, yawning, sleepiness, faintness.

2. It can cause infection, the most frequent being the common cold, tonsillitis and bronchitis. Pulmonary tuberculosis is an air-borne infection and you will keep adding to this list as further experience comes your way.

3. Bronchitis can become chronic; then in its wake over several years of poor health comes chronic heart failure.

4. Industrial pollution can cause pneumoconiosis or silicosis, and this also leads, over years of impaired health, to chronic heart failure.

5. The smoke blanket robs us of some direct sunshine with its cheering rays, its disinfectant property, and its ability to produce vitamin D in the skin. Rickets results from lack of this vitamin.

6. 'Fall out' in the atmosphere has undoubtedly led to an increase in strontium in children's bones; there is undoubtedly an increase in the incidence of leukaemia, and it is thought that there is a connection between these two.

7. The inhalation of large quantities of tobacco smoke, particularly that from cigarettes, does carry with it an increased risk of lung cancer, emphysema and heart disease.

Nature attempts to keep the composition of the atmosphere constant by:

Plant Life. Plants absorb carbon dioxide during the day and liberate oxygen, and in thickly forested areas this helps to replenish the oxygen which human beings and animals have removed from the atmosphere.

Winds which dilute and mix the atmospheric gases.

Rain which washes down suspended matter; the bacterial count is always less after a shower of rain.

Sun which dries and warms the air, prevents the growth of some germs (bacteriostasis) and kills others (bacteriocidal).

Frost which kills many germs, but some can survive, for example *Salmonella typhi.*

Aids to natural ventilation in a city. Wide streets, preferably lined by trees with sticky leaves to which dust can adhere.

Open spaces and gardens wherever possible.

All these allow free circulation of fresh air, to help to counteract the extra pollution which is bound to occur in the atmosphere of a city.

Aids to natural ventilation in a building

Window Contrivances. These are now many and varied and you are advised to examine the ones you encounter and notice their advantages and disadvantages.

Wall Devices. Perforated bricks and iron gratings can be used to aid natural ventilation in a building. You are advised to observe in which rooms and at what heights these are placed to gain the best advantage.

Chimneys. These form a good means of promoting exchange of gases between inside and outside air. When a fire is burning a strong upward current is caused by the warm air rising up the chimney, cool air coming in through the windows, cracks and crevices to take its place. Chimneys also act as ventilators when the fire is not burning due to the aspirating effect of the wind. When a fireplace that is not going to be used is boarded in, a means of ventilation of the chimney should be incorporated, otherwise dampness penetrates the chimney stack.

Artificial Ventilation or Air Conditioning

This is achieved by *mechanical* means which may consist of forcing in air, sucking out air or a combination of the two activities.

Plenum or propulsion method. Air is forced in by electric fans at the base of a building. It may be drawn through screens of hemp or other fabric and kept moist by a flow of water; this filters the air of dust and moistens it. This filtered, moistened air can be warmed by passing over heated coils before it is allowed to enter the building.

345

This method is essential in operating theatres, in immunosuppressive (organ transplantation) units and in units carrying out cytotoxic therapy (administration of drugs that kill cancer cells), as a dust- and germ-free atmosphere must be provided. In the latter two instances the germ-free atmosphere is necessary because the treatment lowers the patient's resistance to infection.

Vacuum or extraction method. Sometimes it is not necessary to install the expensive Plenum plant as it is sufficient to remove the used air by electric fans, thus drawing fresh air in by any inlets.

Balance method. As this term implies it is a combination of the Plenum and vacuum method, the air being forced in at the base and sucked out at the top.

DISADVANTAGES OF ARTIFICIAL VENTILATION

1. The sun's rays do not pass through the windows unless they are of vita-glass.

2. It is more expensive than natural ventilation.

The Respiratory System

So far we have built up Jimmy's skeleton and covered it with muscle and skin; we have given the muscles a blood and nerve supply, and brought nourishment to them, but still they are inactive, for they require oxygen to spark off the metabolic process that releases the energy for movement. This brings us to a consideration of the structures and mechanism whereby the body is able to absorb the atmospheric oxygen and rid itself of excess carbon dioxide and water.

The Nose

This funnel-shaped cavity has the horizontal, perforated plate of the ethmoid bone as its superior boundary. The perforations allow transmission of the nerve of smell (olfactory), the filaments of which are distributed in the upper nasal mucosa. The two nasal bones form the anterior bridge of the nose, whilst a bar from each malar bone, the lacrimal bones, the ethmoidal air sinuses, the turbinate bones and a portion of the superior maxillae complete the lateral walls. The total cavity is divided into two nostrils by a septum, the bony part of which lies posteriorly and is composed of the perpendicular plate of ethmoid, with the vomer below (see Figs. 23 to 26). This bony septum is lengthened anteriorly by hyaline cartilage. The elaborate contour of the lateral walls gives an increased surface area for the ciliated mucous membrane lining of the nose, thus allowing it maximum function in warming, moistening and filtering the air passing through. The cilia sweep in a backward direction, so that any solids deposited on them find their way into the pharynx and stomach, there to be dealt with by hydrochloric acid. The floor of the nose is simultaneously the roof of the mouth and is composed of the hard palate (superior maxillae and palatal bones, p. 52). The external openings are called the anterior nares, and the ones leading from the nose into the nasopharynx are called the posterior nares or choanae. As well as the meati receiving ducts from the various air sinuses (see Fig. 179), the inferior one receives the nasolacrimal duct, so that excessive tears are discharged via the nose.

The nose is richly supplied with blood from branches of the facial and ophthalmic arteries. The venous drainage is unusual in that part of it goes within the skull to the venous sinuses before being returned to the jugular vein. Infection of these venous sinuses can arise as a complication of nasal sepsis.

When observing a patient with respiratory distress you may be the first person to notice dilation of the anterior nares by contraction of the tiny muscles in the inferior lateral nasal walls. This observa-

347

tion of the patient's attempt to get more air is important and should be reported to the doctor without delay.

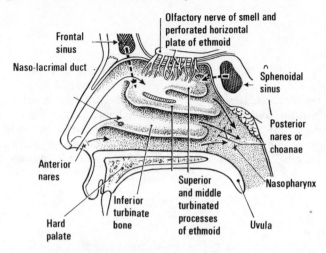

Fig. 179   Diagram of right lateral nasal wall.

THE FUNCTIONS OF THE NOSE

1. It warms, moistens and filters the air passing through it.
2. It is the organ of smell (olfaction).
3. The nasal sinuses give resonance to the voice. Absence of this function is evident when one is suffering from a common cold.

The Paranasal Sinuses

Included in this term are all those air sinuses, viz. sphenoidal, frontal, ethmoidal and maxillary, that have communication with the nose. They are lined with a continuation of the nasal mucosa, so that infection can easily spread. Deviation of the nasal septum can interfere with drainage from these sinuses and can result in sinusitis.

The Pharynx

This tubular cavity extends from underneath the sphenoid bone to the base of the epiglottis. It lies behind the posterior nares, the mouth and the epiglottis, and in front of the cervical vertebrae.

The pharynx is lined by respiratory mucous membrane, supported on a submucosa of areolar tissue, which is adherent to a layer of muscle tissue arranged mainly in circular style with a few longitudinal fibres, so that peristalsis can be initated in the lower portion, to pass food from the mouth to the oesophagus. Scattered in the

348

areolar tissue is a considerable amount of reticular tissue, especially in the roof, enlargement of this tissue having been popularly spoken of as 'adenoids' throughout the years. The correct name for this reticular tissue in the roof of the pharynx is pharyngeal tonsil. Opening from the upper lateral walls are the pharyngo-tympanic tubes, which during swallowing equalize the air pressure on either side of the eardrum. Enlargement of the pharyngeal tonsil can press upon these openings with consequent impairment to hearing. The

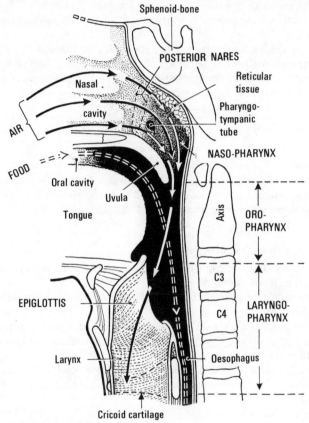

Fig. 180   Sagittal section of the pharynx.

anterior pharyngeal wall is not complete, for it contains the openings from the nose and mouth. Opening out of the posterior base of the pharynx is the oesophagus, and opening out of the anterior base is the larynx or voice box. Can you now name the seven openings in the pharyngeal walls?

BLOOD SUPPLY

Pharyngeal artery.

VENOUS DRAINAGE

Pharyngeal vein.

349

White motor neurones from the central nervous system to the muscles. White sensory neurones from the central nervous system. Grey motor neurones from the autonomic nervous system to the blood vessels.

FUNCTIONS

1. The pharynx provides a passage for food and air.
2. It continues to moisten, warm and filter the air passing through it.
3. It continues lubrication of the food passing through it.
4. During swallowing it allows air to pass along the pharyngotympanic tube to the middle ear to equalize the pressure on either side of the drum.

(See p. 274 for the mechanism of swallowing.)

The Larynx

Commonly called the voice box, this organ forms the prominence known as Adam's apple, more marked in the male, for his larynx is wider anteroposteriorly, than that of the female.

POSITION AND RELATIONSHIPS

Anteriorly. The neck muscles, subcutaneous tissue and skin are to be found in front of the larynx.

Posteriorly. One glance at Figure 180 confirms that the oesophagus lies behind the larynx.

Laterally. On either side lies a common carotid artery, a jugular vein, a nerve plexus, lymphatic nodes and vessels, a sternocleidomastoid muscle, subcutaneous tissue and skin.

Inferiorly. The larynx is continuous with the trachea below.

Superiorly. The larynx opens from the anterior base of the laryngopharynx.

STRUCTURE

In the main this complicated but roughly tubular organ is lined with respiratory mucous membrane, supported on a submucosa of areolar tissue carrying the nerves and the blood and lymphatic vessels.

In a recessed area in each lateral wall, the lining becomes specialized to form the vocal cords. The laryngeal muscles are short and numerous; they belong to the striped, skeletal, voluntary classification and are supplied with white motor neurones from Broca's area in the cerebrum, best developed on the left side in right-handed people, and vice versa. On contraction the laryngeal muscles move the laryngeal cartilages whereby the vocal cords are slackened and tightened in the complicated process of voice production.

The thyroid cartilage. This is the largest; it is shield-shaped with a midline prominence known as the Adam's apple. This prominence gives attachment to the anterior border, of the vocal cords.

The cricoid cartilage. This lies below the thyroid; it is the shape of a signet ring and is placed with the signet plate at the back, the upper portion rising between the lateral borders of the thyroid cartilage. The cricoid is the first complete cartilaginous ring of the respiratory tract and lies immediately above the trachea.

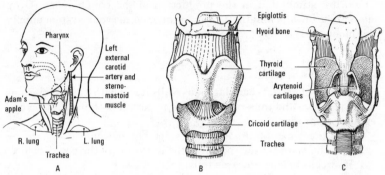

Fig. 181 The larynx. A, Position and relationships. B, Anterior view. C, Posterior view.

The arytenoid caltilages. These are two triangular-shaped structures perched on the posterior surface of the cricoid signet. They give attachment to the posterior border of the vocal cords, and as they move they shorten and tighten the cords to produce high pitch, or they lengthen and loosen the cords to produce low pitch.

The epiglottis. This is a leaf-shaped piece of hyaline cartilage rising at an angle from the upper thyroid cartilages. The remaining larynx rises and shelters underneath this 'hood' during the act of swallowing (Fig. 136).

BLOOD SUPPLY

Laryngeal artery.

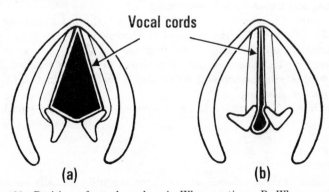

Fig. 182 Position of vocal cords. A. When resting. B. When speaking.

VENOUS DRAINAGE

Via various branches which open into the internal jugular vein and another set that open into the brachiocephalic veins.

NERVE SUPPLY

White motor neurones from the central nervous system supply the muscles. This area is well supplied with white sensory neurones which are stimulated in the production of the cough reflex. Grey motor neurones belonging to the autonomic nervous system supply the blood vessels.

FUNCTIONS OF THE LARYNX

1. Equipped with its cartilaginous walls it can withstand variations in air pressure and thus remain patent to permit both inspiration and expiration of air.
2. It is the organ of phonation or voice production.
3. It is the main sensory organ in the initiation of the cough reflex.
4. It sweeps mucus upwards to be swallowed.

The Trachea and Bronchi

POSITION AND RELATIONSHIPS

The upper portion of the trachea lies in the cervical region and it is only in this part that the operation of tracheostomy can be performed. The lower portion descends into the upper central area of the thoracic cavity, ending at the midsternal level. The space between the lungs which these tubes occupy is called the mediastinum.

Anteriorly. The isthmus of the thyroid gland lies across the front of the cervical trachea, this area being covered with muscles, subcutaneous tissue and skin. The sternum lies in front of the thoracic trachea, and at midsternal level the trachea divides into two bronchi, the costal cartilages lying in front of these two bronchi.

Posteriorly. The oesophagus separates the trachea from the lower cervical and upper dorsal vertebrae.

Laterally. The right and left lobes of the thyroid gland are in intimate contact with the cervical trachea, whereas the right and left lung lie on either side of the thoracic trachea.

Inferiorly. The heart is the main organ below the trachea and bronchi. Distally the bronchi divide into bronchioles which actually form part of the lung.

Superiorly. The trachea is continuous with the cricoid cartilage of the larynx.

STRUCTURE

The lumen of these tubes is widest at the top and gradually becomes smaller as the bronchi divide into bronchioles. They are lined with ciliated mucous membrane, the cilia sweeping any unwanted material upwards, so that it can be expectorated or swallowed. Poisonous gases can paralyze the cilia. Without their protective function a person is predisposed to pneumonia. This mucous membrane lining is supported on a submucosa of areolar tissue carrying nerves and blood and lymphatic vessels. Outside this the walls are of fibromuscular tissue, interspersed with C-shaped rings of hyaline cartilage, so that the tube can remain patent during changes in air pressure. The incomplete portion of the C is placed at the back where the trachea is in contact with the oesophagus, to accommodate a bolus of food passing down this latter tube. The muscle fibres are of the internal plain, involuntary category and are supplied with grey motor neurones belonging to the autonomic nervous system. The left bronchus is longer and more horizontal than the right to accommodate the heart. In consequence inhaled foreign bodies find their way more easily into the right bronchus.

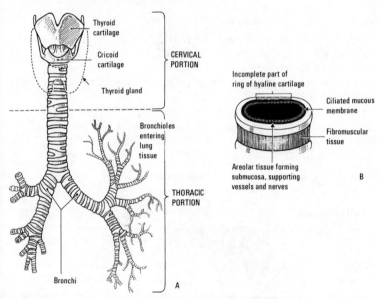

Fig. 183   A, Diagram of trachea and bronchi.   B, Cross section of trachea.

BLOOD SUPPLY

Tracheal and bronchial arteries from the thoracic aorta.

VENOUS DRAINAGE. Part of the blood is returned directly to the right atrium of the heart in the pulmonary veins, the remainder is returned to the superior vena cava.

NERVE SUPPLY

Grey motor neurones from the autonomic system supply the plain

353

muscle, the sympathetic portion dilating to give a greater air intake, and the parasympathetic portion restricting air intake.

FUNCTIONS

1. To remain patent and allow inflowing of air during inspiration and outflowing of air during expiration.
2. To continue to warm, moisten and filter inflowing air.
3. To sweep mucus upwards so that it can be expectorated or swallowed.

The Lungs

These two spongy, cone-shaped organs occupy the major portion of the thoracic cavity.

POSITION AND RELATIONSHIPS

Anteriorly. The sternum and the first five pairs of ribs with their costal cartilages and intercostal muscles are to be found at the front of the lungs.

Posteriorly. The upper ribs with their intercostal muscles lie behind the lungs.

Laterally. More than five of the upper ribs plus their intercostal muscles are in contact with the sides of the lungs, because of the dome of the diaphragm in this area (Fig. 188).

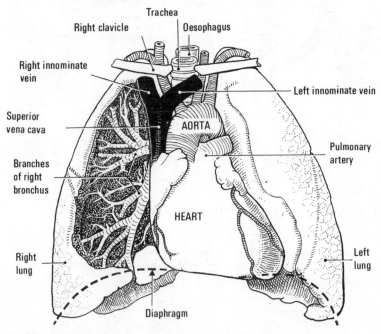

Fig. 184   Diagram showing organs associated with the lungs.

Medially. The heart and large blood vessels occupy the major portion of the space between the lungs (mediastinum), and all the tubes passing through the thorax to the abdomen (oesophagus, aorta, inferior vena cava, thoracic duct) plus lymphatic nodes and nerves, are to be found here.

Inferiorly. The base of each lung rests on the dome of the diaphragm, below which is the liver on the right side, and the stomach, pancreas and spleen on the left (Fig. 188).

Superiorly. The apex of the lung has the first and second ribs and their intercostal cartilages above them, as these structures form the 'roof' of the thorax (Fig. 188).

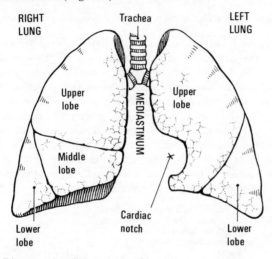

Fig. 185   Diagram showing division of lungs.

STRUCTURE

Since the heart lies between the lower medial lung borders, that area in contact with the heart is called the cardiac notch. A bronchus enters the middle medial aspect (root or hilum) of each lung, and breaks up into many similar but ever smaller tubes called bronchioles. These spread throughout the lung and at the end each breaks up into several alveolar ducts, leading into a cavity rather like a balloon, the walls of which present the appearance of a bunch of grapes (Fig. 186).

The walls of the terminal bronchioles are of mucous membrane, submucosa and internal, plain, involuntary muscle fibres. It is these muscle fibres that go into spasm in asthma, and throughout your training you will learn of drugs that relax these muscle fibres, and can therefore be used to relieve respiratory distress. Each alveolus is like a tiny balloon at the end of a tiny bronchiole. Alveolar walls are a single layer of simple epithelium. As the alveoli inflate when air is drawn into the lungs, the cells in the walls are stretched. Part of the effort of breathing goes into stretching the alveolar cells—overcoming the surface tension of the fluid on the cells. It is made easier

355

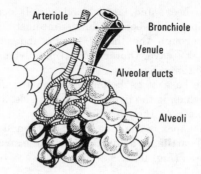

Fig. 186   Diagram of alveolar ducts with alveoli opening from them.

by a detergent-like substance secreted by the cells, called surfactant, that lowers the surface tension. It is said that, if the alveolar membrane could be spread out, it would cover a tennis court. Tiny capillaries from the two pulmonary arteries (which enter each lung at the hilum) are in direct contact with the alveolar membrane, so that air in the alveolus is separated from blood in the capillary by one layer of respiratory epithelium and one layer of squamous (vessel) epithelium. This facilitates interchange of gases. By diffusion, blood loses carbon dioxide and gains oxygen (Fig. 187). These capillaries become venules, and unite to form the two pulmonary veins that leave each lung at the hilum.

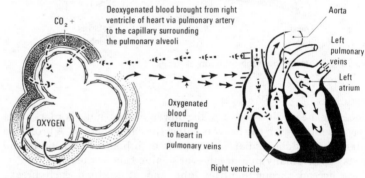

Fig. 187   Diagram of the blood supply to the alveoli.

All this mass of bronchioles, alveoli, blood vessels, lymphatic vessels and nerves is supported by elastic connective tissue and enclosed by a serous membrane called *the pleura*. That which is adherent to the external surface of the lung is called the *visceral pleura*. It dips in between the fissures that divide the right lung into three lobes, and the left lung into two lobes. At the hilum this pleura is reflected to form a sac, so that there is a double layer; this reflected layer is adherent to the chest wall and the upper diaphragmatic surface and is called the *parietal pleura*. Being of serous membrane these layers glide smoothly on each other, the potential space between them being the pleural cavity. (*N.B.*—Anything abnormal

occupying this space, e.g. fluid, air, blood or pus, will compress the
lung tissue and produce difficulty in breathing—dyspnoea.)

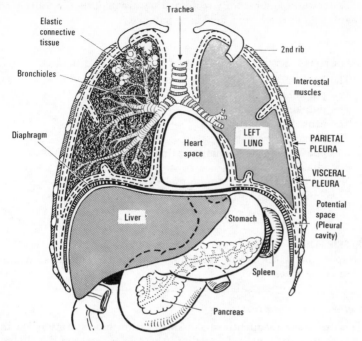

Fig. 188   Diagram of lungs showing disposition of pleura.

BLOOD SUPPLY

The bronchial arteries from the thoracic aorta bring oxygenated
blood to nourish the lung tissue. The pulmonary arteries from the
right ventricle bring deoxygenated (venous) blood for oxygenation.

VENOUS DRAINAGE

Some blood from the bronchial arteries and all blood from the
pulmonary arteries is returned in the pulmonary veins which enter
the left atrium of the heart. The remaining blood is drained away in
the bronchial veins which eventually connect up with the superior
vena cava.

NERVE SUPPLY

Grey motor neurones from the autonomic system supply the plain
muscle fibres, those from the sympathetic portion widening the tubes
and causing a greater air intake, whilst those from the parasympathetic (vagus nerve) restrict the air intake. There are grey sensory
neurones which respond to distension in the alveolar walls, and
these connect to the respiratory centre in the medulla oblongata.

357

To supply the body with oxygen and help it to get rid of excess carbon dioxide, water and heat. By so doing they are important organs in the regulation of the body's acid-base balance.

Man, his Health and Environment

STRUCTURES ENTERING AND LEAVING THE HILUM OF THE LUNG

The bronchi
The pulmonary arteries
The pulmonary veins
The bronchial arteries
The bronchial veins
Lymphatic vessels communicating with lymphatic (hilar and mediastinal) nodes
Nerves from the sympathetic and parasympathetic (vagus nerve) portions of the autonomic system.

RESPIRATION

Respiration is the act of breathing, and the term implies an interchange of gases.

EXTERNAL RESPIRATION

This implies the sequential acts of inspiration and expiration. During inspiration, air is sucked down into the deep recesses of the pulmonary alveoli, there to be in close contact with deoxygenated (venous) blood in the capillaries. The changes in the air (Table VII) take place, the gaseous ones due to diffusion, the humidity changes due to evaporation from the moist membrane and the temperature changes by conduction of heat from the highly vascular alveoli and tubes. During expiration air is squeezed out of the lungs.

Some air (2 to 3 litres) is always held within the lung, so that interchange of gases is a continuous process and not confined to the

Table VII. Contrasting the properties of inspired and expired air.

CONSTITUENT	SYMBOL	INSPIRED AIR	EXPIRED AIR
Oxygen	O_2	20%	16%
Carbon dioxide	CO_2	0·04%	4·04%
Water vapour / Humidity		Variable	Saturated
Temperature		Variable, rarely 21°C, (70°F) in Britain	Body temperature 37°C, (98·4°F)
Bacteria		Variable	Increased in respiratory infection

period of inspiration. Half a litre of air is squeezed out during the expiration of ordinary quiet breathing, and this is replaced by entry of another ½ litre during inspiration. Tidal air is the name given to this changing ½ litre.

The amount of air that can be expired after the deepest possible inspiration can be measured by a spirometer, and is known as the vital capacity of the lung. A simple test is to ask the patient to count as he exhales, and you may be asked to record this on the case notes. Another test is to ask him to blow out a lighted match held 10 cm away from his fully open mouth. The vital capacity averages 3 to 4 litres, and is of clinical importance in diagnosis and prognosis, for the lower it is, the greater has been the destruction of lung tissue, or impairment of the mechanism of respiration.

INTERNAL RESPIRATION

As distinct from pulmonary ventilation this implies the reverse interchange of gases that takes place between oxygenated blood and tissue cells. As the tissue cells use up oxygen they produce carbon dioxide. By diffusion this carbon dioxide passes into the oxygenated blood, which contains less, whilst simultaneously the oxygen which is more plentiful in the blood diffuses into the tissue cells (Fig. 96). The 'reduced' or deoxygenated blood then returns to the heart to be sent to the lungs, where, through external respiration, it will lose its excess carbon dioxide and gain some oxygen.

The Muscles of Respiration

These muscles belong to the striped, skeletal category and though they are supplied with white upper motor neurones from the pre-Rolandic area of the brain to the appropriate level of the spinal cord, and white lower motor neurones from the spinal cord to each muscle, and *can* therefore be controlled by the will, they also function under a powerful automatic control, for there is a maximum period during which the breath can be 'held,' in spite of 'will-power'. Do use this opportunity for revision of muscle tissue on pages 92 and 105.

THE DIAPHRAGM

This is the largest muscle of respiration.

Origin. (1) Two pillars or crura arising from each side of the lumbar vertebrae. (2) The inner surface of the lower ribs. (3) The base of the sternum.

Insertion. The muscle fibres form a dome, and are inserted into an aponeurosis which makes the top of the dome.

The inferior vena cava, oesophagus, aorta and thoracic duct pass through the diaphragm.

Action. As the fibres contract they pull the aponeurosis downwards; this increases the vertical dimension of the thoracic cavity

359

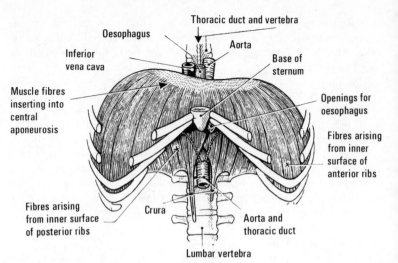

Fig. 189　Anterior view of the diaphragm.

(inspiration), and decreases the vertical dimension of the abdominal cavity. As the fibres relax the aponeurosis ascends to its resting position; this decreases the vertical dimension of the thoracic cavity (expiration) and increases the vertical dimension of the abdominal cavity. The diaphragm is concerned in the acts of sneezing, coughing, vomiting, defaecation, micturition and parturition.

Blood supply. The phrenic arteries which are the first pair to leave the abdominal aorta.

Venous drainage. This is accomplished by the phrenic veins pouring their contents into the abdominal inferior vena cava.

Nerve supply. White upper motor neurones from the pre-Rolandic area of the cerebral cortex to the cervical region of the spinal cord and white lower motor neurones in the form of the phrenic nerves. It

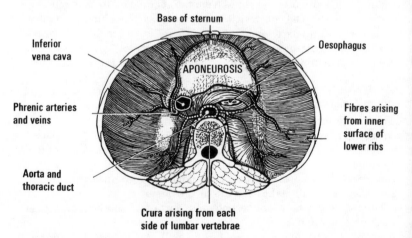

Fig. 190　Interior surface of the diaphragm.

must be remembered that the phrenic nerve is innervated by impulses from the respiratory centre in the medulla oblongata most of the time.

EXTERNAL INTERCOSTAL MUSCLES

Origin. Lower border of the rib above (see Fig. 46).
Insertion. Upper border of the rib below.

INTERNAL INTERCOSTAL MUSCLES

Origin. Upper border of the rib below.
Insertion. Lower border of the rib above.
Action. Their concerted action draws the rib-cage forward and upwards, often likened to the action of a bucket handle, and as the ribs are attached to the sternum, this bone is also pushed forward and upward, thus giving considerable increase in dimension from side to side and from front to back (see Fig. 32).
Blood supply. The intercostal arteries which are branches from the thoracic aorta, and are transmitted in a groove on the under surface of each rib.
Venous drainage. The intercostal veins which run in the same groove as the arteries (see Fig. 47).
Nerve supply. White upper motor neurones from the pre-Rolandic area of the cerebral cortex to the appropriate level of the spinal cord, and lower motor neurones (intercostal nerves) from the spinal cord to each muscle. These intercostal muscles are innervated by impulses from the respiratory centre in the medulla oblongata most of the time.

The Automatic Control of Respiration

Breathing takes place anything from 12 to 20 times a minute in an ordinary healthy person, without the individual necessarily being aware of it. When the tissues require more oxygen as in exercise we breathe more deeply and more quickly to supply it.

The Chemical Initiation of Inspiration

Cells in the arch of the aorta and at the bifurcation of the carotid arteries are highly sensitive to increased acid content of blood. As blood collects the acid carbon dioxide from the tissues, these cells are stimulated to send impulses to the respiratory centre in the medulla, which in turn sends impulses to the diaphragm and intercostal muscles causing them to contract and produce inspiration. (This will rid the body of some of its excess carbon dioxide.) With the enlargement of the volume of the thoracic cavity there is consequent decrease in the pressure of the contained air. Atmospheric air therefore rushes in to equalize these pressures.

The Nervous Stimulus for Expiration

As the alveoli distend to accommodate this inrushing air, the sensory nerve endings in their walls are stimulated and this impulse passes to the respiratory centre in the medulla, and cuts out the chemically stimulated inspiratory impulses. Mainly by elastic recoil and relaxation of muscle, expiration is produced, the decrease in the thoracic volume increasing the pressure of the contained air, so that it rushes out into the atmosphere which is at a lower pressure.

Via the white upper motor neurones, the rate and depth of respiration can be brought under voluntary control. Emotion also affects respiration as you will discover by close observation of your patients who have pain or are very frightened.

There are many deviations from normal breathing, and a nurse's training develops her powers of observation so that any deviations are noted immediately, and reported to the doctor for confirmation and interpretation.

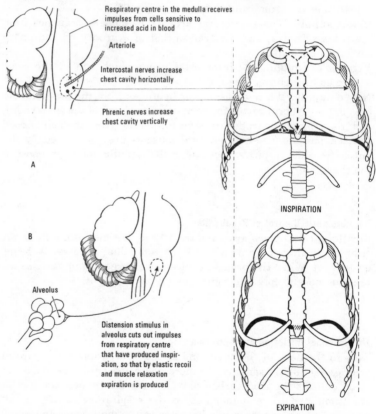

Respiratory centre in the medulla receives impulses from cells sensitive to increased acid in blood

Arteriole

Intercostal nerves increase chest cavity horizontally

Phrenic nerves increase chest cavity vertically

A

INSPIRATION

B

Alveolus

Distension stimulus in alveolus cuts out impulses from respiratory centre that have produced inspiration, so that by elastic recoil and muscle relaxation expiration is produced

EXPIRATION

Fig. 191   A, Diagram of chemical stimulus for inspiration.   B, Diagram of nerve stimulus of alveolar distension for expiration.

CARE OF RESPIRATORY SYSTEM IN THE MAINTENANCE OF HEALTH

Care needs to be taken so that the young, small baby does not suffocate. There are special pillows available for babies. One type is illustrated in Figure 192. We have previously mentioned the inadvisability of taking a baby into the parents' bed. Each year a number of children are accidentally strangled with straps. When procuring pram and chair harnesses, as well as walking reins, they should be of the British Standards Institution specification.

When children can co-operate they should be taught to breathe through their noses and how to use a handkerchief properly (they tend to nip the end of the nose and if there is any infection present this can have disastrous results). Some part of every day should be spent out in the open air, preferably away from the industrial smoke of a city. A few deep breaths daily help to ventilate the bases of the lungs (and suck blood up the inferior vena cava thus preventing venous stasis in lower limbs), and schools do help to inculcate healthy breathing by good posture, and exercises in the physical-training programme. Good nutrition goes a long way towards building up resistance to infection, so that a well-balanced diet is a prerequisite to the health of any system. Avoidance of excessive dust is desirable, and here the Councils help in the summer-time by spraying the streets. People should not go out unnecessarily in the fog, and legislation is in progress with regard to smoke abatement areas which will help to cut down the formation of fog. Covering the nose and mouth with a paper handkerchief when coughing and sneezing will protect other people s respiratory system.

The Respiratory System

Fig. 192   The 'Safesleeper' pillow.

Smoking[165,166]

Countries make laws about smoking, even if it is just to tax tobacco as an import and export. In our country the *sale* of tobacco is subjected to *limited* control by law. Cigarette packets must exhibit a label to the effect that smoking can endanger health. Tobacco is heavily taxed and may not be sold to children. There is no advertising on television. Here are two examples of what other countries do:

In 1971, Singapore enforced a total ban on all advertisements relating to smoking in broadcasts, newspapers and magazines, and on bill-boards and neon signs. Foreign magazines are exempt. A Bill was introduced in the Oregon State Legislature that would designate nicotine and nicotine tars as dangerous drugs. If passed into law it would put tobacco in the same category as marijuana. All countries realize that smoking is not just a health and behavioural problem, complex though the behavioural aspects are; they have to be viewed in the light of economic and political aspects of the problem.

It is now customary to speak of the *cigarette-induced diseases.* They include bronchitis, lung cancer, emphysema and heart disease. In Britain over 20,000 deaths in men aged between 35 and 64 are caused annually by smoking.[167,168] The chances are that two out of every five heavy smokers, but only one in five non-smokers will die before 65. The figures for women are less than for men. Dr Russell, research worker, Addiction Research Unit, London University Institute of Psychiatry, said that, '69 per cent of men and 43 per cent of women in the United Kingdom are regular smokers. Despite the established link between smoking and ill health, there has been little change in smoking prevalence in the last 10 years. Seventy-seven per cent of smokers wish to stop smoking,[169] or have tried unsuccessfully to do so, but less than one in four ever manages to break himself of the habit permanently. It is far easier to become *dependent on cigarettes* than on alcohol or barbiturates. While the *majority* of alcohol or drug users indulge only *intermittently,* no more than 2 per cent of smokers succeed in remaining *occasional* smokers. Of those who smoke more than a single cigarette during adolescence, 70 per cent will continue smoking for the next 40 years. Only the teenager who has never attempted to smoke, or who has attempted it only once and decided he disliked it, has much chance of beco.ning a *non-smoking adult.* The individual who reaches the age of 20 without smoking is unlikely to become addicted'. An association has been found to exist between the ability to stop smoking and the age of weaning. In one study it was found that those who stopped easily were weaned at an average age of 8 months, whereas those unable to stop were weaned at an average age of 4·7 months.

Action on Smoking and Health Ltd. (ASH) was set up by the Royal College of Physicians as an independent body to combat the hazards of smoking. Dr John Dunwoody, Director General, wants to make the diseases associated with cigarette smoking 'as socially unacceptable as V.D.'.

Women who *smoke during pregnancy*[170,171] tend to have smaller babies, and have two to three times as many premature babies as non-smokers. Of the children followed up at 7 years old from one week's births in 1958, those born to smoking mothers could read less well, were smaller and were less well adjusted to school.

The Report of the Royal College of Physicians, *Smoking and Health,* which appeared in March 1962, contains a careful examination of the evidence produced by research in a number of countries.

364

Many doctors were so impressed with this evidence that they gave up smoking. Another report, *Cigarette Smoking and Health Character- istics; a Public Health Service Review,* from America, published in 1967, discusses the relationship between smoking and death from *coronary disease.* It found that smokers were absent from work more often and for longer periods than the non-smokers. One way of getting carbon monoxide into the lungs is by smoking. The blood of a 20-a-day smoker had double the carbon monoxide content of that of traffic policemen and others exposed to high concentrations of exhaust fumes.

Countries report different ages at which children start smoking. In several surveys the two main causes for children starting to smoke are given as social pressure from peers and nervousness. In one sur- vey it was found that the smoking children had less O levels. The money spent on Health Education to prevent people starting to smoke, and to stop people smoking is infinitesimal compared with the 8 million pounds' worth of advertising by the cigarette companies. Expert persuasion on this scale needs more resistance than many people can muster, and they might be helped by equally expert persuasion to stop smoking. The sale of cigarettes provides that £8 million, and the tax claimed by the Government, and the money for the coupon schemes and gifts, and a profit for the firm! Many people are pressing for Government legislation against cigarette coupons and gifts. Two articles[167,172] suggest what we as nurses can do about it. 'The psychological dynamics of smoking'[173] dis- cusses the complexities associated with smoking. Another article[174] has thoughts on smoking and eating.

CONDITIONS THAT CAN INTERFERE WITH THE HEALTH OF THE RESPIRATORY SYSTEM

These are legion and I only propose to discuss their application from a general point of view.

The Common Cold (coryza)

Several different viruses are responsible for this condition, so that one attack does not produce immunity. Many working days are lost because of this infection, therefore each human being is morally obliged to do all in his power to avoid spreading it. It should be con- sidered an offence to appear in any public place, including trans- port, whilst suffering from a cold. Coughs and sneezes spread disease; trap them in your handkerchief, preferably a paper one and burn it immediately.

Adenoids

This is enlargement of the reticular tissue in the roof of the pharynx.

If it enlarges anteriorly and blocks the posterior nares it causes persistent mouth breathing with its attendant susceptibility to the common cold and other upper respiratory infections, as the initial warming, moistening and filtering function of the nose is lost. Mouthbreathing also alters the shape of the hard palate and you will learn to recognize the typical adenoid facies. With their difficulty in breathing such children have difficulty in eating and are often less well-nourished than they should be. If the reticular tissue enlarges laterally it can press on the opening of the pharyngotympanic tube and interfere with hearing. This can give rise to poor learning, hence the importance of audiometric tests for school children.

Infections Producing Sputum

Most people at some time or other have slight infections which produce some sputum. It is the disposal of this which constitutes a hygiene problem. The best method is the use of a paper handkerchief with immediate burning or disposal into a covered bin. To be condemned is expectoration on to the streets. In some towns this is a finable offence, and this is in the best interests of the nation's health. Germs contained in the sputum are dried by the air and can then be blown about in the wind and inhaled by some unsuspecting person. Tuberculosis patients are taught their responsibility with regard to sputum disposal.

Pneumoconioses

This is a group of conditions that arise in workers as a result of the dust hazard to which they are exposed. Everything possible is done to reduce this hazard to a minimum, and it is our duty to encourage people to use the protective measures supplied for them whilst they are at work.

Pulmonary Tuberculosis

Great credit is due to the Public Health Authorities for the part they have played in the decreasing incidence of this disease over the past years. Vigilance is still necessary to decrease this still further. Early detection is so important that the public needs encouragement to use the mass radiography unit when it visits locally. Children at 13 years of age are offered a Heaf test and if they show no resistance to the tubercle bacillus they are offered a B.C.G. vaccination to give them resistance. School leavers are encouraged to visit the mass radiography unit. Sanitoria are provided where those suffering from the disease can be nursed back to health, without endangering other citizens. There is also an extensive follow-up programme for these people when they leave hospital. (All people who have been in contact with a newly diagnosed tuberculosis patient are traced and encouraged to have a chest X-ray and medical examination.)

Cancer of the Lung

The incidence of this disease is on the increase, and much publicity is being given to the part played by cigarette smoking in the production of this condition. Deaths from lung cancer each year were two for every 5 000 of the population two decades ago; currently the numbers are five for every 5 000. Only one-fifth of the patients seen by doctors are in a sufficiently early stage to benefit from surgery. Dependence on tobacco is discussed on page 364.

Local authorities and hospitals vary in their response to the Minister's plea to discourage smoking. Some of them conduct smoke-free meetings; some display in public: (1) No Smoking notices, and (2) Posters illustrating the dangers of smoking; some have withdrawn the sale of cigarettes from their premises, and some conduct clinics to help those people who want to rid themselves of their addiction to cigarettes. The Central Council for Health Education sends two mobile vans out into the country to help local authorities inform the public about the dangers to health involved in smoking. Young university graduates who are trained to talk to school children, young people and adults drive the vans, which are equipped with film and filmstrip projectors, tape recorders, and exhibition material.

SUMMARY OF THE RESPIRATORY SYSTEM

THE NOSE

This external organ is lined by an extensive mucous membrane that warms, moistens and filters the air passing through. It is the organ of smell and gives resonance to the voice. It receives ducts from the paranasal sinuses, and the nasolacrimal duct. Posteriorly it leads into the pharynx.

THE PHARYNX

Placed at the back of the nose, mouth and epiglottis this cylindrical tube allows passage of food and air. It is richly supplied with reticular tissue, especially in the roof. There are seven openings on its walls. Can you name them? What is the function of the pharyngotympanic tube? Can you describe a child with 'adenoids'?

THE LARYNX

This anterior structure in the neck is the voice box. With its cartilaginous walls it can remain patent during inspiration and expiration. It is the main sensory area for initiation of the cough reflex. It sweeps mucus upwards.

367

These tubes continue to be lined with ciliated mucous membrane and have C-shaped rings of cartilage to maintain patency. Part of the trachea is in the neck and in close contact with the thyroid gland. It is in this portion that tracheostomy can be performed. The remainder is in the upper thorax, with a lung on either side. At midsternal level the trachea divides into two bronchi, one going to each lung. The left bronchus is longer and more horizontal than the right to accommodate the heart; consequently inhaled foreign bodies find their way more easily into the right bronchus.

THE LUNGS

These two spongy cone-shaped organs occupy the major portion of the thoracic cavity. The space between them is called the mediastinum, and contains the heart and great blood vessels and all tubes passing to and from the abdomen. The lungs are made up of ever-branching bronchioles; at the end of each there is an alveolar duct, from which many alveoli open rather like a balloon and in the shape of a bunch of grapes. These have a rich blood supply from the pulmonary arteries, the blood and the air being in close contact, thus permitting the air to pass 4 per cent oxygen into the blood and gain 4 per cent carbon dioxide from the blood. The bronchioles, alveoli and blood vessels are supported on elastic connective tissue, which together with nerves, lymphatic vessels and nodes, form the substance of the lungs. Each lung is closely invested with visceral pleura (serous membrane) which is reflected at the root or hilum to form a sac. The reflected portion (parietal pleura) is adherent to the chest wall, the potential space between the two layers being the pleural cavity. The vital capacity of the lungs is 3 to 4 litres of air, but only $\frac{1}{2}$ a litre is changed with each quiet respiration. As well as the gaseous exchange in the lungs, heat and moisture are lost from the body. Bacteria are expelled and can be dangerous if infection is present. The stimulus for inspiration is of chemical origin from the acidity of the blood passing over cells in the aortic arch and carotid artery, the impulse being transmitted to the respiratory centre in the medulla. The stimulus for expiration is of nervous origin and arises from stretching of the nerve endings in the alveolar wall; this cuts out the impulses which produced inspiration and, by elastic recoil and relaxation of muscle, expiration is produced. The main muscles of respiration are the external and internal intercostals, and the diaphragm, the latter assisting in the expulsive actions of the body, viz. sneezing, coughing, vomiting, micturition, defaecation and parturition.

CARE OF RESPIRATORY SYSTEM IN THE MAINTENANCE OF HEALTH

Safety for a baby's breathing must be considered in relation to the type of pillow, pram and chair harness and walking reins. Good habits of deep breathing and exercise, nose blowing and covering the nose and mouth during sneezing should be established during childhood.

Smoking is now known to play a multifactorial role in the production of disease so that abstinence is to be desired, especially during pregnancy.

Several conditions can interfere with the respiratory tract such as the common cold, enlarged adenoids, any infections producing sputum, pneumoconioses, pulmonary tuberculosis and cancer of the lung.

The Urinary System

We have already discussed some of the organs via which the body gets rid of waste products of metabolism, namely the skin and the lungs. We have discussed the colon which gets rid of indigestible food that the body cannot utilize, so that we are ready to discuss the urinary system which excretes the waste products of protein metabolism, amongst its other functions.

The Kidneys

These two bean-shaped organs lie at the back of the abdominal cavity with their hili facing the lower thoracic and the upper lumbar vertebrae. The kidneys are supported in a pad of fat—perirenal fat—that from animals being purchased as suet. The right kidney is slightly lower than the left, probably owing to the large space occupied by the liver on the right side.

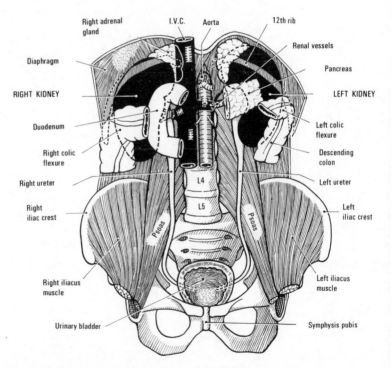

Fig. 193 Position and relationships of the kidneys.

Anteriorly. A portion of liver, duodenum and the right colic flexure lie in front of the right kidney. A portion of the pancreas touches the mid-anterior surface of the left kidney, and then the left part of the transverse colon, the left colic flexure, and the upper part of the descending colon lies in front of the rest of the left kidney (Fig. 193).

Posteriorly. The strong loin muscles, especially the iliopsoas, lie behind the kidneys. The upper part of both kidneys lie on the inner surface of the respective twelfth ribs, which fact you will need to remember when getting the instruments ready for the operation of removal of the kidney.

Superiorly. An adrenal gland lies on top of each kidney.

Inferiorly. Coils of small intestine supported on their mesentery lie below each kidney.

Medially. The vertebral column lies between the two kidneys. Immediately in front of it are the two great vessels, viz. the inferior vena cava on the right and the aorta on the left. These give and receive the arteries and veins to and from the kidneys. The receptaculum chyli also lies in front of the lumbar vertebrae.

The Urinary System

STRUCTURE

The kidney is enclosed within a capsule of white fibrous tissue. Next comes the cortex, granular in appearance to the naked eye, then a portion, the medulla, striped in appearance to the naked eye, and then a space, from the base of which a tube (the ureter) leads to the urinary bladder. This space is called the kidney pelvis, and its lateral border is bounded by the medial surface of the medulla, which presents five or more major openings (major calyces), into each of which open five or more minor calyces. (A glove can be used for illustration,

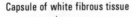
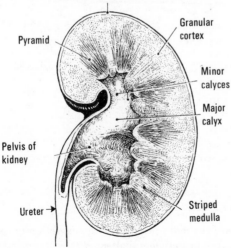

Fig. 194   Naked eye appearance of a sagittal section of the left kidney.

371

the cuff being likened to the major calyx, and the fingers and thumb to the minor calyces.) A cone (pyramid) made up of straight collecting tubules, converges on each minor calyx, and so urine is poured into the pelvis to be transmitted to the ureter and thence to the bladder.

As the brick is the basic unit of the house, so the nephron is the basic unit from which the kidney is constructed. There are millions of these nephrons and they are supported on a connective, interstitial tissue. This connective tissue also serves to support the elaborate network of blood vessels, which are necessary for the functioning of the kidney.

A nephron. This starts as an invaginated cup called Bowman's capsule. Entering this invagination is a wide afferent arteriole from the renal artery. Within this Bowman's capsule the arteriole breaks up into a capillary network called a glomerulus. The Bowman's capsule together with the glomerulus is known as a Malpighian corpuscle (Fig. 195). The network of capillaries reunite to form an efferent arteriole, which leaves the Bowman's capsule and breaks up into a second set of capillaries along the urinary tubule. This efferent vessel is of much narrower bore than the afferent vessel, so that tremendous *pressure* is created in the blood passing through the capillary plexus, sufficient to *force* the liquid part of the blood out into the urinary tubule. The Bowman's capsule is made of one layer of epithelial cells, and this is the pattern throughout the nephron, so that exchange of fluid can take place easily. Leaving the base of the Bowman's capsule (rather like the stem of a wine glass) is the first convoluted part of the tubule; a loop of Henle then dips down into the medulla, returns to the cortex to become the second convoluted tubule, which drains into a straight collecting tubule, which in turn drips the urine into a minor calyx, to pass through a major calyx into the kidney pelvis and down the ureter into the urinary bladder.

BLOOD SUPPLY

Renal artery, a branch from the abdominal aorta.

VENOUS DRAINAGE

Via renal vein into the inferior vena cava.

NERVE SUPPLY

Grey motor neurones from the autonomic nervous system to the blood vessels.

FUNCTION

The kidneys are the main organs for the excretion of urea, which is the end product of protein metabolism. You will remember that there is ·02 per cent urea dissolved in blood plasma, and whatever the

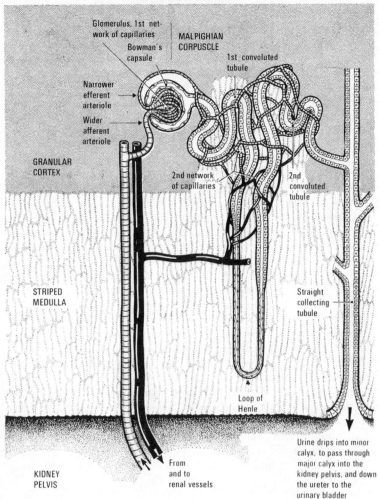

WHITE FIBROUS CAPSULE

Glomerulus, 1st net-
work of capillaries
Bowman's
capsule

MALPIGHIAN
CORPUSCLE

1st convoluted
tubule

Narrower
efferent
arteriole

Wider
afferent
arteriole

GRANULAR
CORTEX

2nd network
of capillaries

2nd
convoluted
tubule

STRIPED
MEDULLA

Straight
collecting
tubule

Loop of
Henle

Urine drips into minor
calyx, to pass through
major calyx into the
kidney pelvis, and down
the ureter to the
urinary bladder

KIDNEY
PELVIS

From
and to
renal vessels

The Urinary
System

Fig. 195 Diagram of a nephron.

protein intake there is a minimum of 2 per cent urea dissolved in urine. This is attributed to the fact that 100 litres of fluid are forced out into the Bowman's capsules each day, and only one to one and a half litres of urine are passed daily; there must be tremendous reabsorption of fluid between the rest of the kidney tubule and the second set of capillaries, without reabsorption of urea; this could account for the increased percentage in the urine. There are other waste products from protein metabolism, urates, creatinine and uric acid which are excreted in the urine.

Other substances normally present in the blood can be removed when they are in excess, for example glucose, sodium, chloride, potassium, calcium, and magnesium. This is the balancing mechanism whereby the body's electrolytes are kept at a constant level.

373

The kidneys help to regulate the water balance of the body. In hot weather when more perspiration is being secreted so that the body can be cooled by its evaporation, the kidneys secrete much less urine. Conversely, in cold weather when perspiration is at its minimum the kidneys secrete much more urine. An excessive fluid intake leads to the passing of a large amount of dilute urine. Diuretics open up more glomeruli and cause an increased flow of urine.

The blood must be kept at a very slightly alkaline reaction and the kidneys play a parge part in the maintenance of the body's acid-base balance by excreting excess acids and alkalis.

BLOOD PRESSURE IN THE KIDNEY

About one litre of blood passes through the kidneys each minute. Over half of this is plasma and most of the plasma—minus the plasma proteins—is *forced* from the glomeruli into the Bowman's capsules. It is necessary for reabsorption to take place along the tubules, leaving the waste products behind. If anything should happen to cause a general lowering of blood pressure throughout the body, or a local lowering of blood pressure in the kidney, this first *forcing* step in the formation of urine cannot take place. Insufficient filtrate is presented to the nephrons for treatment, and the waste products accumulate in the blood. To a limited extent the kidneys are able to correct a fall in general blood pressure, and so ensure an adequate flow of blood through the glomeruli. When the blood pressure within the kidney falls below a certain level, the kidneys secrete an enzyme called *renin,* which activates a substance in blood—called *angiotensin,* which causes constriction of arterial muscle walls. This raises the blood pressure. Angiotensin also increases the output of *aldosterone*, from the adrenal cortex, which in turn encourages reabsorption of salt from the nephrons. Hence the osmotic force exerted by the blood is raised, tissue fluid enters the circulatory system in greater quantities and further raises the blood pressure.

HORMONES INFLUENCING URINE FORMATION

Antidiuretic hormone. This is secreted by the posterior pituitary gland whenever tissue fluid becomes too dilute. Secretion of antidiuretic hormone stops whenever tissue fluid becomes too concentrated. Secretion is controlled by osmoreceptors in the midbrain and hypothalamus. Shrinkage of these osmoreceptors occurs when tissue fluid is too concentrated and this sets up impulses resulting in the greater secretion of antidiuretic hormone. The hormone is then carried to the kidneys causing greater absorption of water from the tubules. As the amount of water in the blood increases, and later the amount in tissue fluid, the osmoreceptors swell up and nerve impulses cause diminished secretion of antidiuretic hormone.

Aldosterone. This, secreted by the adrenal cortex, causes reabsorption of sodium from nephrons. It is made whenever concentration of salt in tissue fluid begins to fall. As salt is reabsorbed and concentration in tissue fluid returns to normal, aldosterone is reduced.

Cortisol. This causes renal excretion of calcium and phosphate.

RENAL ERYTHROPOIETIN

Kidneys make this substance which influences formation of red blood cells. In some diseases the tissue which makes this substance becomes diseased, and insufficient is produced, resulting in anaemia. In other diseases, secretion of erythropoietin is increased and there are far too many red blood cells—polycythaemia.

URINE

Composed of 96 per cent water in which is dissolved 2 per cent urea and 2 per cent other salts. The pH varies with the needs of the body. Between the kidneys and the lungs, the body's pH is kept constant. The amount varies according to fluid intake, the weather and muscular activity.

The Ureters

These are two long tubes leading from the kidney pelvis on the posterior abdominal wall, descending on either side of the vertebral column and passing forward along the pelvic brim, to enter obliquely the posterior base of the urinary bladder, which lies immediately behind the symphysis pubis. The ureters are constructed so that the urine passes along them by peristaltic action.

There is an inner lining of mucous membrane supported on a submucosa. The next layer is of plain, involuntary muscle tissue, the inner fibres being arranged in circular fashion so that on contraction they constrict the lumen and force the urine onwards. Relaxation of these fibres widens the tube. The outer fibres are arranged in a longitudinal fashion so that on contraction they shorten the tube. The outer layer is of white fibrous tissue.

BLOOD SUPPLY

Ureteric arteries.

VENOUS DRAINAGE

Via ureteric veins into the inferior vena cava.

NERVE SUPPLY

Grey motor neurones from the autonomic nervous system supply the plain muscle.

FUNCTIONS

To conduct urine from the kidney pelvis to the bladder.

The Urinary Bladder

This distensible organ is, when empty, a content of the pelvic cavity. It has a fixed triangular base on the pelvic floor. The base of the triangle lies posteriorly and is marked by the entrance of a ureter on either side. The apex of the triangle lies anteriorly, and is marked by the exit of the urethra, the tube conducting urine from the bladder to the exterior. The remaining portion of the bladder is called the fundus or apex, and it rises and falls according to its contents. When full it protrudes into the abdominal cavity and lies against the anterior abdominal wall (Fig. 196B). A surgeon can incise the full bladder via the abdominal wall, without risk of contamination of the peritoneal cavity. Conversely, it is essential to empty the bladder before any other abdominal operation. In the male the bladder lies in front of the rectum, in the female the uterus separates the bladder from the rectum.

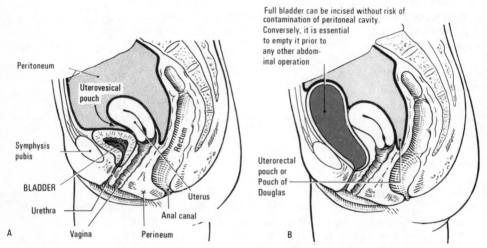

Fig. 196 Position and relationships of the female urinary bladder. A, When empty. B, When full.

STRUCTURE

The bladder is lined with transitional epithelium to allow distension and protect the underlying tissues against penetration of urine. The lining is supported on a submucosa of areolar tissue carrying blood and lymphatic vessels and nerves. Next comes a layer of plain, involuntary muscle fibres (detrusor muscle), which relax to accommodate incoming urine, and contract to expel urine. The outer coat of peritoneum is incomplete and only present on the posterior and superior surfaces. Figure 196 illustrates that the rising bladder pushes the peritoneum upwards, whilst the bladder's anterior surface is in contact with the anterior abdominal wall.

BLOOD SUPPLY

The vesical arteries from the internal iliac artery.

376

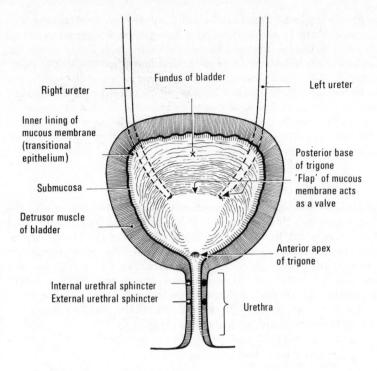

Fundus of bladder

Right ureter

Left ureter

Inner lining of
mucous membrane
(transitional
epithelium)

Posterior base
of trigone

'Flap' of mucous
membrane acts
as a valve

Submucosa

Detrusor muscle
of bladder

Anterior apex
of trigone

Internal urethral sphincter
External urethral sphincter

Urethra

Fig. 197   Diagram of the urinary bladder.

VENOUS DRAINAGE

Via the vesical veins into the internal iliac, which joins the external
iliac to form the common iliac, the two common iliacs joining to
form the inferior vena cava.

NERVE SUPPLY

Sensory neurones which respond to distension or pressure, connect
with the micturition centre in the lower spinal cord. Motor neurones
from this micturition centre supply the muscle in the bladder wall
and urethral sphincters.

FUNCTION

To act as a reservoir for incoming urine until there is sufficient to
produce a stimulus for micturition.

The Urethra

This tube conducts urine from the bladder to the exterior, and in the
female is only 4 cm (1 to 1½ in) long, whilst in the male it is 15 to 18 cm
(6 to 7 in) long, since it has to travel through the penis. The urethra

377

is lined with transitional epithelium supported on a submucosa. The circular fibres in the muscle layer are gathered together into two sphincters, the internal one being of plain fibres at the exit from the bladder, and the external one of skeletal fibres just below. The outer covering is of white fibrous tissue. The functioning of the sphincters is to remain tightly closed whilst urine is collecting in the bladder, and to relax in response to a sensory stimulus, so that urine can be voided. It must be remembered that in the male the urethra is also the organ for transmission of semen, but this will be discussed more fully with the reproductive system.

REFLEX ACT OF MICTURITION

In animals and babies the procedure is purely reflex, but in man the desire can be inhibited until it is convenient to void the bladder. Collection of 210 to 300 ml (7 to 10 oz) of urine in the bladder stimulates the autonomic sensory nerve endings and the impulse passes to the micturition centre of the spinal cord. From here impulses are transmitted in motor neurones, and these cause the urethral sphincters to relax whilst the muscle in the bladder wall contracts to force the urine out. The act is assisted by the descent of the diaphragm increasing intra-abdominal pressure.

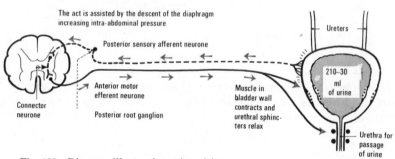

Fig. 198 Diagram illustrating micturition.

CARE OF THE URINARY SYSTEM IN THE MAINTENANCE OF HEALTH

Correct nourishment is essential for the health of all tissues, and it is especially important to partake of adequate fluids, to decrease the possibility of any of the substances in solution crystallizing in the urine. It is important that 'keeping dry' is not demanded of a child before his neuromuscular mechanism is sufficiently well developed for him to be physically capable of complying without mental stress. Sometimes a child wets the bed as an unconscious manifestation of insecurity, and this should never be treated as 'bad' behaviour, but conscious measures should be sought of building up the child's security. Children should be taught their responsibility to respond to the desire to pass urine, as overdistension of the bladder can lead to back pressure in the kidneys. It is for this reason that children

378

should be encouraged to be independent when visiting the lavatory before they go to school. Also it is rank unkindness to expose a child to the ridicule of his school mates, because he has not been allowed to.acquire independence in this direction.

Little girls should be taught to perform post-defaecation toilet in a posterior direction to avoid soiling of the urethral orifice. (Inflammation of the urinary bladder (cystitis) is much more common in women, because the urethra is only 2½ to 4 cm (1 to 1½ in) long.)

Drugs are excreted by the kidney. Self-medication is *not* recommended. Take only those drugs ordered by the doctor.

Kidney failure in young adult life can be due to a long, insidious infection of the kidney. The National Kidney Research Fund suggest that all children up to 5 years old should have routine urine examinations one month after birth, at 2 years and again on starting school. The Fund considers that this would be a great advance in preventive medicine.

SUMMARY OF THE URINARY SYSTEM
The urinary system comprises two kidneys, two ureters, one bladder and one urethra.

The Kidneys
These lie on the posterior abdominal wall, one on either side of the vertebral column. They are bean-shaped with a medial depression where the arteries enter and the veins and ureters leave. They are surrounded by perirenal fat, and are enclosed within a fibrous capsule, below which is a granular cortex, from which extends a striped medulla which borders the space called the pelvis of the kidney. Kidneys are composed of millions of nephrons supported on interstitial connective tissue. Each nephron is composed of an invaginated Bowman's capsule into which fits a glomerulus of blood capillaries, the two together being known as a Malpighian corpuscle. Leaving the Bowman's capsule is a first convoluted tubule, a loop of Henle, and a second convoluted tubule entering a straight collecting tubule, which pours its contents into a minor calyx, to pass through a major calyx and the pelvis on its way to the ureter. A wide afferent arteriole enters the glomerulus, and a narrow efferent arteriole leaves the glomerulus to break up into a second capillary network around the rest of the urinary tubule, for purposes of reabsorption. The difference in bore of the entering and leaving arteriole is responsible for creating the pressure in the blood whereby the fluid part is driven into the Bowman's capsule. One hundred litres arrive in the capsules in this way, and yet only 1 to 1½ litres of urine are excreted daily. Along the tubules much of this fluid is reabsorbed, so that in health the urine contains only those substances that the body wants to get rid of. Secretion of urine is controlled

by the antidiuretic hormone and aldosterone. The enzyme renin is the controller of the amount of the hormone aldosterone. Renin rises in conditions of low sodium (and low blood pressure—and they are often associated). Renin acts on angiotensin, a vessel constrictor in the blood. Vasoconstriction raises blood pressure. Renin also increases aldosterone, which conserves sodium, so that by this twin mechanism, sodium and blood pressure are raised to normal levels. Cortisol causes renal excretion of calcium and phosphate.

FUNCTIONS OF THE KIDNEYS

1. They produce the filtrate—urine, from the blood.

2. They help to regulate the body's water balance.

3. They assist in the adjustment of the electrolyte content of the plasma.

4. They assist in the adjustment of the acidity and alkalinity of the blood, so that only a slight alkalinity is maintained.

5. They eliminate urea, urates, creatinine and uric acid from protein metabolism.

6. They excrete the end products from drugs and toxins, and for this reason fluid intake should be increased in fevers.

7. They help in the maintenance of normal blood pressure.

The Ureters

Two long, muscular tubes that convey urine by peristaltic action to the posterior base of the bladder.

The Bladder

A distensible organ which, when empty, is a content of the pelvic cavity. It has a fixed triangular base on the pelvic floor, and is lined by transitional epithelium, supported on a submucosa. The next layer is of plain, involuntary muscle. The serous membrane, peritoneum, is only present on the posterior and superior surfaces, so that as the bladder rises it pushes this peritoneum upwards and the anterior bladder comes to rest against the anterior abdominal wall, from whence it can be pierced; thus it is necessary to send a patient who has to have an abdominal operation to the operating theatre with an empty bladder.

The Urethra

This tube conducts urine from the bladder to the exterior. In the female it is $2\frac{1}{2}$ to 4 cm (1 to $1\frac{1}{2}$ in) long, and in the male 15 to 18 cm (6 to 7 in) long. The circular muscle fibres are arranged in an internal and an external urethral sphincter. A sensory stimulus from the

380

bladder causes these to relax whilst the bladder muscle contracts to void urine, in the reflex act of micturition which can be controlled in man.

The Endocrine System

The ability of the human subject to take in nourishment and oxygen, and to excrete waste products has been described. The mechanisms involved are, however, extremely complex and so a highly efficient and yet readily adaptable control system is required. Such a function is provided by the endocrine glands acting with the nervous system.

The word 'endocrine' means secreting internally, and therefore an alternative name for an endocrine gland is a ductless gland. The secretions produced by these glands are absorbed directly into the bloodstream and are called hormones. These hormones are substances which have highly specific effects on nearby or distant tissues and organs and so are often referred to as 'chemical messengers'.

The endocrine glands in the human body are the pituitary, thyroid, parathyroids, pancreas, adrenals, ovaries, testes, thymus and pineal. The pancreas has both an internal (endocrine) and an external (exocrine) secretion (p. 291).

The last two decades have brought greatly increased knowledge of the endocrine system and the nervous control of it. The latter is exerted by cells in the *hypothalamus,* just above the pituitary gland. The cells are sensitive to the level of the different hormones in the blood passing through them. *Below* a particular level, the cells secrete a *releasing factor* for that particular hormone. The releasing factor passes in special blood vessels (called the pituitary portal system) down to the anterior lobe, where the deficient hormone is secreted (Fig. 199). The two posterior lobe hormones, are actually secreted in the hypothalamus and pass down to be stored in the posterior lobe. When the level of the hormone in the general circulation rises above the optimum, secretion of the releasing factor, and the two posterior hormones, ceases temporarily. Conforming with modern terminology, this is called a *feed-back* mechanism, and it serves to prevent inappropriate secretion of any one hormone.

The Pituitary Gland (Hypophysis)

POSITION AND RELATIONSHIPS

The pituitary is already familiar to you. This gland rests in the pituitary fossa of the sphenoid bone and is bounded superiorly by the tough diaphragma sellae, a lamina of the dura mater, which separates the gland from the hypothalamus at the base of the brain. The pituitary gland should, therefore, be added to the list of organs contained in the cranial cavity.

382

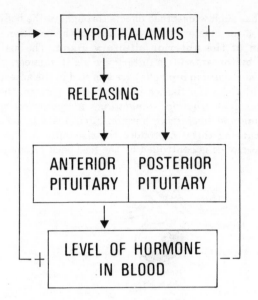

Fig. 199   Feedback mechanism.

STRUCTURE

The adult pituitary gland weighs approximately 0·6 grams, and is
ovoid in shape. It is clearly divided both anatomically and functionally into a larger anterior and a smaller posterior lobe. The former is
derived embryonically from the posterior pharynx and the cells
comprising it are of glandular type. The posterior lobe, on the other
hand, is of nervous origin and results from a downgrowth from the
third ventricle of the brain. The gland is connected to the hypothalamus just behind the optic chiasma by the pituitary stalk or
infundibulum which contains important nerve fibres from the hypothalamus to the posterior lobe and vascular communications
(pituitary portal system) between the hypothalamus and the anterior
lobe.

BLOOD SUPPLY

Hypothalamo-hypophyseal branches from the internal carotid artery
(part of the circle of Willis).

VENOUS DRAINAGE

Via hypophyseal branches into the cavernous sinus, leaving the
skull by the jugular vein.

FUNCTIONS

To supply the body with many hormones as illustrated in Figure 200.

383

Since it has such widespread effects throughout the body, no wonder it is called the master gland, or 'leader of the endocrine orchestra'. **Function of the anterior pituitary gland.** The main hormones secreted by the anterior pituitary are *six* in number. Two of these (*growth hormone* and *prolactin*) are secreted in the anterior pituitary, and circulate to the tissues in which they perform their function. The other four (*thyroid stimulating, adrenocorticotrophic, follicle stimulating* and *luteinizing* hormones) circulate to other endocrine glands, causing them to produce their secretions. The secretion of all six hormones is controlled by the feed-back mechanism.

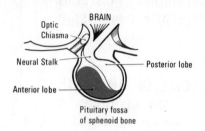

HYPOTHALAMUS

RELEASING FACTOR

GROWTH HORMONE	PROLACTIN	THYROID STIMULATING HORMONE	ACTH	FOLLICLE STIMULATING HORMONE	LUTEINIZING HORMONE	ANTIDIURETIC VASOPRESSIN	OXYTOCIN
All tissues		Thyroid gland	Adrenal	Ovaries and testes	Ovaries	Kidneys	Uterus and breast
Raises blood sugar. Stimulates fat break-down. Stimulates protein synthesis.	Prepares pregnant woman's breasts for milk production. Suckling causes milk to be secreted.	Stimulates production of thyroxine.	Stimulates production of cortisol.	Stimulates ovum-containing follicles in females and production of sperms in males.	Stimulates ovary to secrete oestrogen and progesterone. Forms corpus luteum.	Effects water balance by regulating absorption in kidney tubules.	Causes contraction of uterus and expulsion of milk from breast in response to suckling.

ANTERIOR LOBE OF PITUITARY   POSTERIOR LOBE

Fig. 200   Pituitary gland and hormones.

Growth hormone. This has an effect in all tissues of the body. It not only raises blood sugar as described below, but it stimulates fat breakdown and protein synthesis. If the feed-back mechanism is such that there is deficiency during childhood, a dwarf results. If there is too much during childhood, a giant results. If the over-secretion occurs in adult life, the soft tissues and the bones of the hands, feet and face enlarge. The tongue enlarges and may protrude from the mouth. The condition is called acromegaly (Fig. 201).

The growth hormone is also important in the regulation of blood sugar. Some people identify this function separately and talk about a diabetogenic hormone. The action is *antagonistic to that of insulin.* Its action is therefore necessary *after a meal has been absorbed,*

384

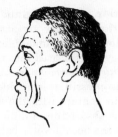
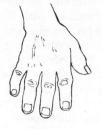

A. EXCESSIVE GROWTH OF BONES OF FACE AND THICKENING OF SKIN

B. SPADE-LIKE HANDS

Fig. 201   Acromegaly.

when metabolism changes from mainly carbohydrate oxidation to mainly fat oxidation. When the growth hormone level is persistently high, as in acromegaly, diabetes frequently occurs. The islets in the pancreas become exhausted by trying to secrete sufficient insulin to keep the increased blood sugar from the extra growth hormone, down to normal level.

Prolactin. Prolactin is the second hormone secreted in the anterior pituitary. It acts only on the pregnant woman's breasts. It prepares them for milk production. Once birth has taken place, suckling stimulates sensory nerves in the nipples, impulses are transmitted to the hypothalamus, and milk is secreted. When the baby is no longer put to the breast, secretion of milk stops.

The thyroid stimulating hormone (TSH). This hormone is secreted in response to a thyroid releasing factor (TSF), reaching the anterior pituitary in blood from the hypothalamus. Thyroid stimulating hormone is trophic and has its effects on the thyroid, stimulating it to secrete its own hormone, thyroxine, which has an effect on every cell in the body. If thyroxine rises above an *optimum* level, secretion of the releasing factor is inhibited.

The adrenocorticothrophic hormone (ACTH). The adrenocorticothrophic hormone is secreted in response to its releasing factor (ACTF) from the hypothalamus. Adrenocorticotrophic hormone is trophic and has its effect on the adrenal cortex causing it to secrete cortisol. Again the feed-back mechanism inhibits the releasing factor when the level of cortisol is above the optimum.

The follicle stimulating hormone (FSH). This is secreted in response to its releasing factor (FSF), and has the same feed-back mechanism to control the supply in the blood. It is trophic to the ovaries in the female, where it is essential for the development of the ovum-containing follicles; and to the testes in the male, where it is responsible for production of sperm.

The luteinizing hormone (LH). This hormone follows a similar pattern of secretion control. It is also trophic to the ovary in the female, where it controls the secretion of oestrogen and progesterone —responsible for the formation of the corpus luteum.

385

Functions of the posterior pituitary gland. The two hormones associated with the posterior pituitary are actually secreted in the hypothalamus and pass down the infundibulum into the posterior pituitary, from whence they are released into the blood stream.

First there is the *antidiuretic hormone* of which we have spoken (p. 374). It is also called vasopressin. It is produced in response to the concentration of plasma in contact with the secreting cells in the hypothalamus. It regulates absorption of fluid in the renal tubules, and thus helps to regulate the body's water balance. Absence of this secretion causes passage of a large amount of pale, dilute urine and complaints of thirst—diabetes insipidus.

The second hormone is *oxytocin,* and it too is secreted in the hypothalamus, passes down the infundibulum into the posterior pituitary, to be released into the blood stream as needed. Oxytocin is thought to be responsible for the sensation of orgasm, which in the female is accompanied by contractions of the uterus, during sexual intercourse. Though oxytocin is present in the male, its function is uncertain. During pregnancy in response to sensory impulses from the uterus, there is increased secretion of oxytocin which causes more vigorous contraction of the uterus thus helping to expel the baby. Oxytocin also allows expulsion of milk from the breast in response to sucking of the nipple. The sucking, in fact, causes the secretion of oxytocin, which is why the baby does not get milk at the first few sucks.

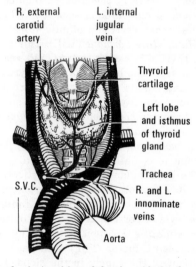

Fig. 202   Position and relationships of the thyroid gland.

The Thyroid Gland

POSITION AND RELATIONSHIPS

This gland has three well-defined parts, two lobes which are joined across the midline by an isthmus. The thyroid therefore clasps the upper part of the trachea and extends upwards on either side of the larynx (Fig. 202).

The thyroid is composed of many closed vesicles, which are lined by cuboidal epithelial secretory cells. The vesicles are filled with a semi-fluid colloid substance which acts as a reservoir for the thyroid hormone, thyroxine, which is produced by the secretory cells in the presence of iodine. The vesicles are held together by a delicate fibrous network, the stroma of the thyroid. Throughout the thyroid tissue in a parafollicular position (Fig. 203) there are clumps of large cells with clear cytoplasm—'C' cells.

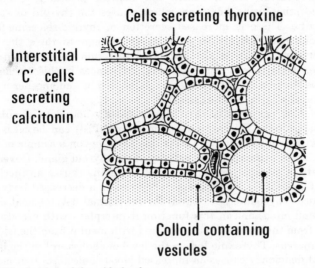

Cells secreting thyroxine

Interstitial 'C' cells secreting calcitonin

Colloid containing vesicles

Fig. 203   Section of thyroid gland.

BLOOD SUPPLY

Two pairs of arteries are involved. The superior thyroid arteries which arise from the external carotid arteries and the inferior thyroid arteries from the subclavian arteries.

VENOUS DRAINAGE

Via branches into the jugular and brachiocephalic veins.

NERVE SUPPLY

Grey fibres from the cervical ganglia of the sympathetic trunk and from the vagus nerves.

FUNCTIONS

A recently identified hormone called *calcitonin,* is thought to be secreted by both the thyroid and the parathyroid glands. Its exact place in health and disease is, as yet, incompletely understood. It has two possible roles: control of serum calcium levels and regulation of bone mineralization. Calcitonin, therefore, may be of considerable importance in the aetiology and treatment of some types of degenerative bone disease and also in the essential repair mechanisms of bone following trauma.

The thyroid gland secretes *thyroxine*, an iodine-containing compound that regulates the *rate* of metabolism. It is stored in the colloid and released to the blood as needed. In response to lowered thyroxine in the blood, nuclei in the hypothalamus secrete the thyroid-releasing-factor, which causes the anterior pituitary to secrete thyroid stimulating hormone, which causes the thyroid to secrete thyroxine. There is increased secretion of thyroid-releasing-factor in prolonged exposure to cold, in an attempt to raise the body temperature. There is also increased secretion in prolonged food intake. It is thought that individual differences in this mechanism may be the reason why one person gains weight and another merely maintains weight, on the same diet.

About 90 per cent of thyroxine is closely bound to circulating plasma protein. This protein-bound iodine (PBI) can be measured. You will meet patients who have this test done on a sample of their blood, as an estimate of the activity of the thyroid gland. Governing the rate of metabolism, an excess of thyroxine causes an increased body temperature, while a deficiency causes a decreased body temperature. Thyroxine is essential for the normal development of the *body* and *mind,* and in this function it interplays with growth hormone from the anterior pituitary, and with insulin from the islets of the pancreas. Thyroxine lowers the level of cholesterol in the blood, so that deficiency causes an increased blood cholesterol that may be a disadvantage for the cardiovascular system. Estimation of blood cholesterol is a useful diagnostic tool. Thyroxine is needed for the conversion of carotene into vitamin A. This provides another useful diagnostic criterion, i.e. a yellow tinge to the skin of people who are deficient in thyroxine. Thyroxine potentiates the effectiveness of adrenaline and the sympathetic nervous system, so that there is a rapid heart beat with increased thyroxine, and a slow heart beat with decreased thyroxine.

If an undersecreting thyroid is present at birth, the child is called a cretin (Fig. 204). The hair is sparse and coarse, the nasal bridge is depressed, and the nostrils are wide. The tongue seems to be too big for the mouth. The skin is coarse and the child is overweight. There is a pot belly and often an umbilical hernia.

It is important to recognize that the child may appear normal at birth, for the fetus has been nourished with maternal blood. The characteristics of cretinism usually become manifest after weaning. The health visitor, either visiting in the home or seeing the baby at the clinic, will note any lack of development, and will advise the mother to take the baby to her general practitioner. If the diagnosis is made sufficiently early and adequate thyroxine treatment given, normal physical and mental development can be attained.

In an adult, lack of thyroxine causes the hair to become thin, the

388

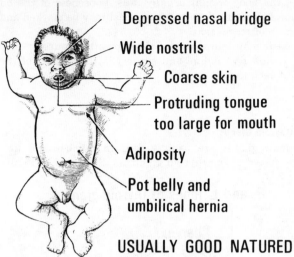

Sparse, coarse hair

Depressed nasal bridge

Wide nostrils

Coarse skin

Protruding tongue
too large for mouth

Adiposity

Pot belly and
umbilical hernia

USUALLY GOOD NATURED

Fig. 204   Infantile cretinism.

skin to coarsen and become dry. Lethargy with slowness of movement, speech and thought is usually marked. Intolerance of cold,
constipation and a tendency to put on weight are also characteristic
features. Hoarseness of the voice may occur and a few patients
accumulate a jelly-like material in the subcutaneous tissues particularly of the face and lower legs which is referred to as myxoedema
and, if present, is pathognomonic of hypothyroidism. These changes
are reversible with adequate thyroxine replacement therapy.

Oversecretion (Hypersecretion; Hyperthyroidism). This condition
is spoken of as hyperthyroidism, toxic goitre, thyrotoxicosis, Grave's
disease or exophthalmic goitre. It presents a picture of excessive
metabolic activity throughout the body (Fig. 205). The patient is

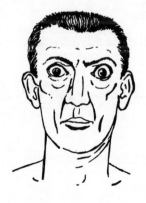

Prominent
eyes—
proptosis

There may
be visible
swelling
in neck

Fig. 205   Hyperthyroidism.

389

tremulous, nervous and cannot be still. In spite of an increased
appetite, the body weight usually falls. Looseness of the bowels is
a common complaint. Intolerance of heat may be marked and undue
sweating is often present.

Treatment may follow one of three possible courses; antithyroid
tablets, radioactive iodine and surgery with removal of the major
portion of each lobe.

The Parathyroid Glands

These small glands are usually four in number and lie on the
posterior aspect of the thyroid gland, as illustrated in Figure 206.

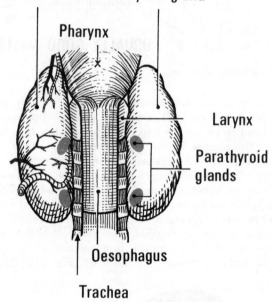

R. and L. lobes of thyroid gland

Pharynx

Larynx

Parathyroid
glands

Oesophagus

Trachea

Fig. 206　Position and relationships of the parathyroid glands.

BLOOD SUPPLY

Branches from the inferior thyroid arteries supply all four glands.

VENOUS DRAINAGE

Via branches into the jugular and brachiocephalic veins.

NERVE SUPPLY

This is similar to that of the thyroid gland.

390

To secrete *parathormone*. A high blood calcium inhibits secretion of the releasing factor in the hypothalamus, and a low blood calcium stimulates secretion. Parathormone releases calcium from bone into blood, at the same time reducing the excretion of calcium via the kidney. Simultaneously it causes renal excretion of phosphate, important because when both calcium and phosphates are present in considerable amounts an insoluble complex is formed. This can damage the tissues, especially arterial walls. Some *calcitonin* is also secreted by the parathyroid glands (p. 387). Should blood become alkaline as from excessive vomiting of acid gastric juice, or when excessive acid carbon dioxide is breathed out as in overbreathing, more calcium is bound to the blood protein and is not available to the tissues for normal neuromuscular activity, so that the condition of tetany ensues. *Tetany* can also result from *hyposecretion* of parathormone. It is characterized by muscular twitching and in severe cases painful spasm of the hands and feet—carpo-pedal spasm. The condition is quickly relieved by the administration of calcium. *Overactivity* of the parathyroid glands raises the blood calcium, and if prolonged, stones may form in the kidney and salivary glands. Calcium may also be deposited in the arterial walls. The extra calcium is drawn from the bone stores, leaving the bones weak and cystic in appearance. The condition is called osteitis fibrosa cystica.

The Endocrine System

The Islets of Langerhans

Scattered throughout the *pancreas* are tiny islands of special cells that secrete hormones directly into the bloodstream. The beta cells produce *insulin*, and the alpha cells produce *glucagon*. These islet cells have nothing whatever to do with the production of pancreatic juice, which is of great importance for normal digestion of food within the intestine; pancreatic juice is conveyed from the pancreas to the duodenum by the pancreatic duct. Revision of the position and relationships of the pancreas can be accomplished by studying Figure 142.

BLOOD SUPPLY

From the splenic, hepatic and superior mesenteric arteries.

VENOUS DRAINAGE

Via many branches into the portal vein.

NERVE SUPPLY

Grey neurones from the autonomic nervous system.

As a direct result of an increased amount of glucose in the blood perfusing the pancreas (as occurs after a meal), the beta islet cells secrete extra insulin. *Insulin* converts soluble glucose into glycogen in the liver and muscles thus lowering blood sugar. Insulin also converts some glucose into fat. Both these substances are storage forms of nutrients. Insulin also stimulates the synthesis of amino acids into protein. As the blood glucose falls some time after a meal, insulin secretion decreases. This helps to stop the cells using glucose, except those of the liver and brain, which must have glucose for energy. Cells other than those of the liver and brain start using fat for energy. Fats are broken down into fatty acids by the growth hormone, and it further helps to stop the cells using glucose by

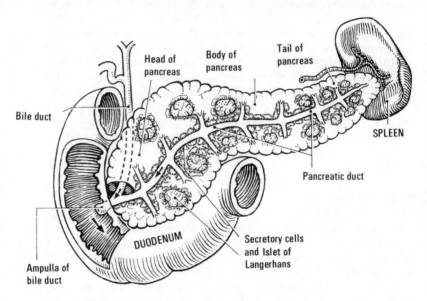

Fig. 207   The pancreas showing the islets of Langerhans.

rendering them impermeable to glucose. Blood glucose level helps to control the amount of growth hormone in the blood. Reciprocal to insulin, growth hormone is increased by a low blood sugar and decreased by a raised blood sugar. Adrenaline can mobilize blood glucose in an emergency, to ensure sufficient glucose for the brain (p. 223). *Glucagon* secreted by the alpha cells of the islets, has the opposite effect to insulin. It raises the blood sugar and facilitates the breakdown of protein into glucose. Relatively little is known about when and why this happens, but it is thought that it occurs in starvation. With this complex interplay of hormones, the normal blood glucose level is maintained at 80 to 120 mg per 100 ml, whether one eats one or four meals a day. Failure of the islets to produce sufficient insulin results in *diabetes mellitus*.

The Adrenal Glands

POSITION AND RELATIONSHIPS

These two small, conical glands lie perched on top of each kidney
(Fig. 208). Their posterior surface is in contact with the muscle fibres
arising from the inner surface of the ribs to form the diaphragm.

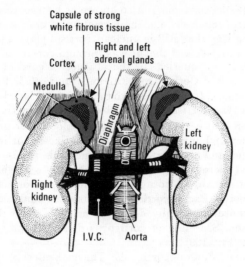

Fig. 208   The adrenal glands.

STRUCTURE

Each gland is encapsulated and shows a distinct outer cortex and
inner medulla, each secreting their own hormones. The medulla is
much smaller than the cortex.

BLOOD SUPPLY

Each gland is supplied by at least three arteries, one from the renal
artery, one from the phrenic artery and finally one from the aorta.

VENOUS DRAINAGE

Each gland has only one vein leaving it. The right adrenal vein opens
into the inferior vena cava, and the left adrenal vein opens into the
left renal vein which empties into the inferior vena cava.

NERVE SUPPLY

To both adrenal glands there is a rich supply of sympathetic fibres
which are derived from the coeliac plexus, but they contain also many
preganglionic fibres, which pass through the coeliac plexus from
the splanchnic nerves and end in relation to the cells of the adrenal
medulla.

393

To secrete corticosteroid hormones. These used to be divided into three groups—the glucocorticoids, mineralocorticoids and the androgens. Now that it is known that their functions overlap, particularly the first two groups, it is customary to use the name of the main hormone in each of the first two groups.

Cortisol (Glucocorticoids). Also known as hydrocortisone. It takes part in gluconeogenesis, i.e. the formation of glucose from protein and fat. In a way that still is unclear, cortisol plays a part in water balance. It was believed that it was through control of sodium content but this is now queried. In the absence of cortisol, water is excreted only slowly. Cortisol plays an important role in the regulation of blood pressure. It influences production of red blood cells. In conditions of stress, cortisol level can be increased five times. At levels higher than those normally found in blood, cortisol has anti-inflammatory and anti-allergic properties. Excess cortisol interferes with collagen and muscles, making them weak. Excess cortisol also causes renal excretion of calcium and phosphate, the blood calcium being maintained by withdrawal of calcium from bone, thus weakening the skeleton. Excess cortisol also raises the blood sugar, much of which is characteristically deposited as fat around the shoulders and abdomen. Secretion of cortisol is dependent on the feed-back mechanism (Fig. 199).

Aldosterone (Mineralocorticoids). The main effect of aldosterone is to increase the renal excretion of potassium, and conserve sodium and chloride. It is thought that the sodium content of the body controls the level of aldosterone. When sodium is low, secretion of the hormone is high. There is controversy as to whether the adrenocorticotrophic hormone plays any part in the secretion of aldosterone. If it does, it is thought to be a very small part. *Renin*, an enzyme secreted by the kidney is a controller of the amount of aldosterone. The amount of renin rises in conditions of low sodium (and low blood pressure—and these are often associated). Renin acts on angiotensin in the blood, a blood vessel constrictor. Vasoconstriction raises the blood pressure. Renin also increases the output of aldosterone, which conserves sodium, so that by this twin mechanism both sodium and blood pressure are raised to normal levels. Conversely, high sodium and blood pressure decrease the secretion of aldosterone by this renin-angiotensin mechanism, thus lowering sodium and blood pressure.

Androgens. Though this word is usually associated with male sex hormones, the androgens are present in females as well. They are responsible for the growth of pubic and axillary hair; the secretion of sebum in the skin; the development of muscle and the production of sexual desire in both sexes.

FUNCTIONS OF THE MEDULLA

To secrete two substances adrenaline and noradrenaline. The latter is also the chemical transmitter of the postganglionic sympathetic nerve fibres. Under usual circumstances the secretion of adrenaline and noradrenaline by the adrenal medulla is too small to have any significant physiological effects. Both of these hormones are, however, potent agents, adrenaline being on the whole more effective than noradrenaline. Both substances increase tissue oxygen consumption and raise the body temperature, the blood sugar rises as a result of increased breakdown of liver glycogen to glucose, and the mobilisation of free fatty acids from adipose tissue is increased.

There are important qualitative differences, however, in their effects on the cardiovascular system. Although adrenaline when applied locally is a potent vasoconstrictor, when given systemically it causes a net vasodilatation, as well as a marked increase in heart rate and cardiac output. The diastolic blood pressure does not rise.

In contrast, noradrenaline causes a net vasoconstriction and as it also has a potent stimulatory effect on the heart, there is a sharp rise in systemic blood pressure. Adrenaline, but not noradrenaline, is a powerful bronchodilator.

The secretion of these substances is increased in any emotion, so that they have been described as preparing the body for flight in response to fear, and fight in response to anger.

It is conventional to include the *thymus* and the *pineal glands* in any discussion of the endocrine glands. The functional importance of these two structures remains far from clear and it is possible that they do not behave as true endocrine glands. There is, however, some indirect evidence that they do secrete hormones although their nature is as yet unknown. It is therefore appropriate to describe these glands in this chapter. Quite apart from this, the thymus is of importance in its recently identified immunological role.

The thymus gland is now considered to be part of the *central lymphatic system*. It is believed to be necessary for the development of the peripheral lymphatic system. In its absence, gammaglobulins are not produced, the tonsils and adenoids do not develop, and death results from lack of immunological response.

The Thymus

POSITION AND RELATIONSHIPS

This organ lies in the upper mediastinum of the thoracic cavity (Fig. 209).

Anteriorly. The upper sternum lies in front of the thymus.

Posteriorly. The lower trachea and its bifurcation into two bronchi together with the great vessels from the heart lie behind the thymus.

Inferiorly. The heart lies below the thymus.

Superiorly. The large vessels at the thoracic inlet are in close appropriation to the upper pole of the thymus.

Laterally. The lungs lie on either side.

STRUCTURE

The thymus when fully developed is divided into an outer cortex and an inner medulla. The cortex is very similar to lymphatic or reticular tissue. The thymus is almost as large as the heart in an infant; it increases in size until puberty, and then the lymphatic tissue is replaced by fibrous fatty tissue in the adult.

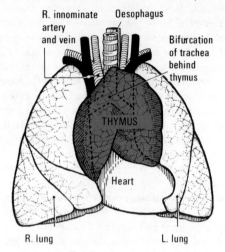

Fig. 209   Position and relationships of the thymus gland in an infant.

BLOOD SUPPLY

Scanty, a few branches from the mammary and thyroid arteries.

VENOUS DRAINAGE

Mainly into the left brachiocephalic vein.

NERVE SUPPLY

A few grey neurones from the autonomic nervous system.

FUNCTIONS

In early life it is essential for adequate development of lymphocytes in the lymph nodes, the spleen and the lymphoid aggregations of the gastro-intestinal tract. As a consequence of diminished lymphocyte formation there is a deficiency of antibody formation and the individual is much more liable to external hazards, particularly infections.

There is also some evidence that in certain diseases, such as, myasthenia gravis, which is characterized by severe muscle weakness, there may be pathological overactivity and enlargement of the thymus with overproduction of abnormal antibodies, harmful to the body's own tissues. This is referred to as a state of auto-immunity. The exact place of the thymus in the causation of this group of disorders is still poorly understood, but in myasthenia gravis, surgical removal of the thymus is not infrequently followed by improvement in the patient's symptoms. It is also possible that the increased susceptibility to carcinoma in human ageing, may depend ultimately on waning thymic function.

The Pineal Body

POSITION AND RELATIONSHIPS

This structure lies in the groove between the superior quadrigemina just above the most posterior part of the third ventricle. It is attached to the wall of the third ventricle by a stalk.

STRUCTURE

It is a small cone-shaped organ and is composed of two types of cells, 'parenchymal' cells which resemble epithelial tissue and neural cells most of which are astrocytes.

BLOOD SUPPLY

The arterial supply is sparse and consists of only a few small branches from the posterior cerebral artery.

VENOUS DRAINAGE

The venous drainage runs to the internal cerebral venous system.

NERVE SUPPLY

The pineal receives a nerve from each superior cervical ganglion of the sympathetic nervous system.

FUNCTIONS

In mammals, recent studies have suggested that the pineal is concerned in the regulation of pituitary-ovarian function. It is thought that this effect is due to inhibition of secretion of *pituitary gonadotrophin*. There is also evidence that the pineal contains a substance *adrenoglomerulotrophin* which stimulates release of aldosterone by the adrenal cortex. In fish and frogs the pineal produces a substance *melatonin,* which lightens the pigmentary cells of the skin.

The physiological significance of these functions in man is controversial. The pineal body normally undergoes calcification in adults and there are no observed effects. On the other hand, absence or hypoplasia of the pineal has been reported in association with upsets of sexual development. Tumours of the pineal have been recognised to be associated with similar upsets. Sexual precocity is associated with 25 to 30 per cent of pineal tumours. This has been found only in boys. Hypogonadism is also found, but it is debatable whether the upset results from abnormal function of the pineal itself or damage to surrounding structures in the brain.

Many of the claims that the pineal is a functionally active organ depend on indirect evidence. Further information is required before a final decision can be made about this enigmatic organ.

The ovaries and testes are, of course, important endocrine glands but they will be discussed with the reproductive system on page 436.

CARE OF THE ENDOCRINE SYSTEM IN THE MAINTENANCE OF HEALTH

The great majority of diseases resulting from abnormality of function of the endocrine system are of unknown cause and so there are few precautions which can be taken to prevent these disorders. One factor, which requires emphasis, however, is the irrefutable association between obesity and diabetes mellitus. The problem of over-weight has now become a major hazard of life in all modern developed countries. Unfortunately preventive measures, and particularly education of the public regarding the serious dangers of obesity, have received all too little attention. In areas where pre-diabetic surveys are offered it is up to the public to take advantage of them and thus preserve their health.

The hormone-dosing of livestock animals intended for human consumption has been questioned. The Food and Agriculture Organization (FAO) of the United Nations lays down guide lines for all countries to follow. It is a dilemma of society that one might unknowingly imbibe hormones in meat—imported as well as home-produced—because of decisions and actions taken by agriculturalists.

Jimmy is almost complete, but he needs eyes with which he can become aware of his environment.

PERCEPTION BY SIGHT

'Seeing' is much more than projection of an image on the retina. It involves interpretation by the person of that image. The interpretation of what one sees can vary. Look at Figure 210 and you will find that your interpretation keeps changing. Sometimes the white will be foreground when you will see a vase-like object; sometimes the white will be background when you will see two profiles. When seeing the vase you are unaware of the profiles and vice versa. Turn to any psychology book and you will find a wealth of information about how a child learns perception by sight. This complicated process continues to be a highly individual thing. Several different lists will be

Fig. 210   Perception by sight.

written if several people, having been in the same room, are asked to make a list of the things that they saw in that room. What each person saw had special relevance for that person, in other words, there was selectivity in his seeing. Similarly, if each person were asked to make a list of the things that he observed about the other people in the room, the lists would be different, and again, they would manifest the selectivity of the observer. All health workers need to be aware of the selectivity of their seeing, so that they can train themselves to compensate for this natural process of selectivity, by consciously developing their powers of observation of patients. Highly developed powers of observation should be part of the expertise of health workers.

Though one can only learn one thing at a time, when each area of knowledge is mastered, one needs to develop a functional approach to the human body. Without co-ordination of the sensations received from the eye and the ear, the brain will inadequately control the balancing mechanism. Lack of co-ordination of stimuli from the eye, the ear and the digestive tract, can, in predisposing circumstances, produce travel sickness. This is yet another example of the *interdependence* of all parts of the body.

There is a minimal time for 'correct' perception, though subliminal perception occurs without our awareness. Subliminal perception can be used to make people more suggestible.

Advertisers are aware of the subtle process of perception. One needs to develop a critical attitude to their expertise, so that one is not unduly influenced by it.

THE EYE
POSITION AND RELATIONSHIPS

The eye, a 23 mm (1 in) sphere, is set anteriorly on a perfect horizontal and vertical plane, as it lies within the bony conical socket, called the orbital cavity, which has a medial slant. To accomplish this the eye rests on a pad of connective tissue containing a considerable amount of fat. When depleted this gives rise to the 'sunken' appearance of the eyes which you will learn to recognize clinically. There are four guiding reins, in the form of four strap-like muscles, having their origin at the back of the cone, and coming forward, one to be inserted on the top of the eyeball, one on the bottom and one on each side. These are the recti (rectus, sing.) muscles and are designated superior, inferior, medial and lateral. There are also two oblique muscles within the orbital cavity, the superior one arising from the back of the orbit, and travelling along the roof until it reaches the inner angle at the front; here it passes through a ligamentous sling and then travels backwards to be inserted into the eyeball on its superior, slightly posterior surface. The inferior oblique is a very short, but wide strap of muscle that arises from the lower, inner angle of the orbit, and is inserted into the inferior eyeball. The lacrimal gland is also tucked into the anterior orbital cavity, at the superior, lateral border.

STRUCTURE OF THE EYE

The outer coat. The posterior five-sixths is covered by sclera and the anterior one-sixth by cornea.

The Sclera. This is made of strong connective tissue and forms the white of the eye. At its anterior circular border it is continuous with the cornea, and this union is called the corneoscleral junction. The sclera does not allow any light rays to pass through, so that the 'window' of the eye is merely the anterior one-sixth. The muscles contained within the socket are inserted on to this scleral coat. It is pierced posteriorly by the optic nerve.

The Cornea. This transparent, avascular structure completes the anterior one-sixth of the eyeball. It is convex anteriorly and has often been likened to a watchglass superimposed on the globe. It is highly sensitive, and any irritant approaching the cornea sets off the blinking reflex, so that the eye has an automatic protection. It is because of corneal sensitivity that you will be instructed to instil eye-drops on to the white of the eye.

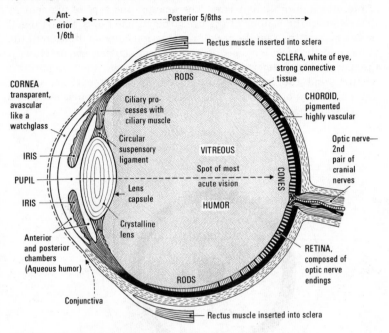

Fig. 211   Horizontal section of the eye-ball.

Any scar tissue from previous corneal injury or inflammation will prevent the passage of light rays and thus interfere with vision, which can be restored by corneal grafting from a donor eye.

Anteriorly the conjunctiva is adherent to the cornea (Fig. 217), and its posterior surface is continuously washed and nourished by aqueous humor.

The middle coat. This is again made up of two portions, the choroid being present in the posterior five-sixths and the iris in the anterior one-sixth.

The Choroid. This lines the sclerotic coat, and is highly vascular and pigmented, so that it absorbs the light rays and can be likened to the black paint on the inside of a camera. As it nears its anterior, circular border it is thrown into folds or processes known as the ciliary processes. These have two important functions: they give attachment to the suspensory ligament of the lens capsule, and they secrete aqueous humor into the posterior chamber. On the anterior surface of these processes there is some plain muscle, arranged in a circular fashion and called the ciliary muscle. Contraction of these fibres lessens the pull on the suspensory ligament and the lens

401

capsule, so that the latter becomes shorter and thicker or more convex. These two structures, the ciliary processes and the ciliary muscle, constitute the ciliary body.

The Iris. This fibromuscular, pigmented curtain is continuous with the choroid anteriorly, but it is incomplete centrally and this 'hole' is the pupil. It appears black merely because the choroid at the back of the globe absorbs light rays. The amount and type of pigment in the iris gives the 'colour of the eye'. The muscle fibres are arranged in two planes, the radial ones on contraction causing dilation of the pupil, and the circular ones on contraction causing constriction of the pupil. The former fibres are innervated by the sympathetic portion, and the latter by the parasympathetic portion, of the autonomic nervous system. The iris is bathed and nourished on its anterior and posterior surfaces by aqueous humor.

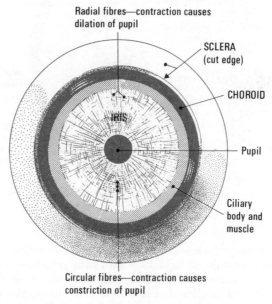

Fig. 212   The iris showing radial and circular muscle fibres.

The function of the iris is therefore to regulate the amount of light entering the eye. In dim light the radial fibres contract and enlarge the pupil thus allowing maximum light to enter; in bright light the circular fibres contract and lessen the pupil thus preventing excessive light from entering. This 'light reflex' can be tested by shining a torch on the eye and gives useful information about the nervous system.

The inner coat. The posterior portion of the globe is lined by the retina, which is composed of the sensory endings of the optic nerves —the second pair of cranial nerves. These sensory nerve endings are of two types, the rods found on the outer circumference functioning in dim light, the cones situated centrally functioning in bright light. The cones are most dense at the centre, which is in fact immediately

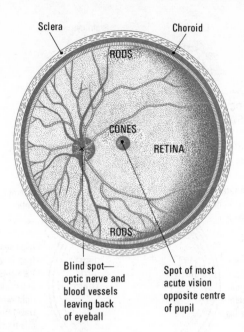

Sclera

Choroid

RODS

CONES

RETINA

RODS

Blind spot—
optic nerve and
blood vessels
leaving back
of eyeball

Spot of most
acute vision
opposite centre
of pupil

Fig. 213 Anterior view of the retina.

opposite the centre of the pupil, and is the spot of most acute vision. The fibres from both types of nerve ending leave the back of the eyeball as the optic nerve, below the horizontal axis, slightly to the nasal side. The optic nerve leaves the back of the bony orbit, and enters the skull via the optic foramen, which is in front of the pituitary gland. Here in front of the pituitary gland some of the optic nerve fibres cross (Figs 24 and 214) in the optic chiasma, before becoming the optic tracts. The optic tracts pass backwards through the cerebrum to the lateral geniculate bodies. These bodies act as relay stations and from them the nerve fibres continue backwards to the cells in the occipital grey matter, where the light impulses are translated into sight.

The rods require vitamin A in the performance of their function of seeing in dim light. Absence of vitamin A gives rise to night blindness.

There is a point on the retina, (where all the fibres from the rods and cones are converging to form the optic nerve), which is devoid of rods and cones, and is therefore incapable of being stimulated by light rays, and it is referred to as the blind spot.

The retina is pressed against the choroid by the vitreous humor, a jelly-like material stretching from the posterior surface of the lens to the retina behind.

The lens. This is a circular, biconvex structure, enclosed in a capsule, which around its circumference is attached to the circular suspensory ligament, which in its outer circumference is attached to the anterior circumference of the ciliary processes of the choroid.

403

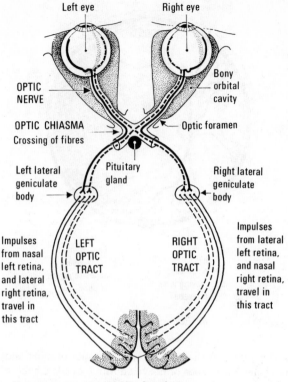

Fig. 214   The optic chiasma.

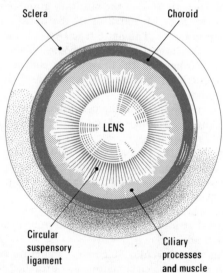

Fig. 215   Anterior view of the lens suspended from the ciliary muscle by the suspensory ligament.

Interspersed in the ciliary processes are the ciliary muscles, which on contraction lessen the pull of the suspensory ligament on the lens capsule, allowing the lens to become more convex anteriorly and posteriorly—in other words to become shorter and fatter. This is necessary in near vision when the light rays from the object being viewed are more divergent, and need more bending to bring them into focus at the spot of most acute vision. When the ciliary muscles are relaxed, the suspensory ligament is taut and the lens therefore less convex anteriorly and posteriorly—in other words long and thin. This is sufficient to bend the light rays, travelling from a distant object and almost parallel as they reach the eye, so that they focus on the spot of most acute vision.

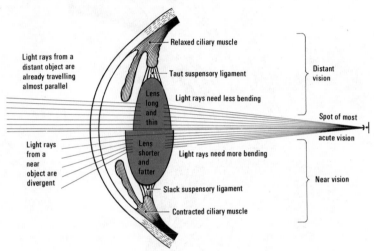

Fig. 216   Section of the lens in near and distant vision.

Aqueous humor. This fluid is secreted by special glands in the ciliary processes and is poured into the posterior chamber, between the lens and the iris. It circulates through the pupil, and is absorbed into veins around the angle where the posterior surface of the cornea joins the iris. Increased pressure in this intra-ocular fluid causes intense pain and threatens loss of vision.

Refracting media of the eye. These are structures which take part in bending the light rays—(more or less according to whether they come from near or distant objects respectively), the cornea, aqueous humor, lens, vitreous humor.

The eyebrows. These are thickened portions of skin and subcutaneous tissue lying over the orbital ridge of the frontal bone. The skin bears specially strong hairs which help to prevent sweat from the forehead dropping into the eyes. The brows are associated with facial expression.

The eyelids. These consist of two loose folds of skin which, when their free edges are in apposition, cover the eyeball. A circular muscle brings about this apposition or closing of the eyes. The eye-

405

lids are strengthened by tarsal plates made of strong fibrous tissue. The muscles which help in blinking, etc., are inserted into these plates. The inner surface of the lids is lined with mucous membrane known as conjunctiva, which is illustrated in Figure 217.

At the junction of the skin and mucous membrane on each lid, there are projecting eyelashes and special glands—Meibomian glands. It is necessary to remember this name, so that you will be able to learn about Meibomian cysts.

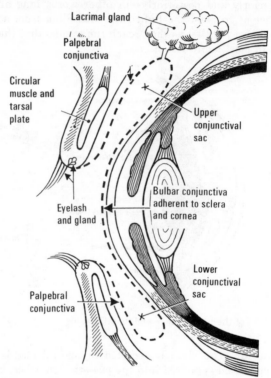

Fig. 217   Disposition of the conjunctiva.

The lacrimal apparatus. The lacrimal gland is tucked into the upper, outer portion of the orbit, and it pours its secretion of tears into the upper conjunctival sac via several small ducts. These tears are assisted in their passage across the eyeball by blinking, and are eventually drained into two tiny canals at the inner angle of the eye, via which they reach the lacrimal sac, from whence they are conveyed by the nasolacrimal duct to the nose. Tears contain an enzyme called lysozyme, capable of destroying bacteria (bacteriolytic).

CARE OF THE EYES IN THE MAINTENANCE OF HEALTH[175,176]
The first hazard to which the eyes are exposed as the baby descends the birth canal is that of infection from the mother, so that one of

the duties of the attending midwife, in the home or hospital, is to look carefully for any redness or discharge in the newborn baby's eyes (p. 133). This is reported immediately to the doctor and vigorous treatment instituted to prevent damage.

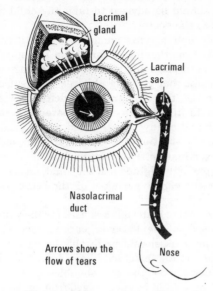

Lacrimal gland

Lacrimal sac

Nasolacrimal duct

Arrows show the flow of tears

Nose

Fig. 218   The lacrimal apparatus.

Because of shock or asphyxia some newborn babies require the administration of oxygen. This is performed under carefully controlled conditions, as excessive oxygen at this stage can cause the formation of fibrous tissue in the vitreous humor, which can cause blindness.

Each individual should be provided with a personal face cloth and towel throughout life. Eye infections can be spread via these media.

At all stages throughout life the eyes require to be well-nourished, and since they present a moist membrane to the air for the greatest part of 24 hours, it is important that this epithelial tissue is kept in good condition by adequate vitamin A, with its added anti-infective properties. Vitamin A is also necessary for the formation of visual purple via which vision in dim light is accomplished. A British diet is unlikely to be lacking in this vitamin, for it is present in all dairy produce, and as the pigment carotene in yellow fruits and vegetables.

In early infancy a baby may appear to squint. It takes time for a baby to learn to use both eyes together, but children do not 'grow out of' a true squint. They may appear to be seeing all right because they suppress the picture received from the squinting eye. Should sight be lost from the 'good' eye for any reason, the poverty of sight in the squinting eye will be evident. A squint is when a pair of eyes do not look in the same direction. The squint need not be present all the time. This is one of the reasons that parents disregard a squint. On the other hand, *one* eye may persistently 'turn'—but the eyes can squint alternately. A squint is most obvious when a child is tired, fractious or ill. In a long-sighted child a squint can develop at between 3 and 6 years. One should not wait for increasing age to cure it. Parents should be advised to take any child in whom they suspect a squint to the doctor.

An outdoor life and playing games are good activities for the development and strengthening of far vision in young children. As the eye muscles, both plain and striped, get a chance of rest during sleep, it is important that growing children have sufficient sleep.

It is important also that any suspected defect in vision, whether noticed by the midwife, health visitor, parents, school nurse or school teacher should be investigated, and in any case sight testing should take place by the time the child is 6 years old—at his first school medical examination. There are various arrangements for the education of the partially sighted and the blind, and the sooner these special arrangements can be partaken of, the better for the afflicted child.

Large print and coarse materials are used for the young until their eyes have developed sufficient convergence to cope with finer print and materials. The book should preferably be on a steady surface, with the lines horizontal, and should have a matt finish to the page and the light falling over the left shoulder. A good posture for reading and writing should be encouraged from an early age, as this evens out the work of the eye muscles, and allows them to remain healthy and in balance.

As children have more bookwork in their formal education, they should realize that glancing from the book temporarily gives the ciliary muscles a rest, and prevents a feeling of eye fatigue.

If occupation is sought in a factory where arc-light welding takes place, then the eyes must be protected from sparks and excessive glare by the wearing of special goggles. Whenever a saw or drill is in use it is wise to protect the eyes with goggles, visors, spectacles or face screens. The British Safety Council, whose slogan is, Safety is *my* business, has an annual 'Eye Safety' week.

Some hobbies require the wearing of protective goggles, for example motor-cycling, open-car touring, speed boating, deep-sea diving. Consideration for the prevention of accidents should be shown by vehicle drivers who wear spectacles. The type of frame with wide, heavy sides restricts lateral vision. In the United States plastic lens in spectacles is mandatory, because it is estimated that 100 000 adults and 20 000 children are injured every year in accidents with glass lenses that shatter. In Britain, for an extra £2 lenses can be made in plastic instead of glass. This greatly reduces the risk of damage to eyes, nose and forehead. In all countries where traffic is on the increase, there is an increase in eye injuries, caused by passengers being thrown against the windscreen. An attempt can be made to reduce this danger by using a material other than glass for windscreens or by the more general use of safety belts.

CONDITIONS WHICH CAN INTERFERE WITH THE HEALTH OF THE EYES
FOREIGN BODIES
Most of the time the blinking reflex works efficiently, but flies, dust, sand or even an eyelash can find its way into the eye. Sometimes the immediate reaction of excessive watering is sufficient to wash out the foreign body. Sometimes it can be washed out by using plain tap water, or it can be removed by the corner of a clean handkerchief. A particle in the upper conjunctival sac is less easily removable than one in the lower conjunctival sac. When first-aid has been rendered a doctor must inspect the eye for corneal abrasions; these can later interfere with sight if there is any formation of fibrous tissue.

SQUINT (Strabismus)
Imbalance in the muscles that move the eye, and can be due to incorrect length of muscles, or to weakness of same. Treatment may strengthen weak muscles, and such children need support and encouragement to persevere. Extra love and kindness may be needed to off-set any unkind remarks made by other children.

Surgery may be necessary and the child should be protected from adverse stories of hospital life, and should be given a simple, true picture of what he might expect as a patient having an eye operation. His parents and relatives should visit frequently during the hospital period.

MYOPIA (Shortsightedness)
Because the eyeball is too long from front to back, the light rays come to a focus *before* they reach the retina, and thus a blurred image is seen. The child discovers that by holding the object nearer, the light rays are more divergent and need more bending, and thus they focus nearer to the retina, and give a clearer image. The treatment is to wear a concave lens which will diverge the light rays, so that the environment can be seen and accidents avoided.

HYPERMETROPIA (Longsightedness)
Because the eyeball is too short from front to back, the light rays come to a focus *beyond* the retina. Light rays coming from a distant object may reach their focus at the retina, so that distant vision is good, but near vision blurred. The treatment is to wear a convex lens which will converge the light rays from near objects so that they will be brought to a focus on the retina.

STYES
These result from infection of an eyelash follicle with a pus-producing germ, which may get there from dirty hands, infected face cloth or towel. Good routine personal hygiene will help to prevent styes.

409

The normal healthy body can cope with a few germs placed on the eyelid, but when the tissues are slightly below par, and/or the number of assaulting germs large, the battle ensues in the form of a stye.

BLEPHARITIS

Inflammation, usually of a low grade, occurring round the eyelids. This condition needs long and patient treatment as ordered by the doctor; it also requires surveillance of personal hygiene so that there is no danger of spreading the condition. Prevention of unsightly red eyes is important from a psychological point of view; not only is the child robbed of his confidence because he knows his red eyes are not very nice to look at, but he is robbed of his playmates as they become frightened of 'catching' sore eyes.

PINK EYE

This is a highly infectious inflammation of the conjunctiva, occurring most often in school children, hence in an outbreak the infected children should be excluded from school and treated by their own doctors. Strict hygiene measures must be inculcated into the remaining children and they should be inspected daily by the school nurse for signs and symptoms of early infection.

TRACHOMA (Granular conjunctivitis)

This virus infection is responsible for much blindness, and has been the subject for extensive research work by the World Health Organization. The slogan for a World Health Day was 'Preserve sight; prevent blindness'. Since then there has been a gradually declining incidence of trachoma, especially in those areas serviced by a mobile eye clinic.

Trachoma is rife in many tropical and subtropical countries, especially where sanitation and flies have not been brought under control, and where high standards of personal hygiene are not practiced. The Medical Research Council's Trachoma Research Unit is trying out a vaccine against trachoma.

CONGENITAL CATARACT

The cause of opacity of the lens at birth is unknown, but we now know that if a woman contracts German measles early in her pregnancy, the possibility of her producing a baby with a congenital defect is increased; it is therefore important to prevent pregnant women from coming into contact with those suffering from German measles. Immunization is now offered to girls who have not had the infection, on leaving school.

410

Any inflammation of the cornea that results in scarring can prevent the entry of light rays, sufficiently to produce blindness. Corneal grafting from a *donor* eye can remedy this. Full information, and corneal-grafting-donation-forms can be obtained from The Director-General, The Royal National Institute for the Blind, 224 Great Portland Street, London, W1. Neither age nor the wearing of spectacles is a deterring factor.

There is a growing body of opinion, that poor sight might be a contributory factor in road accidents. In 1935 when seeing a number plate at a distance of 25 yards in daylight became necessary for a driving licence, this standard of vision was probably all right. But with today's increased numbers of, and faster cars, more coloured cars and red and green traffic lights, many people think that peripheral vision and degree of colour-blindness should be tested—as part of the driving test. Under Section 42 of the Road Traffic Act, 1962, police have the power to check vision of drivers, but they rarely exercise this power. The Optical Information Council states that eye-sight is the neglected factor in road safety.

Concerning Blind People

There are many societies in existence throughout the world to give assistance to blind people. In this island there is the *Scottish National Institute for the War-Blinded* (which among other things provides visiting by an after-care officer), the *Royal National Institute for the Blind* and *St Dunstans*. These are probably the best known, but there is a *National Federation of the Blind*, a '*Mobility of the Blind*' *Association* and a *Royal Commonwealth Society for the Blind*.

In our country *education services* for the blind and partially sighted comes under the jurisdiction of the *Department of Education and Science*. Blind children usually attend special schools for the blind. The teachers are specially trained. Partially sighted children may go to such schools, or may go to a special class in an ordinary school, or an ordinary class in an ordinary school, with special facilities available to them, e.g. a desk near the blackboard. According to their intelligence level they follow the same curriculum as sighted children. With improved methods of educating blind people there has been a rise in their entries to universities and now they attend classes for the Open University. A plastic 'paper' enables blind people to read and write without learning braille. A special pen causes the words to rise a few thousandths of an inch above the surface. In a test of blind, partially sighted and sighted but blindfolded people, almost all could make out half-inch letters. Pictures can now be incorporated into books for the blind. A machine made by Rolls-Royce apprentices produces a master drawing on a zinc plate and then presses it out on to thick manilla card. The Argonne braille machine[177] converts symbols recorded on magnetic tape into upraised dots on a plastic belt that moves past the fingertips. On the

411

revolving belt the dots are erased. One such belt can be used for
many months. There is a possibility that there will be electronic
artificial eyes[178] that can be used in conjunction with miniature
television cameras. Ultrasonic spectacles[179] are being evaluated in
Britain, Australia and the United States. Measures taken by another
country to educate its blind population are given in an article by
Ross.[180]

In a more general educational sense, in some areas, volunteers
prepare casettes of talking news and deliver them through the letter-
boxes of blind people. The Royal National Institute for the Blind
in collaboration with the British Medical Association have produced a braille edition of the Family Doctor publication, 'You and
Your Baby'. It also carries the advertisements in braille, so that
blind mothers can have choice about baby clothes, equipment and
foods. The Library Association and the National Association for the
Education of the Partially Sighted, published a book, *Print for the
Visually Handicapped Reader.*

Mobility for the blind is achieved through sensory and other aids.
Probably the two best known to the public are the white walking
stick and the guide dog. Another version of the white stick is fitted
with a global type of castor that wheels in front of the blind person,
and some find this preferable and more information-giving than the
tap-tapping of the stick. There is a long cane available that increases the range of mobility of the user, but a long training course
in its use is necessary and there are only a few training centres in
England and as yet there is not one in Scotland though one is being
considered. Other problems are lack of instructors, and the availability of money for the blind person to take the long-cane course.
There is also a sonic aid—a transmitter gives a warning signal if
there is anything in front of the blind person. The most recent addition is a cane equipped with laser beams[181] which warn the blind
person by a low-pitched bleep or finger signal, if there is a step or
obstacle up to 12 feet ahead. Some pedestrian crossings are fitted
with an audible signal when it is safe to cross.

Registration as a blind person depends on the degree of blindness,
and on the cause and whether the condition is likely to worsen or
not. Registered people are visited by the welfare social committee
and social workers trained in the welfare of the blind. If the blind
person receives a social security benefit, he is eligible for a supple-
mentary benefit of £1·22½. He may be able to claim income tax
relief. He is given priority with 'talking book' equipment. The
rental of this is £3 a year, but some local authorities undertake to
pay this. The 'books' and postage are free. Radio licences are free to
registered people and there is a concession for television licences.

In the *domestic sphere* there are many difficulties to be overcome
by the blind person. There are many gadgets available and I can
mention only a few. Eating is a big problem. Taking what one thinks
is a fork full of potato and finding that it is carrot is disconcerting.
It helps if the server always follows the same pattern (Fig. 219). It
is difficult for a blind person to know if there is food on the fork.

412

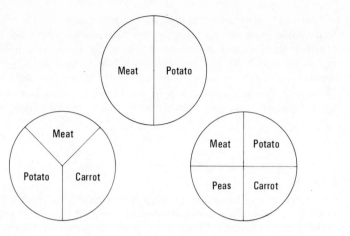

Fig. 219   Arrangement of food on plate for blind person.

There is a special one that has a tiny electrical circuit built into the handle and it is connected to the prongs. When there is food on the fork, a buzzer vibrates on the handle. It can also be used to find out if there is food on the plate. Knowing when sufficient liquid has been poured into a vessel is accomplished by distinguishing a change in sound or by putting a cleanly washed finger over the edge of the vessel and allowing the fluid to reach the finger. This method precludes the use of boiling liquids. There is a warning device for when clothes are hanging out and it rains. It is marketed at £5·62½ and is sold to registered blind people at £1·87½. These examples are given to help you to think about the many things that we sighted people do without conscious thought, but which require adaptation by and for the blind person.

Recreation is as necessary for blind people as for sighted people. There is a weekly braille magazine for teenagers and young persons produced by the Royal Blind Asylum and School in Edinburgh. Blind people can play bowls. One stands at the far side of the green, and claps his hands slowly and clearly. By listening carefully to the sound, the other players can judge the direction in which to roll their bowls down the green. Blind people have taken part in cross channel relay swims, guided by music from radio on the escort boat. Some parks have a special garden in which scented plants are grown. Some nature trails have plates in braille explaining points of interest, and a knotted nylon guide rope warning of the obstacles ahead. Several card games are adapted and have braille dots identifying the cards. In a more challenging sphere, nine blind Africans in association with the Outward Bound Trust, climbed the highest mountain, Kilimanjaro, 19 430 feet. Their leader said that it was intended as a demonstration to all Africa that trained blind people have the mental and physical stamina to achieve exacting goals, and to justify their place in the economic and social life of modern Africa. In our country, a blind man climbed Ben Nevis.

413

There are *sheltered workshops* for the blind in which they are able to do daily work and earn an income. Some blind people are able to compete on the open labour market at such things as typewriting, operating a switchboard, gardening—where things are adapted for them, for example tape measure in braille. There are blind physiotherapists. The Viking 1010 sewing machine has been adapted so that a blind person can do everything on it that a sighted person can do.

LIGHTING

The town planning, and buildings which have been and are being erected in the middle twentieth century, are all tribute to this era's 'light consciousness'. There is a gradual movement away from narrow streets of tall houses, inhabited cul-de-sacs, small barred windows hung with dark, heavy curtains and narrow dark passages.

There are spaces in the cities for a little greenery, and the maxim in buildings seems to be as much window space as possible. A change in the pattern of living is also to be observed, for people no longer sit at home, but go out into the light and sunshine to benefit from the stimulating rays of the sun, the production of vitamin D in their skins and the bactericidal effect of the ultra-violet rays.

Natural Lighting

All rooms and corridors should have sufficient window space to make maximum use of daylight, which by showing up the corners encourages cleanliness, discourages accidents, and gives a feeling of well-being. The windows must be kept clean to allow maximum light entry, and the rays of light should strike a light-coloured, glossy wall and ceiling to allow maximum reflection of light. In classrooms light-coloured furniture is used for the same reason. For the right-handed person light should, where possible, fall over the worker's left shoulder, and vice versa, so that shadow is not cast on to the immediate area of work.

414

Artificial Lighting

The candle provided the first chapter in the story of artificial lighting, and as tribute to its usefulness light is still measured in candlepower. The following are the characteristics of 'good' artificial light:

1. Sufficient intensity to show detail clearly.
2. Supply constant, i.e. without flicker.
3. It does not shine into the eyes, nor is it reflected from any bright object into the eyes.
4. Uniform illumination of the object being viewed without the interference of shadows.

There are therefore three simple rules to follow: 1. Make sure there is sufficient general lighting in the room; 2. Use local lighting for special activities, in addition to general lighting; 3. Do not place lights where they will cause glare. These criteria are best fulfilled by electric light, and since it is used most in our country we will discuss it next.

ELECTRICITY

This is now the main source of lighting, and into the bargain it is clean, labour-saving, immediately available on moving the switch, does not require a means of ignition (matches were scarce in World War II), has no naked flame as a fire hazard, does not use up oxygen from the air and does not add any products of combustion to the air which it slightly warms.

When considering any lighting scheme the function of the room must be taken into consideration. Diffused wall lighting has become increasingly popular for rooms used for relaxation, and for dining-rooms. A central pendant of 100 to 150 watts, either from a single or multiple socket, is useful in an all-purpose living-room; it is usual to supply one or two table or standard lamps, which can be moved to the area in which one person might choose to do closer work. For kitchen and bathroom where steam is a hazard, lights with the globe fitting directly on to the ceiling or wall are considered safest; and operation by a pull cord is safer than the more usual wall switch.

Fluorescent strip lighting has been extensively used wherever good, shadowless light is called for. If set into the ceiling it does not become a dust trap. It is economical in running costs, but the incessant slight noise can become tiresome.

Electric light can be used for outside lighting, and entrances should be well lit to assist those looking for names and numbers, and to prevent accidents from fumbling with gates and steps. Street lighting is now of major importance in the prevention of accidents and there is a lighting section at the Industrial Health and Safety Centre, Horseferry Rd., Westminster, SW1. Dr Kraft Ehricke, a top space scientist says that it would be possible to put in orbit a sun reflector to light up city areas with five times the brightness of a full moon. He says that it would be an important deterrent to crime. Light from battery-operated torches is extremely useful as they are safe and portable, and should always be included in an emergency camping and car kit.

GAS

Gas is produced from the distillation of coal. Natural gas is found within the earth's strata and is being used increasingly in this country. When burned within an incandescent mantle it glows to give a reasonably effective light. Unfortunately it takes oxygen from the atmosphere, and gives out waste products of combustion (mainly carbon products). By this process it spoils interior decorations. The switch is on the light fitting, which makes it inconvenient as one has to penetrate the dark room, before one can make the light available. A means of ignition is required which carries a fire hazard, and the housewife must never allow the supply of matches to run out. The welding of gas pipes must be perfect. When gas starts to flow, flaws would be disastrous. Each weld is X-rayed, then the pipes are sealed off and subjected to double the pressure they will be under when gas flows. There is a danger of leakage from old gas pipes, and carbon monoxide poisoning can result. Gas lights are permanent fixtures, so that the person desiring to do close work must sit in the area of good illumination from the existing light, and has not the pleasure of choosing a quiet corner of the room. Some council houses still have gas lighting, and in some towns it is still used for street lighting.

Calor gas from cylinders has been used increasingly in recent years for the provision of light in caravans. The instructions on the cylinder must be carefully carried out to avoid explosion.

OIL LAMPS

From a reservoir, paraffin oil seeps up a wick, and on combustion produces a flame. Considerable labour is involved in tending the wick and reservoir, the particular smell is evident in the room, suitable storage place must be found for the inflammable oil, oxygen is used from the atmosphere, and waste products of combustion are added to the atmosphere, a means of ignition is required and fire hazard is always present, though the newer models have an automatic extinguisher in case of overturning. They can also burn the paraffin under pressure, and can have an incandescent mantle, which greatly improves the illuminating power, but has a hissing noise which some people find disturbing.

Hurricane oil lamps still provide a useful standby during electricity cuts, etc., and to fulfil their emergency function they must be kept clean and filled.

CANDLES

With all their disadvantages a packet of candles should still be standard household equipment for emergency. They should be kept in a tin box to minimize fire hazard. Care should be taken when using them for table and festive decoration, as the fire hazard from draught on a naked flame is considerable.

CHEMICAL LIGHT

Tetrakisdimethylaminoethylene (mercifully shortened to TMAE), with the addition of air gives a slow, almost heatless oxidation that produces a high light output. Emergency chemical lighting systems are being fitted in helicopters and aircraft escape slides. Chemical light distress-signals are being incorporated into life-jackets and life-rafts. This emergency light source is ideal for campers, outdoor people and drivers. For car use there is a chemical light strip that fastens to the back bumper. A tab is pulled which lets air into the sealed transparent envelope—and the light warns other motorists.

Hospital Lighting

CORRIDORS

Construction should be such that adequate natural lighting satisfies daytime needs. At night artificial lighting should have sufficient illumination to allow adequate observation of any change occurring in a patient returning from the operating theatre. For the same reason lifts must have adequate lighting—and an emergency light, should the electric current fail.

WARDS

Many styles and patterns of natural and artificial lighting are to be seen and you will be able to discuss their advantages and disadvantages better at the end of three years, when you have had experience of working in several differently lit wards. It is only recently that natural lighting of buildings has been scientifically studied, enabling the design of windows to be improved. It has been found that, while old hospitals were sometimes underlit, modern hospitals are often overlit. Very large areas of glass can produce severe discomfort from glare.

Adjustable lights are useful over the beds, and there should be a socket at each bedside for a bell lamp, when treatment requires a good light. There should be some form of dimming for night time, and arrangements for adequate illumination for continuous observation of an ill patient, without disturbance of the other patients. All light switches should be silent in operation, and shuttered so that walls can be washed without danger.

OPERATING THEATRES

Strong shadowless lighting that does not overheat the atmosphere must be provided, and it must be capable of being fed from an emergency battery system. The fittings must all be safe for the frequent wall washing which must be done in a theatre.

Possible Effects of Inadequate Lighting on Health

1. Gloominess, which interferes with mental well being.
2. 'Dark corners' are not conducive to cleanliness, dirt is dangerous.
3. Badly lighted areas are accident-prone.
4. The pupils have to be dilated and this constant work fatigues muscles, with consequent headache.
5. Adoption of a bad posture, to catch a gleam of light, especially for close work.
6. Exclusion of ultra violet rays of sun, thus loss of: (a) their disinfectant properties; (b) their ability to activate the skin ergosterol to produce vitamin D.

SUMMARY OF THE EYE AND LIGHTING

The spherical eye is supported in a bony conical socket by a pad of fat, four recti and two oblique muscles. It is protected by eyelashes, eyelids and eyebrows.

The sclera. This covers the posterior five-sixths of the eyeball, and forms the 'white' of the eye. At its circular, anterior border it meets the transparent, watchglass *cornea* covering the anterior one-sixth of the eyeball.

The choroid. The choroid forms the middle coat of the posterior five-sixths, and nearing its anterior border it is thrown into folds— the ciliary processes. These secrete aqueous humor which is poured into the posterior chamber; they also give attachment to the circular suspensory ligament of the lens capsule. At its circular, anterior border it meets the *iris*, the coloured, muscular curtain that relaxes and contracts to regulate the amount of light entering the eye. It is incomplete centrally at the pupil.

The retina. This is composed of the highly specialized endings of the optic, the second pair of cranial nerves, which leave the eyeball below the horizontal axis, slightly to the nasal side. The nerve endings are rods and cones, rods being found on the periphery for appreciation of dim light, and cones being most numerous centrally for the appreciation of bright light. The spot of most acute vision is directly opposite the centre of the pupil. The spot at which the optic nerve leaves is called the blind spot.

The lens capsule. This is attached circularly to the suspensory ligament. It contains the elastic lens, which becomes more convex to focus divergent light rays from a near object, or less convex to focus parallel light rays from a distant object.

Aqueous humor. Aqueous humor is present from the anterior surface of the lens to the posterior surface of the cornea, that is in the anterior and posterior chambers.

Vitreous humor. A jelly-like substance stretching from the posterior surface of the lens to the retina at the back of the eyeball.

Tears. These are secreted by the *lacrimal gland* in the outer, upper orbit, and are assisted in their passage across the eyeball by blinking. At the inner corner they pass into two tiny canals, thence into the lacrimal sac, and via the nasolacrimal duct to the nose. They contain a bactericidal substance lysozyme.

In the care of the eyes the following points are important:

Pregnant women should not be exposed to the risk of contracting German measles. Girls who have not had German measles are offered immunization against this disease, on leaving school. Very young babies should not be given excessive oxygen. Vitamin A is necessary for the prevention of night blindness and the health of the covering epithelial tissue (conjunctiva). The earlier any defect in vision is noted, the sooner can appropriate education be instituted for that child. Sight testing is done annually throughout school life. Protective glasses should be worn where necessary, whether it be for work or play. Good standards of personal hygiene prevent infectious eye conditions from spreading.

Apply the following questions to the following forms of lighting:

Is it constant?
Does it flicker?
Is the intensity sufficient to show detail clearly?
Does it shine into the eyes?
Is it reflected into the eyes from a bright object?
Does it cast shadows?
Is it clean? Electric.
Is it labour saving? Gas.
Is it immediately available? Oil lamp.
Does it require a means of ignition? Candle.
Is there a naked flame or other fire hazard? Chemical.
Does it use up oxygen from the air?
Does it add waste products of combustion to the air?
Is it noisy?
Do the fittings form a dust trap?
Is it portable?

The Ear

Jimmy now needs to be endowed with hearing, so that together with his sight, touch and smell, he can become fully aware of his environment.

AURAL PERCEPTION

'Hearing' is much more than stimulation of the sensory apparatus in the ear by sound waves. It involves interpretation of that stimulation by the person, and this can vary, as witnessed by a guessing game. The contestants listen to recorded sound and are invited to identify it. There are no visual clues. This reinforces that visual and aural perception can each be used to complement the other in interpretation of the environment. Light waves travel faster than sound waves, so that one can visually perceive an action in the distance, for example dropping of a heavy structure and its arrival on the ground, seconds before one 'hears' the impact on the ground. There is the natural phenomenon of seeing lightning before hearing thunder. Many clues can be used by a person in the detection of the source of sound. People vary as to what sounds, including intensity of same, they find pleasurable or the reverse. Sometimes a person wants to attend exclusively to what he is hearing, and therefore closes his eyes to exclude interference from visual perception.

There is selectivity in aural perception. At an orchestral concert, a budding violinist in the audience can selectively listen to the sound from the violins; a mother hears the *first* cry of her 1 year old child, when she has not paid attention to sound of similar intensity from the same direction as her sleeping child. Many jobs involve training in aural perception, for example many people know when an engine is working properly from the sound it makes. Nurses need a well-developed aural acuity so that they can immediately recognize change, for example in a patient's breathing. A well-practised total process of aural perception is essential for *'listening'* to patients/clients—*one of the most important functions of all health workers.*

The sensations received from the ear have to be co-ordinated with those received from the eye, in control of the balancing mechanism. Travel sickness results from lack of co-ordination of stimuli from the eye, ear and digestive tract—we cannot get away from that *interdependence*!

For purposes of description and function the ear is divided into three parts—the external, middle and internal ear.

The External Ear

This is composed of the pinna and the external auditory meatus.

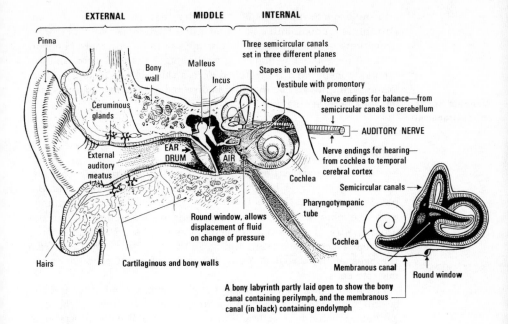

Pinna

Three semicircular canals set in three different planes

Bony wall

Malleus

Stapes in oval window

Incus

Vestibule with promontory

Ceruminous glands

Nerve endings for balance—from semicircular canals to cerebellum

AUDITORY NERVE

External auditory meatus

EAR DRUM

AIR

Nerve endings for hearing— from cochlea to temporal cerebral cortex

Cochlea

Semicircular canals →

Round window, allows displacement of fluid on change of pressure

Pharyngotympanic tube

Cochlea

Hairs

Cartilaginous and bony walls

Membranous canal

Round window

A bony labyrinth partly laid open to show the bony canal containing perilymph, and the membranous canal (in black) containing endolymph

Fig. 220   Diagram of the ear.

THE PINNA

This is commonly called the ear lobe. It is a delicately folded cartilaginous structure, with a few small muscles around the cartilage, covered by subcutaneous tissue and skin. Many animals retain voluntary control of the ear muscles and can arrange the pinna in an erect funnel, thus collecting maximum sound waves. In man the pinna is exposed on the side of the skull, and has an excellent blood supply in an attempt to keep it warm. Especially marked in males are the hairs at the entrance to the canal leading from the pinna.

THE EXTERNAL AUDITORY MEATUS

Leading from the pinna to the tympanic membrane or ear drum, this canal is 3 to 4 cm (1 to $1\frac{1}{2}$ in) long. It rises upward and backward and then falls downward and forward. Since the inner half of the canal has bony walls and the outer half has cartilaginous walls, this 'hillock' can be modified by grasping the lower half of the pinna, and pulling it in a backward and slightly upward direction—a necessary procedure when examining the ear.

The canal is lined throughout by skin into which is incorporated special glands (ceruminous glands) that secrete wax (cerumen). Wax[182] contains lysozymes and immunoglobulins—that act locally. Caucasians and Negroes secrete a 'wet' honey-coloured wax. Mongoloid peoples secrete a 'dry', grey, brittle kind of wax. Lysozymes

421

are found much more frequently in dry than in wet wax. One important immunoglobulin is found almost excluvisely in dry wax. (Infections of the ear canal do not appear to be any more common with either wax.) The floor is slightly longer than the roof, and the tympanic membrane is stretched across this oblique medial border. When syringing an ear the fluid is directed along the roof and space left for the returning fluid.

The Middle Ear

Medial to the external ear and lateral to the internal ear, this air-containing cavity is hollowed out of the temporal bone. Via the pharyngotympanic (Eustachian) tube, the middle ear is continuous with the pharynx anteriorly and medially. Posteriorly it is continuous with the mastoid air cells, and all these structures share the same mucous membrane lining, so that infection can travel from the back of the throat along to the middle ear and through to the mastoid cells.

The lateral wall is composed of the tympanic membrane, and the medial wall contains four important structures, namely the *oval window,* covered by membrane to which is attached the footpiece of the stapes, one of the small bones strung across the middle ear. Below the oval window is a slight prominence—the *promontory*— and in close association with this is the *facial,* the seventh pair of cranial nerves. Disease of, or operation on, the middle ear can disturb this nerve, for which reason the nurse must continually observe such patients for weakness or paralysis of the face muscles. Below the promontory is the *round window,* again covered by membrane which acts as a safety valve when there are changes in the pressure of fluid on its medial surface.

Stretching from the tympanic membrane to the oval window are three small bones (ossicles) in the shape of a hammer, an anvil and a stirrup, and known as the malleus, incus and stapes respectively. They vibrate in response to sound waves which they conduct to the internal ear.

The Internal Ear

This comprises a complicated series of canals hollowed out of the petrous portion of the temporal bone. Immediately within the medial wall of the middle ear lies the vestibule, and from its inferior border the cochlea is hollowed out in the form of a snail's shell, that is two and a half turns round a central pillar (the modiolus). From the superior border of the vestibule, three semicircular canals set in three different planes of space are hollowed out. It must be understood that each semicircular canal leads from, and returns to, the vestibule.

Supported within this complicated bony structure, that is the three-semicircular canals, the vestibule and the cochlea, is an exact replica made in a tubular membrane. This membrane floats in fluid, and this fluid separating it from the bony canal is called perilymph. Completely contained within the membranous canal (as ink in a fountain-pen filler) is another fluid called endolymph. Stretching out into this endolymph are the fine endings of the auditory, the eighth pair of cranial nerves. Those from the cochlea join up with cells in the temporal cerebral cortex, where the stimulus is interpreted as sound. Those from the semicircular canals join up with cells in the cerebellum, as well as the cerebral cortex, so that in the main, interpretation of position in space is carried out at an automatic level, that is in the cerebellum.

Hearing[183] and Balance

Sound waves are collected by the pinna and transmitted along the external auditory meatus to vibrate the eardrum, which in turn vibrates the three ossicles stretched across the air-containing cavity of the middle ear. The footpiece of the stapes acts as a plunger and disturbs the perilymph, which disturbs the membranous labyrinth and the contained endolymph. The nerve endings projecting from the membranous labyrinth into the endolymph are stimulated by these vibrations, and transmit the stimuli to the temporal cerebral cortex, where they are interpreted as sound. Most audiometers test hearing over the *tone range* from the frequencies 250 to 4 000 cycles per second and over the *intensity range* of 20 to 85 decibels. A person able to hear sound throughout the tone range 250 to 4 000 c.p.s. at an intensity of 20 db has normal hearing.

Any movement of the head disturbs the endolymph in the semicircular canals, thus stimulating the nerve endings that connect with the cerebellum for the automatic maintenance of balance and equilibrium, though this can be brought through to the conscious mind (cerebrum) when required.

CARE OF THE EARS IN THE MAINTENANCE OF HEALTH

At the first opportunity for complete observation of the baby, after the excitement of the birth, the midwife observes the presence of the pinna and the external auditory meatus. As the weeks pass, reaction of the baby to sound is noted by the parents and the health visitor. Testing is done by the health visitor with rattles emitting sounds of three different pitches, and records are kept of the baby's reactions. This is done in an attempt to diagnose hearing defects early, so that the necessary steps can be taken to avoid a deaf child being dumb, for children learn to talk by imitating the sounds heard. There is a risk of the woman who contracts German measles early in pregnancy producing a congenitally deaf child, so that contact with such infection should be avoided where possible. Immunization is now offered, to girls who have not had German measles, on leaving school. Thorough drying of the skin fold behind the pinna is essential in young babies to avoid the condition of cracked ears.

Audiometric tests[184] are increasingly performed annually at school so that any developing defects will be observed and treated early. There is an account of a comprehensive service in Durham.[185] There are various forms of education available according to the age, and degree of deafness of the child. Teachers, after basic training, take a special course to qualify them for teaching deaf children.[186,187]

When there is *discharge from the ear*, the condition is best attended to at the school clinic, for direct light is necessary for any examination or treatment of the ear. Such children must also be advised against swimming until the condition heals.

Since a survey in Salford showed a surprisingly high proportion of children run down by motor vehicles to be suffering from a significant degree of low-tone deafness, several other local authority education committees are currently providing hearing details of children involved in accidents.

Concerning Deaf People

The Royal National Institute for the Deaf is the leading and most comprehensive association on deaf and hard of hearing matters in this country. Advice, and many more concrete services are given free and gladly. It publishes many useful booklets for those with impairment of hearing and for normal people who are in close contact with deaf people. The Institute has a technical department which tests hearing aids before their advertisement is accepted in 'Hearing' a monthly magazine for the deaf and hard of hearing. There are various voluntary societies that also give valuable service in this field. Locally, there is usually a 'Hard of Hearing' club that keeps members informed of the many items that become available to make life easier for them. There are flashing lights on alarm clocks, kettles and phones.[188] But light demands that the person is within seeing distance, so a wailing sound is being tried. A new fire alarm system which works by vibrating the pillows of deaf patients is being developed at a home of the Royal National Institute for the Deaf. It is thought that a small receiver implanted under the scalp and communicating with the brain, may one day enable the deaf to hear. Various societies are campaigning for cheaper television licences for deaf people, and for better programmes for them. Countries vary as to whether deaf people can drive a vehicle. In our country they can drive. Some deaf people can be communicated with by sign language, but the number of hearing people that can communicate in this medium is infinitesimal. Lip-reading is a complicated technique and there are not sufficient skilled teachers. Though some people master the technique of lip-reading, it is limiting in that the speaker's lips have to be visible. It is thought that there is a great need for rehabilitation centres for the deaf, where they can learn the full range of accessories to non-verbal communication.

A hearing aid has two users, the speaker to the person using this apparatus, as well as the person 'listening' through the apparatus. Clarity of speech is much more important than volume. The speaker needs to enunciate all the consonants clearly, and speak a little more deliberately than usual. The voice should be modulated to a single pitch, rather than dramatic loud and soft emphasis. It is wise to stand or sit towards the deaf side—the side he wears his aid, but keep within his view so that he can augment with lip reading. Deaf people tend to complain that others shout at them and generally treat them as if they were stupid. Intelligence level does not change when a person becomes deaf.

For those with naturally 'soft' voices—when talking with a person who is dull of hearing, but not sufficiently so to benefit from a hearing aid, do speak a little louder. It is rank unkindness to continually say, 'She can't hear me'. Remember you can discipline yourself to speak louder, the other person simply can not hear any better.

America pioneered a National Theatre for the Deaf and this is very successful. The company has toured in many countries. Britain is also in the process of developing a Theatre for the Deaf,[189] but as yet it has no financial help from the Government, though the arts for hearing people are subsidized.

Many teachers think that non-hearing children need an extra two years at school in the acquisition of language, and are campaigning for this. Career possibilities are restricted for these people. They are able to register as Disabled Persons, and each employing authority has to employ a proportion of disabled persons by law.

There is a quadrennial World Congress of the Deaf; thousands of deaf people gather to hear medical and scientific experts talk on audiology, education and sociology. In addition to technical and human studies, the Congress affords an opportunity to deaf artists and sportsmen and sports women to participate in artistic and sporting events.

Hearing as it Affects Mental Health

To grow up with normal behaviour patterns in spite of a physical weakness is a great feat, and a tribute to all concerned in this 'growing up' situation. It is easy for the deaf child[190] to become an introvert and to cease making an effort to participate in normal groups. It is equally easy for the deaf child to be over-aggressive in participation with others, as a mask to his insecure feelings because he cannot hear properly.

The deafness accompanying increasing age can bring many changes in behaviour, indeed it may be one of the reasons such people are assessed as 'difficult'. Not hearing their voice properly, they tend to speak loudly, which others can find irritating. Not hearing properly, they fabricate the rest of the story, and repeat this as truth to a third person, which can have dire results. The young must remember that this is not done as malicious gossip, and they must persevere with encouragement to the older member to seek advice about the deafness. Deaf people can also be repeatedly wounded by the double-edged blade—of misunderstanding others, and others misunderstanding them.

People who do not hear properly can imagine that they are being talked about, and this pattern of thinking can grow to the similarity of a persecution complex. Devoid of acute hearing, shopping presents difficulties, and this group are more accident prone.

The National Deaf Children's Society is carrying out a survey of mothers with young deaf children. It will take place over a period of two years and will cost about £5 500. While there is much information about mothers' emotional reactions and the guidance they receive, very little is established about what mothers actually do. It is hoped, from frank discussion, that Miss Susan Treble, a teacher and honours graduate in psychology who is carrying out the project will discover some interesting trends to add to this important aspect of parents' guidance.

NOISE

With the progression of man's inventions his eardrums are bombarded with more and more sound waves, so that now idyllic peace and quietness are extremely difficult to attain. It is thought that we are becoming 'noise tolerant' and less sensitive to preventable noise than is desirable. In three years you will be able to meditate on this, and consider the part that you can play in making the hospital a quieter place. Florence Nightingale said, '*Unnecessary* noise (however slight) injures a sick person much more than *necessary* noise.' Hospitals have not escaped from the noise increase. In May, 1962, the King Edward's Hospital Fund for London started its anti-noise campaign resulting in posters drawn by Fougasse being displayed in prominent places in hospitals throughout the land. Many hospitals have introduced noise-reducing measures after carrying out noise surveys. There is no room for complacency and all personnel need to be ever vigilant to prevent patients being exposed to excessive noise.

The *Noise Abatement Society* formed in 1959 keeps relevant information from all over the world. There is an *International Noise Abatement Society*. The Noise Abatement Act, was passed in 1960. At the same time the Government set up the Wilson Committee to enquire into the subject and the Committee reported in 1963. The Minister of Transport brought out regulations in 1968 prescribing maximum noise levels (87 decibels) which it is an offence for various classes of motor vehicles including motor-cycles and power-assisted bicycles, to exceed when used on the road.

A scientific study of car crashes at the Baylor University of Medicine in Texas has shown that in three out of four crashes, the radio volume was turned up high in at least one of the colliding vehicles.

To preserve hearing the ears should not be subjected to excessive noise. Prolonged noise (for example throughout the working hours) at an intensity exceeding 85 db in the speech-frequency range of 250 to 4 000 cycles per second, may cause permanent damage to hearing (acoustic trauma). Workers in such noisy surroundings should be supplied with, and encouraged to use some form of attenuator, such as ear muffs or wax plugs. Employers should be encouraged to fit any known noise-abating material or apparatus to existing machines and environment. Designers should produce less noisy machines, and soundproof materials should be used more extensively in building.

In his Annual Report (1970) the Chief Inspector of Factories says that 'rehousing in high flats had made noise from adjacent factories —that was hardly audible in two-storey homes, a serious problem for the occupants. If public attitude to noise hardens, pressure may build up to force industry to seek quieter processes'. At a Conference of the Royal Society for Health held in Edinburgh, Dr H. Schmidt said that 'In Germany poor hearing due to noise effects was already the industrial disease with the largest annual rate of increase. Technology can be civilized only if the people are prepared to pay the proper price'. Noise resistance can be built into a house— at an extra cost. Failing this, people will pay for technological progress with their health and there will, one day, be more people wearing hearing aids than there are at present wearing spectacles. For those interested in noise in industry there are two refer- ences.[191,192]

The ways in which sound is generated, radiated and transmitted; methods of reducing noise; the medical effects of excessive noise and the legal position in different countries are outlined in a handbook.[193]

The British Optical Assocation and the Royal Society for the Prevention of Accidents (RoSPA) are of the opinion that excessive noise creates *tensions* and *fatigue*, both of which contribute to *accidents*. Severe noise vibration in the ears causes impaired depth-perception and visual acuity. They say that the matter requires further research and more action from local authorities to diminish noise nuisance.

Occupational hearing loss is now compensable in Norway, Japan, Canada and about half of the American States. They appear to have resolved the intricacies of distinguishing occupational from non-occupational deafness, and determining which, of possibly several employers, is principally responsible for the disability. Britain is in the process of legislating on this matter.[191,192,193,194,195]

Summary

'Hearing' is much more than stimulation of the sensory apparatus in the ear by sound waves. Aural perception is closely allied to visual perception. The ear has the dual function of hearing and balance.

For descriptive purposes the ear is divided into:

the external ear	composed of	the pinna,
		the external auditory meatus
the middle ear	composed of	the oval window
		the promontory
		the round window
the internal ear	composed of	the cochlea
		three semicircular canals

Subjection of the ears to excessive noise can cause loss of hearing. Legislation attempts to define maximal noise levels to which ears should be subjected and makes provision for deaf people.

The Reproductive System

This story started with the discussion of a cell, followed by the union of a male and female cell, so that it is now expedient to discuss the systems wherein these gametes are made.

THE FEMALE REPRODUCTIVE SYSTEM

The organs comprising this system are divided into the external and internal genitals.

The External Genitalia

These are often referred to as the *vulva*, and comprise:

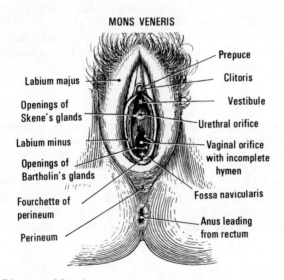

Fig. 221   Diagram of female external genitalia.

THE MONS VENERIS

This is a skin-covered pad of fat overlying the symphysis pubis. It becomes covered with hair at puberty.

THE LABIA MAJORA

This comprises two large 'lips' of skin-covered fatty tissue that form the sides of the vulva, extending backwards from the mons veneris

429

and disappearing into the perineum behind. They develop at puberty and bear hair on the outside; they begin to atrophy after the menopause.

THE PERINEUM

The skin-covered base of a triangular fibromuscular wedge (the perineal body) between the lower posterior wall of the vagina and the anterior wall of the rectum is called the perineum. The apex of the triangle is shown between the vagina and rectum in Figure 68 as a small circle. It will also be seen that the main muscle fibres in the perineal body are those of the levator ani. These can be torn during childbirth, allowing the bladder to fall back into the anterior vaginal wall, or the rectum to fall forward into the posterior vaginal wall, or the womb to descend the birth canal. You will learn to care for patients who are having a repair operation for a perineal tear.

THE LABIA MINORA

Two small 'lips' of highly vascular skin-covered tissue lie inside the labia majora, their anterior union being a hood-like structure called the prepuce, which surrounds and protects the clitoris. Having diverged, the lips meet again posteriorly in a loose fold of skin called the fourchette. There are no hairs in this area, and the modified skin is rich in sebaceous and sweat glands, so that movement can be accomplished without friction.

THE CLITORIS

This is a small piece of sensitive erectile tissue, protected by the prepuce, and forming the anterior apex of the vestibule.

THE VESTIBULE

This is a smooth triangular area, the anterior apex being the clitoris; the sides are formed by the folds of the labia minora, and the posterior base is the orifice of the vagina. The urethra opens into the vestibule.

THE ORIFICE OF THE URETHRA

This projects slightly from the normal surface level of the vestibule. Opening into the base of the urethra are two lubricating glands— Skene's glands—important because they tend to harbour infection, particularly in gonorrhoea.

THE ORIFICE OF THE VAGINA

This orifice occupies the space behind the vestibule and in front of the fourchette, with the folds of the labia minora forming its lateral

430

borders. It is normally a slit from front to back as the side walls of the vagina are normally in contact.

THE HYMEN

The hymen is a thin perforated fold of membrane stretching across the vaginal orifice. It divides the external from the internal genitalia.

THE FOSSA NAVICULARIS

This is a depression between the fourchette and hymen, and it receives the ducts from the Bartholin's glands.

BARTHOLIN'S GLANDS

These are two small structures, one on each side in the posterior portion of the labia majora. They secrete a lubricating fluid and pour it via their ducts into the fossa navicularis, just outside the hymen. Swelling of these glands from any cause gives a typical 'waddling' gait, and pain on sitting down.

The Internal Genitalia

Contained within the pelvic cavity are two ovaries, two utero-ovarian tubes leading to one central structure—the uterus—which is connected to the vulva by the vagina.

THE VAGINA

This is a fibro-muscular tube that opens from the vulva, and passes in an upward and backward direction to receive the lower portion of the uterus (cervix). The cervix descends below the vaginal walls, thus forming a circular groove (the fornix), deeper at the back than the front making the posterior vaginal wall 2 to 3 cm (1 in) longer than the anterior wall. For descriptive purposes the fornix is divided into four quarters termed the anterior, posterior, right and left lateral fornices. The hymen forms the external boundary. There are circular and longitudinal folds in the vaginal lining of modified skin, for not only has this organ to stretch to receive the penis during sexual intercourse, but it has to allow passage of the baby from the womb to the exterior. Even though this modified skin does not contain sweat or sebaceous glands, it is kept moist and free from friction by the secretions from the cervix. Throughout the reproductive phase of life, glycogen is stored in the vaginal cells under the influence of oestrin; Doderlein's bacillus, normally present in the vagina, ferments this glycogen and produces an acid medium in which many bacteria cannot flourish. Before puberty and after the menopause this mechanism does not work so that the vagina is more vulnerable to infection.

431

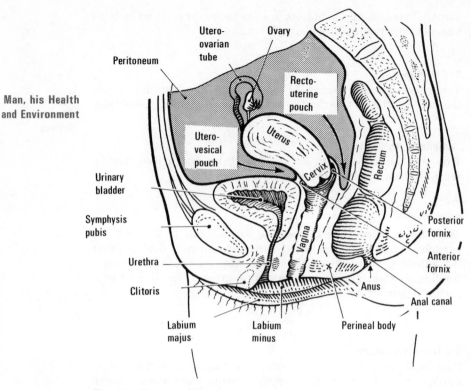

Fig. 222   Section of female internal genitalia.

The superior portion of the vagina passes through the pelvic diaphragm. The bladder and urethra are in close relationship to the anterior vaginal wall, and the rectum to the posterior superior wall; the perineal body separates the lower vaginal wall from the lower rectum and anal canal.

The vagina is supplied with blood by several branches from the internal iliac arteries, including the uterine artery. This blood is drained through the vaginal veins into the internal iliac veins. There are many lymphatic vessels surrounding the vagina. The plain muscle fibres are innervated by grey neurones from the sympathetic and parasympathetic portions of the autonomic nervous system.

THE UTERUS

This highly muscular organ is normally a content of the pelvic cavity. The uterus is separated from the urinary bladder in front by the uterovesical pouch of peritoneum, and from the rectum behind by the recto-uterine pouch of peritoneum.

The uterus is pear-shaped and lies almost at a right angle to the vagina; this position is called anteversion. Superiorly, on either side it receives a utero-ovarian (Fallopian) tube, the portion of uterus

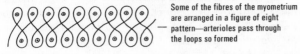

PLAIN INVOLUNTARY MUSCLE FIBRES

Some of the fibres of the myometrium are arranged in a figure of eight pattern—arterioles pass through the loops so formed

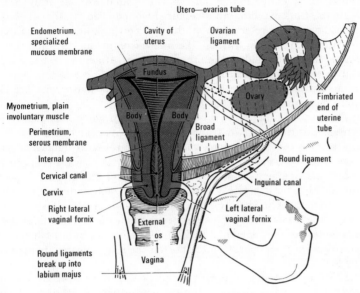

Fig. 223   Diagram of the uterus with its tube and ovary.

above their entrance being called the fundus. The body of the uterus separates the fundus from the cervix. Contained within the body is a triangular cavity, the inferior apex of which communicates with the cervical canal by the internal os. The cervix is a tubular structure enclosing a canal continuous with the cavity of the uterus, and communicating with the vagina by the external os.

The uterus is lined with a highly specialized mucous membrane, which, during the reproductive phase, is in a continual state of change, according to the menstrual cycle. The lining is called endometrium. The middle layer is composed of plain, involuntary muscle fibres, some of them arranged in a figure of eight pattern; arterioles pass through the loops so formed, so that as the uterus enlarges in pregnancy it gets an increased blood supply, and when the loops contract on parturition they act as ligatures and prevent excessive bleeding. Oxytocin, a hormone secreted by the posterior lobe of the pituitary gland, acts upon the uterine muscle to increase contraction at parturition. Grey neurones from the sympathetic and the parasympathetic portions of the autonomic nervous system supply these involuntary muscle fibres. This middle muscular layer is called the myometrium. The serous membrane, peritoneum, forms an incomplete outer layer as it is not adherent to the anterior surface of the uterus; the loose fold formed as it drops from its superior

433

attachment on the uterus to its superior attachment on the urinary bladder is called the uterovesical pouch. The deep fold formed as the peritoneum descends the posterior uterine wall and rises on the anterior rectal wall is called the recto-uterine pouch (pouch of Douglas). A collection of pus in this pouch can be drained per vaginam, thus avoiding an abdominal wound. The peritoneal covering is sometimes referred to as the perimetrium, and it hangs over the uterus and tubes, rather like a sheet on a line, thus forming a double fold at the sides, the broad ligament. It acts as a mesentery, and in it are transmitted blood and lymphatic vessels and nerves. Lying between its folds and attached to the posterior one are the utero-ovarian tubes and the ovaries.

The cervix is supported on the pelvic diaphragm and is kept in its central position by two transverse ligaments which are attached laterally to the bony pelvic walls. Two uterosacral ligaments pass from it to the sacrum. The fundus is kept in its forward position by two round ligaments that pass from it, through the broad ligaments out to the lateral pelvic walls, where they each enter the internal ring, pass through the inguinal canal, and out of it via the external ring, to be inserted into the appropriate pubic bone. Sometimes these ligaments become stretched and allow the uterus to fall backwards (retroversion); an operation to shorten the round ligaments restores the forward tilt.

The uterus is supplied with blood by a pair of ovarian arteries given off from the abdominal aorta, and a pair of uterine arteries given off by the pair of internal iliac arteries. The ovarian arteries descend and the uterine arteries ascend, forming an extensive anastomosis from which the blood is drained by corresponding veins. Oxytocin, secreted in the hypothalamus, stored in the posterior pituitary, and released to the blood as needed, causes contraction of uterine muscle, as occurs in menstruation, orgasm during sexual intercourse, and in parturition.

The function of the uterus throughout reproductive life is to receive an ovum every 28 days, if fertilized to retain and nurture same, and at the end of gestation to expel first the baby, and then the after-birth. If the ovum is unfertilized, then it is expelled in the next menstrual flow. The 28 day cycle whereby the uterus is prepared for, receives and expels an ovum is called the menstrual cycle.

The first four or five days of this 28-day *menstrual cycle* are the days of flow, when the thickened endometrial lining and the unfertilized ovum are shed from the uterus per vaginam. After this there is a short resting phase, and then the endometrium thickens, becomes glandular, and lays down glycogen in its cells, with which the ovum can be nourished should it be fertilized when it arrives in the uterus. This will be discussed in more detail after the tubes and ovaries.

THE UTERO-OVARIAN TUBES (Fallopian tubes)

These are sometimes called oviducts, and are two hollow tubes about

434

10 cm (4 in) long, travelling in the fold of the broad ligament. They are continuous with the upper, outer uterine cavity, and pass above the ovary, so that their fimbriated, flute-shaped ends lie lateral to each ovary. One of the fimbriae is actually attached to the ovary.

The lining is of ciliated mucous membrane which helps to waft the ovum along to the uterine cavity. The lining is supported on a submucosa of areolar tissue carrying blood and lymphatic vessels and nerves. The submucosa is surrounded by plain, involuntary muscle fibres arranged in an inner circular and an outer longitudinal layer to permit peristaltic movement which further assists the passage of the ovum along the tube. The outer covering is of peritoneum which is folded over to form the double layer below, known as the broad ligament.

The female peritoneal cavity is not a closed one, for there is a portal for ascending infection via the vagina, uterus and tubes.

Figure 224 demonstrates the blood supply via the ovarian artery and this is drained by the ovarian vein. The tube is supplied with

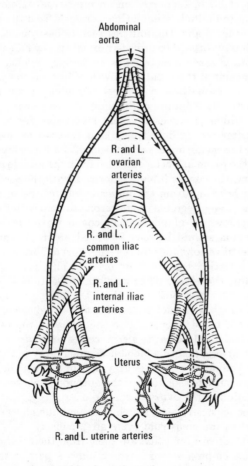

Fig. 224   Diagram showing blood supply to the uterus.

435

grey neurones from the sympathetic and parasympathetic portions of the autonomic nervous system.

The function of these tubes is to permit the passage of ova from the ovaries to the uterus, and of the spermatazoa from the uterus to meet the incoming ova. Fertilization takes place in the outer third of the tube, and the uterine lining is ready to receive the fertilized ovum by the time it reaches the uterus.

THE OVARIES

These are two small glands about the size and shape of an almond. They lie laterally in the pelvic cavity between the flute-shaped end of the uterine tube and the body of the uterus. Each ovary is attached to its corresponding tube by a fimbria, and to the fundus of the uterus by the ovarian ligament. The ovaries lie within the fold of the broad ligament, and are attached to the posterior fold.

Inside this peritoneal lining the ovarian cortex contains millions of follicles at birth, supported on an interstitial stroma. Under the influence of the follicle stimulating hormone from the anterior lobe of the pituitary gland, the follicles secrete oestrogens which allow the reproductive organs to grow to normal size by puberty, and allow the secondary sex characteristics to become evident at puberty, i.e. enlargement of the breasts, growth of hair on the external genitalia and in the axillae, a general rounding of the figure and a characteristically feminine outlook on life. At puberty they initiate ovulation which produces the first menstrual period (menarche), and thereafter one follicle ripens and releases an ovum every 28 days until the menopause, i.e. the end of the reproductive phase of life. After the ovum has been extruded from the follicle at the surface of the ovary, the corpus luteum is formed, with the assistance of the luteinizing hormone from the anterior lobe of the pituitary gland. The corpus luteum secretes progesterone to help with the thickening and laying down of glands and glycogen in the endometrium, in case the ovum should become fertilized on its journey along the tube. If fertilization does occur the corpus luteum persists for 12 weeks: if fertilization does not occur it only persists for 12 to 14 days. Healing then takes place with scar formation, so that whereas the child's ovary is smooth in outline, the adult's ovary becomes increasingly rough with each successive menstrual cycle.

Oxygenated blood is brought to the ovary in the ovarian artery from the abdominal aorta; it also receives an ovarian branch from the uterine artery, and this blood is drained away in the corresponding veins. The nerve supply is grey neurones from the sympathetic and parasympathetic portions of the autonomic nervous system.

The function of the ovary is threefold:

1. To produce oestrogens, which give feminine characteristics and play a part in the menstrual cycle after puberty.

2. To produce ova, and liberate them from their follicles.

3. To produce progesterone, which plays a part in the menstrual cycle and is essential for pregnancy.

436

The Menstrual Cycle

At the beginning of the 28-day cycle, the anterior lobe of the pituitary secretes a follicle stimulating hormone, which passes in the blood to the ovary. Here it causes follicles to undergo a ripening process, whereby they move into the central stroma and gradually secrete fluid. They then move out towards the surface, and when the intracellular pressure is at its maximum, one follicle bursts through the surface of the ovary, releasing the ovum to the peritoneal cavity, to be sucked into the uterine tube by the action of its fimbriae and cilia. All the other ripened follicles then undergo atrophy, continuing

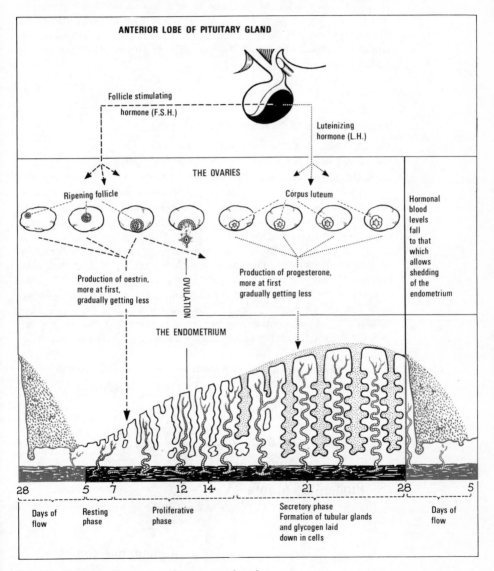

Fig. 225   Diagram illustrating the menstrual cycle.

secretion of oestrogen in the second half of the cycle. The setting free of the ovum is referred to as ovulation and occurs between the tenth and fourteenth days of the cycle. During this ripening process, oestrin is released from the ovary into the blood to be taken to the uterus, where, from the seventh day onwards it causes thickening of the endometrium.

To return to the ovary and the remnant of the burst follicle—it forms a structure known as the corpus luteum, under the influence of the luteinizing hormone from the anterior lobe of the pituitary gland. The corpus luteum produces progesterone, which causes the endometrium to thicken, become glandular and lay down glycogen in its cells. Thus the endometrium is ready to receive and nurture the ovum should it be fertilized when it arrives. If unfertilized, the hormonal blood levels fall to that which allows shedding of the endometrium.

Accessory Organs to the Female Reproductive System

THE BREASTS, MAMMARY GLANDS OR MAMMAE

These organs are present in the male in rudimentary form, but normally are physiologically inactive.

Each breast lies in the superficial tissue of the pectoral muscle (Fig. 60) on the front of the thorax. The weight and size of the breasts vary, they enlarge at puberty, and throughout pregnancy and lactation, and often atrophy in old age. Some females experience a feeling of tension within the breast before the onset of menstruation each month; this is part of a syndrome which was called pre-menstrual, but the term cyclical is now preferred.

The breasts are circular in outline and convex anteriorly, ending in a central prominence known as the nipple. The nipple is surrounded by a pigmented areola. Near the base of the nipple on Montgomery's tubercles are modified sweat glands. They secrete a

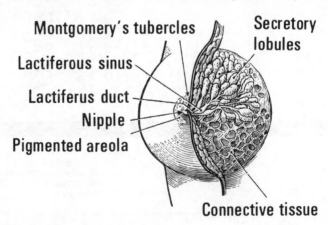

Montgomery's tubercles

Secretory lobules

Lactiferous sinus

Lactiferus duct

Nipple

Pigmented areola

Connective tissue

Fig. 226   Diagram of the female breast.

438

fatty substance for lubrication. The nipple is perforated where it receives the milk ducts (lactiferous ducts). Lymphatic vessels are numerous and they drain into groups of nodes, the axillary, internal mammary and supraclavicular nodes. Any infection or cancer in the breast can be quickly transmitted to these nodes, hence the teaching that any 'lump in the breast' must be examined by a surgeon.

The breasts receive their blood supply from the internal mammary and the intercostal arteries, and blood is drained away in the corresponding veins. White sensory neurones transmit sensation to the post-Rolandic area of the cerebrum for interpretation.

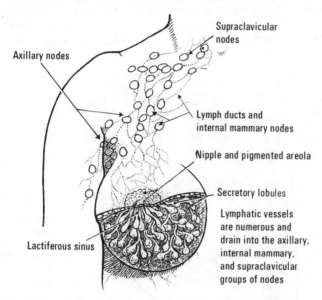

Fig. 227   Diagram illustrating lymphatic drainage from the breast.

Under the influence of the prolactin, a hormone from the anterior lobe of the pituitary gland the breasts secrete a small amount of colostrum during pregnancy and for the first three days after parturition; thereafter they secrete milk, the ideal food for a baby (p. 251). Suckling stimulates sensory nerve endings in the nipples, in response to which, milk is secreted.

THE MALE REPRODUCTIVE SYSTEM

THE TESTES

In these glands the male cells, the spermatozoa, are made. In intra-uterine life the testes form high up in the posterior abdominal cavity near the kidneys, one on each side. Before birth they wend their way downwards, and round the pelvic brim to the lateral anterior pelvic cavity. From this cavity they pass, via the internal inguinal ring into the inguinal canal, pushing a sheath of peritoneum around them. They leave the inguinal canal via the external inguinal ring to drop into the scrotal bag, which is their home thereafter. It is thought that they require exposure to external atmospheric temperatures for functioning. The peritoneum around the testes remains alive and is called the tunica vaginalis; the sheaths of peritoneum in the inguinal canals become obliterated and form part of the spermatic cords. Should they fail to be obliterated two possible sites are left for the escape of a loop of bowel (hernia).

The male-sex hormone, testosterone, is made in the testes and it is responsible for initiating and maintaining the secondary sex characteristics—hair in the axillary and pubic regions, growth of the beard, deepening of the voice due to elongation of the larynx and characteristic mental outlook.

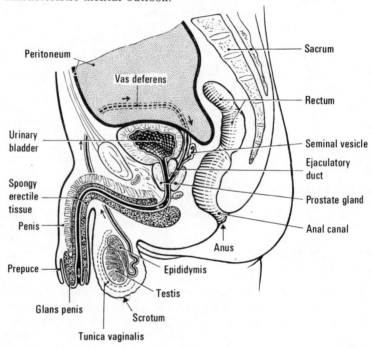

Fig. 228   Diagram of the male reproductive system.

THE EPIDIDYMIS

This is a series of convoluted tubules arranged as an oval body on the posterior superior surface of the testis. It conveys spermatozoa from the testis and deposits them in the vas deferens.

THE VAS DEFERENS

Continuing from the epididymis is the vas deferens. This is a tube which enters the external inguinal ring, to pass through the inguinal canal, and out of it via the internal inguinal ring, thus reaching the

440

pelvic cavity. In the inguinal canal it forms part of the spermatic cord. In the pelvic cavity each vas deferens passes over the top of the urinary bladder and descends behind it, to enter the base of the seminal vesicle lying behind the urinary bladder.

THE SEMINAL VESICLE

This is a pear-shaped, hollow organ that acts as a reservoir for semen, the fluid containing the spermatozoa. The walls contain some plain, involuntary muscle fibres and on contraction the semen is forced along the ejaculatory duct into the urethra.

THE EJACULATORY DUCT

This leaves the base of the seminal vesicle and passes forward into the substance of the prostate gland, where it joins the urethra coming from the urinary bladder.

THE PROSTATE GLAND

This lies at the base of the urinary bladder, and is partly glandular and partly muscular. The ejaculatory duct joins the urethra within the substance of prostate gland. Its secretion is thought to make the spermatozoa motile and ready for their journey through the penis, the uterus and the uterine tubes.

THE PENIS

The urethra traverses this tubular organ, to reach the exterior. The penis is composed of spongy, erectile tissue. At the tip is an enlargement, the glans penis, normally covered with a loose double fold of skin called the prepuce or foreskin. It should be possible to draw this foreskin back over the glans; inability to do so constitutes phimosis, the treatment for which is circumcision.

THE SCROTUM

This is a pouch or bag-like structure, covered with skin, continuous with that of the trunk. The cavity is divided into two by a septum, and each compartment contains a testis with its tunica vaginalis and epididymis, and connecting vas deferens. These structures are supplied with blood by the spermatic artery, and are drained by the spermatic vein, these latter two vessels being part of the spermatic cord.

Man, his Health and Environment

As all tissues need to be nourished so that they can be healthy and grow to adult stature, it is especially important that this should happen in the reproductive system, for perpetuation of the species and satisfaction of man's instincts. As vitamin A plays a part in the health of mucous membranes and prevention of infection it should be included in the diet. Vitamin E is said to be anti-abortive, so that it too should be included. It will be useful revision to write down the foods containing these vitamins.

Meticulous personal cleanliness in these areas is essential, especially in women, and if daily bathing is impracticable then at least a daily wash must be routine. The special cleanliness pertaining to the menstrual period has been mentioned, and it is important that girls develop a healthy attitude to this occurrence, and continue with their normal routine of exercise, fresh air, daily bathing and hair washing.

There are some parts of the world where the women only wear a loin-cloth, so that their breasts are exposed. In western cultures most women wear a brassiere.

The Health Education Council urges women over 24 to help in the *early detection of breast cancer*. A pamphlet which can be obtained from local authority clinics, illustrates how to *examine* the breasts *routinely*. It advises, 'If you find a lump, don't jump to the wrong conclusion and assume the worst. The chances are it's only a cyst. . . . The only way you can really put your mind at rest is to see the doctor.' Breast cancer continues to kill nearly one-fifth of the women who die from cancer each year. If the condition is found sufficiently early, it is curable.

Smear tests in 1970 for the detection of *cervical cancer* numbered 1 850 000, and it is expected that this figure will exceed the 2 million mark by 1976, due to the *Routine Recall Scheme*. Women aged 36 and over, whose previous test result was negative, are invited to have a further test five years after the last examination. Tests are made by general practitioners, local health authority clinics, Family Planning Clinic, and those arranged by some firms for their employees. Dr Helen Pridan did a survey in several countries and found that a woman's chances of contracting cervical cancer increased the younger the age of first intercourse, and with the number of male sexual partners. Social conditions or the number of pregnancies had no significant effect. Virgins had a low incidence and prostitutes a high one.

An Edinburgh study of more than 3 000 mothers found that significantly more of those whose children showed congenital abnormalities had taken aspirin, indigestion tablets, pep pills, and some cough medicines among other common drugs. The doctors state that, during pregnancy, '*self-medication* with common household remedies such as aspirin and antacids should be *avoided*',—although they advise caution in presuming the deformity effects on the basis of the associations shown.

Women who smoke produce smaller babies than those who do not smoke, and most clinics try to help the smoking mothers to *abstain from smoking* during pregnancy.

A Glasgow scientist has won an award for research that shows it is undesirable for pregnant women to be in contact with domestic cats, which can be infected with a protozoan parasite *Toxoplasma gondii.* Disease in human beings as a result of infection is rare, but in England, blood tests have shown that about 30 per cent of the human population are infected. In many instances where a woman has already had an infection the child is completely protected. But there is a possibility of damage to the child in some way, if the *expectant mother* contracts the infection without previous experience of it.

We have already spoken of the possible congenital abnormality of a baby born of a mother who contracted *German measles* in early pregnancy. It is therefore wise for girls of school leaving age who have not had this infection to be protected against this disease by *immunization.*

The reproductive system serves the purpose of *procreation of the species.* Sex education is only one part of the very large subject of Health Education which aims at helping people to develop sensible attitudes to their own bodies, and to treat their bodies with the respect they deserve. Health Education seeks to spread knowledge about the production of a stable physical, social and psychological environment in which a child can pass satisfactorily through each stage of development to become a sensitive, interacting, self-disciplined, *socially responsible adult member of society.* Information about a recently formed group that is concerned about social responsibility can be procured from Graham Heath, Honorary Secretary, The Responsible Society, 1/39 Portland Place, London, W1.

Reproduction is accomplished by *sexual intercourse.* The vagina is normally just a slit from back to front, the side walls being in contact. The penis is normally flaccid over the scrotum. Without erection of the penis and lubrication of the entrance to, and dilation of, the vagina, insertion of the penis into the vagina is impossible. Lubricating fluid is poured, from Skene's glands into the vestibule of the vulva, and from Bartholin's glands into the vaginal orifice, in response to love-play—kissing and mutual fondling of the other's body. Dilation of the vagina occurs with simultaneous erection of the penis, at which time it is inserted. Insertion without adequate lubrication and vaginal dilation can be painful to the woman. (Some women have a lessened secretion of lubricating fluid at, or after the menopause, and are helped by the application of a jelly that is sold at chemists' shops and Family Planning clinics.) After insertion there are rocking movements by the man while the woman bears down synchronously with the inward rock of the man. This phase has been likened to massaging of the penis by the vagina and gives increasing pleasure to both partners, rising to a climax, the orgasm, with ejaculation of the semen. The movements die away gradually and the penis is withdrawn before the vagina returns to its undilated state. Some partners experience orgasm at first intercourse. Other partners take several months or years to acquire the technique of achieving orgasm every time. In some couples each experiences orgasm but not at the same time. Some experience frigidity at times —lubrication of the vulva and dilation of the vagina does not occur in response to overtures. A man can be impotent, i.e. fail to erect the penis, or fail to sustain an erection, at least on occasions. Some men, on occasions, have premature ejaculation, i.e. immediately on erection and before insertion of the penis into the vagina.

Earlier in this book we spoke of the signs and symptoms of early *pregnancy*[196] and the antenatal care that is available to every pregnant woman (p. 36). The fetus grows within a membranous sac containing liquid. The mother experiences a changing body image as she gets bigger. Likewise the father experiences a changing image of the mother-to-be. The mother experiences 'quickening'—the first fetal movements felt by the mother, usually at 16 to 20 weeks' gestation. Some couples have occasional intercourse during the early months of pregnancy, but later there should be abstinence. Although wanting, and looking forward to the baby, some women begin to feel weary towards the end of pregnancy when they feel big, cannot see their feet when standing, have difficulty when bending and cannot sit still for long. A few women experience a greater mood swing than usual. The supporting role of the father is important at this stage. The mother needs encouragement in performing the relaxation exercises taught at the clinic. There is often a feeling of relief at the onset of labour pains (contractions of the uterus), accepted as the beginning of a happy ending. Rupture of the sac surrounding the fetus and escape of liquid can occur just before, or at the onset of contractions when the liquid oozes from the vagina. Or rupture of the sac with explosive expulsion of the liquid can occur when contractions are fully established. When the contractions are fully established and regular, with all the carefully prepared things for herself and her baby, the mother-to-be is admitted to hospital. Some hospitals make arrangements for the father to be present at the birth of his baby. This important decision should be made by the couple during pregnancy. Members of the staff make frequent examinations of the genital tract, recording dilation of the cervix which is the prelude to the baby passing down the tract. The mother is asked to co-operate by bearing down at each contraction of the uterus, thus helping to push the baby out. After expulsion of the baby, there are further contractions to expel the after-birth (placenta). The mother is bathed and made comfortable and again there has to be adaptation to her body image, which is now minus the bulge! Then for both father and mother come all the special feelings associated with holding their baby in their arms. The mother is encouraged to do exercises to strengthen the previously stretched muscles and help the uterus regain its normal size and position.

Whether the baby is the firstborn or subsequent child, the father and mother become involved in the process of *infant and child care*.[197] The new baby has to be integrated into the family unit. He has to be breast or bottle fed and in due time weaned (p. 231). Through the medium of feeding he develops an attitude to food (p. 217). Each time he passes urine and faeces his napkin has to be changed, the soiled napkins have to be rinsed and washed. (There is a reference for napkin care on p. 133, and the process of achieving voluntary control of passing urine and faeces is discussed on p. 293.) The baby has to be suitably clothed at the various stages of babyhood, and he acquires an attitude to clothes and their care (p. 135). His skin has to be cared for, starting with removal of the vernix caseosa by the midwife, and progressing to independent bathing by the child (p. 125). It is normal for the infant and child at various times of the day to be naked and have others look at his body. According to how he learns discretion, about when, where and before whom it is acceptable for him to appear without clothes, he develops an attitude to being naked. This can vary from liking it too much, which can lead to indiscretion. Extremes of indiscretion, such as being naked on the street, are a legal offence in western society. Or he can develop a feeling of guilt about it and this can lead to prudishness. The norm is to feel comfortable when naked in suitable situations. It is normal for infants and children to be fondled and kissed by parents and family. The quality of this handling, and the feeling engendered in the child are variable, and are very likely to have an effect on the amount and type of handling that as an adult he can enjoy giving and receiving. He learns to use discretion about whom he is allowed to touch and kiss, and when this is acceptable behaviour. He also learns when and from whom he should reject touching and kissing. He is continually learning what is, and what is not, acceptable behaviour, and acquiring language which is a complex process. With the ability to walk and talk, his horizon widens to include children from other families. Membership of the first group outside the family is usually nursery or primary school. In the latter formal education is started, and many steps are taken along the long, slow road to independence. (See Normal Development, p. 195.)

Most human reproduction is associated with some form of *marriage*, and *provision of a shelter* (p. 464), in which the young can be nurtured to maturity. The marriage ceremony is celebrated in many different traditional styles throughout the world. Societies differ as to the marital patterns into which the young are socialized. In some primitive societies there are *unwritten laws* and in *developed countries* there are *state laws* about which near relatives are excluded from marrying each other. The disadvantages of near relatives marrying is now known to have a scientific foundation. Genes carrying hereditary disease are *more likely* to be present in *both* parents, this giving rise to a greater incidence of afflicted children. The unwritten law of primitive societies also governs how many wives a man may have at any one time. Though there may be a marriage ceremony, it may not be registered in a legal sense, but it may be honoured by common consent of the community, as may be the material possessions of these families.

In developed and many developing countries, *marriage* is a *legal institution*, and has *to be registered*. The law sanctions the places

at which marriages can be conducted, and the conditions of compliance so that the marriage is legal. Registry Offices, and most of
the churches of most of the denominations have a licence to conduct
'legal' marriage ceremonies in our country. The law also states
that one man can only have one legal wife at any one time. Should a
man marry a second woman while still legally married to one woman,

he is guilty of bigamy, a punishable offence. There are people who
think that the marriage laws will undergo vast changes in the next
30 years. It will be for each country's people to decide whether they
want to continue to promise fidelity until 'death do us part' or for 'as
long as I am able' which might come to be considered more realistic,
since human situations are constantly changing. There are some
clergymen who already use this alternative. Dr Robert McArthur, a
London sociologist is at present researching this subject. It is
thought that of all creatures in the world, human beings are the
most dependent on consistent company, and that if all marriages
were dissolved overnight, most couples would stay together out of
mutual dependence, and that as society progresses marriage and
fidelity may even strengthen.

In our country there are legal rights of a husband and wife, their
treatment of each other, their material possessions and disposition
of these on the death of one or other, or both.

The United Nations Children's Fund (UNICEF) sets out the
rights of children all over the world (p. 31). Transgressors of these
rights can have sanctions applied to them by other members of
society, e.g. neighbours, and if the transgressors continue in their
deviant behaviour, society at large, through the medium of the law,
applies legal sanction. There is at present controversy as to whether
these sanctions should be in the form of punishment or treatment.
In the professional journals you can read about the 'battered baby
syndrome'. Dr Burglass, of the Midland Centre for Forensic
Psychiatry, says that the term is too legalistic and reflects our
preoccupation with the physical injuries done to the child, to the
exclusion of the emotional problems of the parents. He prefers the
Continental term 'child mishandling'. At the same time one cannot
ignore that physical injury to the child will be accompanied by
emotional injury that can be manifested much later in life. Informing
one's elected Member of Parliament of one's attitude to such important social issues, in this instance punishment or treatment of
child mishandlers, allows him or her to know the climate of opinion
among the members of the constituency ready for when these important issues are discussed in Parliament.

Having considered some of the rights of children, we come to the
subject of illegitimacy. When considering numbers of illegitimate
children one has to remember the very different culture patterns to
be found around the world. In some of these, it is the custom for a
man and a woman to co-habit with no thought of getting married.
Yet the children are by definition illegitimate. It has also to be remembered that a country's illegitimacy rates will be related to the
laws on contraception, abortion, minimal age of 'legal' intercourse,

446

and the age of consent for marriage. Also in developed countries, a minority of couples may have children and a stable home, but remain unmarried, because of personal beliefs. In a 1966 to 1968 study of all the countries that have records of illegitimacy, Jamaica had the highest rate with 74 per cent of total live births out of wedlock. It was followed by a list of 25 South American and Caribbean countries, the list broken only by Iceland in sixteenth place (28 per cent) and Sweden in twenty-fifth place (13 per cent). Britain was half way down the list with 7 per cent. Roman Catholic countries rated 6 per cent and Moslem countries 1 per cent. With more widespread knowledge of contraception, availability of contraceptives and more liberal abortion laws, a decrease in the number of illegitimate births throughout the world is expected.

A pregnant unmarried woman often delays seeking medical advice. In such circumstances the growing fetus has inadequate antenatal care and the woman has inadequate preparation for childbirth. If the baby is the result of a casual sex encounter the unmarried person is robbed of male support at a crucial time. If still at school, she has no insurance stamps and cannot claim maternity benefit. If she comes into the category for legal abortion, she is faced with making this decision, otherwise she has to choose whether to bring the baby up herself, or have it fostered or adopted.

In Scotland in 1974, 9·1 per cent of live births were illegitimate compared with 8·8 per cent in the previous year. Birth certificates are now the same in appearance to avoid embarrassment of the illegitimate child on those occasions when one has to produce a birth certificate. It is not necessary to enter the father's name, but in some countries the child can, on attaining the age of majority, ask for this at the National Registry Office. Some countries now allow the illegitimate child to participate in legal inheritance.

The National Council for the Unmarried Mother and her Child prepared a Report on one-parent families in 1971, which was submitted to the Government Committee investigating this problem. It stated that though the Abortion Act has reduced the illegitimacy rate, unmarried mothers still face considerable difficulty in getting proper help from the Government agencies and from the fathers of their babies. It recommends a special child-care allowance of £5 a week to be paid to all mothers who, on their own, support children; additional allowances for each child; and the right to sue the father. A further recommendation is that the word 'bastard' or 'illegitimate' should be outlawed. Instead legal documents should refer to the 'natural child'. Fathers should be encouraged to have their name entered on the baby's birth certificate for the sake of the child. The Report stated, 'It is important that a child should know as much about himself as possible'.

Some countries offer a *Marriage Guidance Counselling Service* to those who find that their marriage is threatened. The trained Counsellors are willing to talk to schoolchildren, members of youth and other clubs, about preventive work in this field. From their vast experience of things that have gone wrong in other peoples' marriages the Counsellors in some areas talk to those contemplating marriage. This is an important contribution to the nation's health, for an unhappy person is not 'at ease' with himself or others and therefore suffers from 'dis-ease'.

Countries have different *divorce* laws, some recognizing only cruelty, adultery and desertion for a specific period, others recognizing irretrievable breakdown of the marriage as a reason for granting a divorce. Irretrievable breakdown underlines the importance of the *quality* of a total relationship. Some people cannot countenance divorce because of their religious beliefs.

The subject of *contraception* is of world wide importance and scarcely a week passes without reference to it in some context in the daily press. Death rates have fallen and birth rates have not been reduced, thus 'people' are polluting the environment! It is referred to as the population explosion. The United Nations Economic Commission for Asia and the Far East cited political hesitancy, admintrative complexities and irrational resistance as three reasons for delays in any significant decline in Asia's population through family planning. Malnutrition is threatening the mental and physical development of millions of children in Asia. Until 1969 in France, users or advertisers of contraceptives were liable to jail sentences. But France has legalized contraception as its contribution to the population crisis. Family Planning projects in many of the developing countries have been carried out by the World Health Organization. In the early projects, the women were fitted with the intra-uterine device (p. 451) as it needed no further precaution. Some of the early loops were less satisfactory than those of today, so some of the later projects offered sterilization to the men. Some Governments pay the men for being sterilized.

The name of Marie Stopes is inextricably associated with the word contraception, for she pioneered the first 'Birth Control' clinics. Now we talk about 'Family Planning'.[198,199,200,201,202,203] Like many other associations, the Family Planning Association started as a voluntary service in this country, but now our Government encourages local authorities to provide clinics for this purpose. In some areas the local authority pays an annual subscription to the Family Planning Association and the clinics continue to be called Family Planning Clinics. The term now has a much broader interpretation and embraces family spacing, family limitation, family postponement and family avoidance. In Britain in 1971 there was a National Family Planning Drive as part of our contribution to the world population crisis. The Health Education Council's poster showing a 'pregnant' man won a British Poster Design Award. Until stronger pressure was put on local authorities to include unmarried boys and girls as clients at Family Planning Clinics, the girls did not have easy access to reliable contraceptives. Provided they knew about them and knew what to ask for at the chemists' shops, they could buy vaginal pessaries, foams and creams and insert them before intercourse, but these are unreliable in killing the thousands of sperms ejaculated into the vagina. The boys had the traditional sources of barbers' and chemists' shops at which they could buy condoms. Some boys procured the pessaries capable of killing sperms and assured the girl that conception would not occur if he inserted one into the vagina before intercourse. The pessaries do contain a substance that will kill sperms *if* the sperms are exposed to it for a sufficient length of time. The pessaries, creams and jellies take time to melt. The motility of sperms is such that they can, unaffected by the substance, swim through the uterus and Fallopian tubes, there to meet an ovum.

The *Brook Advisory Centres* opened to cater for unmarried boys and girls seeking advice about contraception. Clients pay £4 per annum for consultations which vary from 3 to 20 and they pay for any contraceptives procured. Advertising is fraught with difficulties and the Centre's activities are publicized mainly by word of mouth. An appointment can be made, by a boy or girl, by telephone or by calling at the Centre, to see one of the doctors in attendance at clinic sessions. A simple form is completed and the client is given a full medical examination. The services of a consultant gynaecologist are available should any abnormality come to light. The contraceptive pill is prescribed for 85 per cent of the girls. It is suggested to her that her general practitioner should be informed. If she is reluctant to involve the family doctor, the issue is not forced. The seeking of advice is considered to be a *personal responsibility* and the parents are not informed. Opinions and arguments for or against sexual intercourse are not offered. A trained social worker is available to talk with any client who is perplexed and such a discussion may or may not end in the client deciding against contraception. Many people think that the disinhibiting effect of alcohol and some drugs does a great deal more to bring about premarital sex than the availability of contraceptives. There are people who believe, that at a time when the human race is under threat from over-population, it is immoral to deny people the means of avoiding unintended pregnancies. There are others who believe in self-discipline with regard to sex, drugs and alcohol in the avoidance of unintended pregnancies.

Sterilization as a form of contraception has already been mentioned. For the female it means tying the Fallopian tubes, so that it involves an abdominal operation. It is much easier and safer to tie the vas deferens in the male. The Family Planning Association in Britain has as yet, less than 10 male sterilization or 'vasectomy' clinics, but it is their policy to open more. For this service a fee is charged. Availability of the operation on the National Health Service is being 'considered' by our Parliament. At present it is only available for 'medical reasons'. Counselling is important before the operation, because it is irreversible (although there have been claims to the contrary). If a man's children were to die through some tragedy would he be prepared to accept that he could not have another child? If his existing marriage ended for any reason would he be prepared to face the fact that he could not have children by a second wife? (In the United States of America a man about to be sterilized can have his sperm banked so that if he loses his existing children, his

449

wife could replace them after artificial insemination.) In Britain's clinics vasectomy is performed on a Friday night. It involves a small wound at the inner side of each groin, the tying of the vas deferens on each side, then three or four stitches in each wound. The patient goes home and can go to work on Monday. The stitches are removed the following Friday. The experience in Britain is that the men are

mainly over 40. They do not want any more family under any circumstances. They have tried most other forms of contraception and for one reason or another found them unsatisfactory. The Simon Population Project, West Longsight, Crediton, Devon, will provide information about the operation. After couples have discussed their family circumstances and their reasons for wanting a vasectomy, the Project will provide names of surgeons and probable fees (these seem to range from £12 to £50) through the couples' family doctor. Of the annual average of 1 012 men who had vasectomy through the Simon Population Project, 99 per cent would recommend the operation to others.

Attitudes of 18 year olds to sex before marriage have changed significantly in recent years, according to a survey carried out in 1968 and repeated in 1973. In a question intended to test attitudes towards contraception, they were asked if they would risk sexual intercourse without contraceptives. Among the middle class members of the sample, 33 per cent of the men and 68 per cent of the girls said they would. This showed a remarkable swing compared with five years ago, when 67 per cent of the men and 43 per cent of the girls answered 'Yes'. The men had become more cautious, while the girls had become more reckless. Among the working class members of the sample, there was no marked discrepancy between the sexes, and no great change in the five years—77 per cent of the men and 66 per cent of the girls thought the risk was worth taking. But on the whole the girls in both social classes still believed it was the man's job to take precautions. Not that their answers were consistent. When asked to select the most objectionable quality in a husband or wife, nearly all the 18 year olds plumped for 'previous sexual experience with another party'!

There are three *female methods of contraception* approved by the Family Planning Association. Each of them should only be used after medical examination and under medical supervision. First the *oral method*, commonly spoken of as the pill. It is a combination of synthetic hormones which must be taken every day, usually for 20 or 21 days each month. 'Withdrawal bleeding', which is the equivalent of menstruation that the body's hormones normally produce during three to five days of a 28 day cycle, occurs during the twenty-second to the twenty-eighth pill-free days of this *regularly* induced cycle. For the first few months some women experience the side-effects of weight gain, painful breasts and nausea, although in most cases these subside. There is now a 'sequential pill'. A one-hormone pill is taken for 15 or 16 days, and a two-hormone pill for five days and the withdrawal bleeding occurs in the pill-free days. It is said to give rise to fewer side-effects. Discipline is involved in keeping

450

to the regime and the memory has to be relied on, though firms prepare a month's supply with a pill against each number, to make it as safe as possible. It avoids any preliminary fussing before or during the sexual act.

The second method is the *rubber, dome-shaped cap or diaphragm* that should fit snugly round the cervix of the uterus which invaginates into the upper vagina. It forms a mechanical barrier to the sperms. It should always be used in *conjunction with a spermicidal* cream or jelly. Initially the diaphragm is fitted by a doctor who checks the size. He also instructs the client about self-insertion and removal. Most women insert the diaphragm each evening and remove it each morning, except during menstruation when most couples refrain from intercourse. The cap should be left in position for eight hours after intercourse, as sperms can remain active for several hours. Users should keep a check on their weight and see the doctor at a change of half a stone either way, so that he can check if the size of the diaphragm is still effective.

The third approved female method is the *intra-uterine device.* The principle is old, but modern plastics and new designs of the device have renewed interest in this method. The devices are coils, loops, rings and double coils which is the latest and most nearly resembles the shape of the womb. There must be a medical examination and the doctor fits the device in the womb, where it remains, and no further precaution need be taken. Some women are unable to tolerate the device which acts as a foreign body and they may involuntarily expel it, sometimes without their knowledge. Women are advised that there may be some initial discomfort, backache, and increased menstrual bleeding. Occasionally a device has to be removed on medical grounds. It is considered more suitable for women who have borne children than single women. There should be an annual check-up by the doctor.

Some people *erroneously* believe that chemicals alone, in the form of vaginal pessaries, creams and foams inserted into the vagina before intercourse will prevent conception. *Chemicals should only be used in conjunction with another method of contraception.*

The first method of *male contraception* is the protective, *condom* or sheath. It is a very thin but strong covering of latex rubber which is unrolled on to the erect penis before insertion into the vagina. It is available in lubricated and unlubricated form, the former being more popular. At the climax of intercourse, the sperms are ejaculated into this covering. The man must accept the responsibility of seeing that his penis continues to be covered by the sheath during withdrawal. Each protective sheath is used once only. Those that carry the 'Kite mark' are approved by the British Standards Institution.

The second male method is withdrawal of the penis before ejaculation (*coitus interruptus*). Though it is condemned by most doctors as unreliable and capable of causing frigidity, emotional instability, and antagonism between partners, it is still practised. It takes tremendous self-control to withdraw, and there is great risk of frustration in both partners. Whilst the penis is within the vagina, but before the orgasm occurs, a small drop of lubricating fluid is released and this can contain a few sperms.

The Roman Catholic Church countenances the *rhythm method* of contraception. Intercourse can only take place after menstruation and before ovulation which occurs between the tenth and fourteenth days of the menstrual cycle. Some women who have a regular menstrual cycle can find which of the tenth to the fourteenth is her particular day of ovulation because of a slight increase of body temperature on that day.

The cost of contraception differs in different parts of our country. Consultation and contraceptives are free in some local authority clinics. Consultation is free but contraceptives have to be paid for in other clinics. In other clinics, consultation and contraceptives have to be paid for. Countries vary as to the amount of public money spent on keeping down the numbers of world population.

Abortion is *not* a form of contraception, but it can be used for *failed* contraception. In view of the world population crisis, world policy is that it is far better to have family planning clinics than abortion clinics. In this country we are in the curious position of providing free abortions, but charging for contraceptives in many places.

The term *abortion* means expulsion from the uterus of the product of conception before the end of the 28th week. Some women threaten to abort, when there is slight bleeding, but with adequate rest and treatment the pregnancy continues to full term. Other women abort (*spontaneous abortion*) in successive pregnancies and seem to be unable to carry a baby to full term. Then there is the *criminal abortion* which in our country, until 1968 was defined as intentional evacuation of the uterus on *any other than medical* grounds. This may be self- or other-induced. Being a party to a criminal abortion is a legal offence in most countries. Abortion can be carried out as a surgical procedure and the legal reasons for doing this vary from country to country.

Fifty per cent of the world's population now lives in countries with abortion laws that permit termination of pregnancy on *social* as well as medical grounds. India is the most recent addition to the list and their new laws came into effect in January 1972. In this country, the more liberal abortion law, countenancing social grounds, came into effect in 1968. Since then abortion law reform has taken place in Singapore, and South Australia, together with some slight changes in Canada. Some African and Caribbean nations are now giving consideration to amending legislation. German law provides for a maximum of five year's jail for a woman undergoing abortion and anyone helping her to obtain one. A campaign is being waged in West Germany to have the abortion ban removed.

In Scotland the number of recorded abortions per year is 3 544 in 1969, 5 254 in 1970, 6 332 in 1971, 7 600 in 1972, 7 498 in 1973 and 7 545 in 1974. In the last year mentioned 7 413 abortions were carried out in National Health Service Hospitals and 132 in other places; 208 concerned girls under 16 years of age and 1 761 of the women

452

were from 16 to 19 years old. Single women accounted for 43 per cent of the abortions, married women for 45 per cent and widowed, divorced and separated women for 12 per cent. Risk to the physical or mental health of the pregnant woman is the most common statutory ground for termination of pregnancy, accounting for 93 per cent of all recorded abortions in 1974. The 1970 World Health Organization Technical Report series, No. 461, states that 'abortion may precipitate serious psychoneurotic or even psychotic reactions in some women, the risk being greatest in those with previous emotional problems, thus the very women for whom legal abortion is considered to be justified on psychiatric grounds, are those who have the highest risk of post-abortal psychiatric disorders. Careful and competent counselling probably reduces the risk of unfavourable reactions, whether abortion is induced, or the pregnancy carried to term'. The late physical effects of abortion according to the report, are *dominated by* the sequelae of *infection*, but this is associated far more frequently with illegal (including self-induced), than with either spontaneous or legal abortion.

The Reproductive System

In a book *Woman on Woman*, Lady Summerskill says that she used to favour abortion on demand until she visited the Soviet Union in 1931. 'Woman after woman was wheeled into the large operating theatre of the abortion clinic where six women doctors were operating simultaneously. These pregnant women had been exploited by selfish men. It was a relief when I heard that abortion on demand was prohibited in the Soviet Union.' Mr H. Gordon, a consultant gynaecologist in London, said at a Conference that abortion on demand had proved so disastrous in the Iron Curtain countries that some of them had made major changes in their policies over recent years.

This then is a glimpse at the background to the emotive subject of abortion. The last word has not been spoken. Arguments continue to be waged in religious, social and medical circles. Each individual has to develop and fashion his/her own attitudes and beliefs about this, and many other important social issues. It is an essential part of growing into a responsible citizen.

SEXUALLY TRANSMITTED DISEASES[204, 205]

Countries vary as to the diseases which are *legally* defined as *venereal*, i.e. pertaining to, or caused by sexual intercourse. In this country there are three, namely syphilis, gonorrhoea and chancroid (soft sore). There are other diseases that are *medically* but not legally defined as venereal. Of these, those that are *usually* sexually transmitted are non-gonococcal urethritis (NGU, synonym non-specific urethritis, NSU), which occurs in males and is closely associated with thrush (syn. candidiasis) in females; trichomoniasis, genital scabies, lice and warts. Those that are *probably* sexually transmitted are genital herpes and candidiasis (syn. thrush).

In Scotland there were 7 814 new cases of *legally* defined venereal infection in 1950; 6 952 in 1960; 13 073 in 1970 and 21 078 in 1974.

SYPHILIS

The main points are summarized in Table VIII. Further information can be found in the Family Doctor booklet *So Now You Know About VD and Diseases Transmitted Sexually* obtainable, price 10p. (Australia 30 cents, New Zealand 30 cents, South Africa 25 cents, Caribbean 40 cents) at any branch of Boots or from the British Medical Association, 47–51 Chalton Street, London, NW1.

Countries vary as to what tests are done routinely on pregnant women with regard to venereal infection. In this country a *test for syphilis* is done early in pregnancy, so that any infection can be treated and thus an uninfected baby is ensured. In France also, a test for syphilis is obligatory *before marriage*, and during the first three months of the first three pregnancies. All over the world, *blood* taken from *donors* is tested for syphilis. These measures have helped to reduce the number of new cases reported annually, from 1 691 in 1950 to 144 in 1974 (Scotland).

GONORRHOEA[206]

Within two to five days of infecting intercourse, most men know that they have acquired an infection. They experience a burning sensation on micturition, inflammation of, and discharge from, the penis. If this is not treated the inflammation spreads to the epididymis, fine tubules that carry sperms from the testes to the vas deferens. Healing by formation of scar tissue can block these tubes and render a man sterile. *Most women in the early stages of gonorrhoea have no symptoms.* A woman is unaware of the inflammation that may be present in her cervix. This spreads along the uterus and into the Fallopian tubes. Again the scar tissue formed in the process of

Table VIII. Synopsis of syphilis.

	Stage	Time of appearance after infecting intercourse		Signs and Symptoms
An unborn child can be infected	Primary	9–90 days usually around 20	Can be cured by	Ulcer(s) or sore(s) on external genitalia In women it can be in vagina. Within a week or two there is enlargement of glands in groin in an attempt to deal with infection. Ulcer heals, glands subside, so person thinks he/she is cured.
	Secondary	Several months —up to 12		Germs have been multiplying all this time. They invade other tissues and produce a rash and flu-like illness. Again person recovers and thinks he/she is cured.
	Latent	2–30 years		None
	Fourth	Depends on the length of the latent stage		Skin ulcers Heart disease Blindness Insanity

healing can block these tubes and render the woman sterile. Unfortunately there is not a reliable and simple test for gonorrhoea in women. Several different investigations are needed including analysis of cervical and urethral smears. There has been delay in treatment of gonorrhoeal eye infection in newborn babies over recent years, because 'sticky eyes,' caused by non-gonorrhoeal bacteria, are not uncommon in babies. Some people think that pregnant women should be tested for gonorrhoea as a safe-guard to the sight of their babies.

Unlike syphilis, the number of new cases of gonorrhoea reported annually in Scotland is on the increase: 3081 in 1950, 2937 in 1960, 3879 in 1970 and 5110 in 1974. Making up this last figure were 458 boys and 595 girls aged 15 to 19; 1189 men and 708 women aged 20 to 24 and 1186 men and 438 women aged 25 to 34.

CHANCROID

Also called soft sore. It is prevalent in warmer climates. Most of the patients attending clinics in this country are immigrants. There are multiple, ragged ulcers on the penis and vulva, associated with gross enlargement of the inguinal lymphatic glands.

NON-GONNOCOCCAL URETHRITIS

This disease presents a confused picture to the specialists. Its incidence in men now outnumbers that of gonorrhoea. In men the urethral discharge usually appears from a few days to a few weeks after intercourse. There may be no corresponding condition in the sexual partner, but she should be examined and some specialists, even in the absence of any manifestation, prescribe the drug for her that the organism (unknown) causing the male's infection has responded to. The rationale behind this is, that not only will it protect her health, but that of any child she may have later. Stringent tests for cure are advised, because experience shows that 1 in 10 men treated have a recurrence within five years. There can be narrowing of the various tubes in the male genito-urinary tract. About 3 in every 100 infected men get complications, inflammation of the eyes and the joints. Many of the patients are married men who have not had an extramarital relationship. This has led to the speculation that they may be sensitive (allergic) to the vaginal secretion of their wives, especially to *Candida albicans*, which is part of the natural vaginal flora.

TRICHOMONIASIS

This condition occurs mainly in women, and it claimed over 1665 new cases in Scotland in 1974, only 40 of these being among men. It can be present with gonorrhoea which complicates treatment and statistics. It causes a profuse vaginal discharge which is extremely irritating. Many infected men have only a minimal urethral discharge. The treatment fortunately is accomplished orally with Flagyl tablets.

GENITAL SCABIES

Description and treatment of this condition are given on page 73. It has been found more frequently in the genital area during recent years. Being in the genital area it leads infected people to seek advice at the special clinics, but the treatment is the same as for scabies anywhere else in the body.

LICE

This is discussed on page 130 and is only mentioned in this context as infested persons are turning up at the special clinics, fearing that their rash and itchiness is due to venereal disease. The current idiom for infestation is 'a dose of crabs'!

WARTS

The moist areas of the genital region favour rapid growth of these virus-produced growths. They are infectious. They do not threaten life and are easily cured. The sexual partner should always be examined and if necessary treated.

GENITAL HERPES

Blister-like sores around the genitalia, caused by a virus. Similar to herpes anywhere else in the body.

VIRUS HEPATITIS

By the end of 1971 there was speculation that this virus can be sexually transmitted. If you read the daily papers and professional journals you will probably learn more about this in the next few years.

CANDIDA ALBICANS (THRUSH)

This organism is classified as an external vegetable parasite. It has yeast-like cells that form some filaments and is widespread in nature. It can be present in the mouth, on the skin and in the vagina of healthy people. Only in changed circumstances, e.g. during lack of resistance in the vaginal tissues, does it become pathogenic, i.e. produce disease. It causes tiny white patches on the inflamed mucous membrane and produces an irritating discharge. It is difficult to treat infections of the vagina locally, because of the rugae, but an oral drug, nystatin, gives excellent results.

Control of Venereal Disease

Countries are encouraged by the World Health Organization to have a system of *notification* of the Health Authorities of the incidence of *medically defined*, as well as *legally defined* venereal disease, so that all countries can benefit from the *statistics*. Quoting from the World Health Organization Chronicle, 1969, 'Studies show that, contrary to previous experience, the proportion of infected females is now almost the same as that of infected males in some age groups. There is need for better and more up-to-date *national* and *international surveillance* of, and reporting on venereal infections in order to avoid the delays that hinder effective *case-* and *contact-finding'*. The World Health Organization said that 'many countries are reporting an increase in the proportion of young people, including those still at school, among the victims of venereal infection. Despite highly effective drugs, no method has yet been successful in stopping this rising tide of sexually transmitted infections. It has been found that holiday makers and those who in the process of their work constantly travel abroad have increased the incidence of infection on return to their own country. Many people think that with even further rapid expansion of international travel expected in the commercial era of the jumbo jet, this adverse influence on venereal infection control can only be expected to increase. Wars have always increased the incidence of sexually transmitted infections. Servicemen do return to their country of origin and take these infections with them. Currently throughout the world there are pockets of Asian gonorrhoea that is resistant to penicillin. In other countries the rise has occurred during a time of relative peace and against a background of important medical and public health progress. However a

Table IX.   The number of new cases of gonorrhoea reported in Scotland in 1974 by age and sex.

Age in years	Sex Boys	Girls	Total
15 to 19	458	595	1 053
20 to 24	1 189	708	1 897
25 to 34	1 186	438	1 624
Others			536*
Total	2 833	1 741	5 110
	4 574*		

*The sex of the 'others' is not given.

climate of opinion has developed favouring sexual activities thus facilitating the spread. At the same time there is a longer sexual life due to earlier onset of puberty'.

It is not difficult to see that the rising number of infections is an international problem. Arrangements need to be made for infected persons to *continue* their *treatment while out of* their own *country.* But even in the developed countries, where medical services are highly organized, already there is strain on their clinical facilities for investigation and treatment. The difficulties in contact tracing (trying to find consorts of infected persons, encouraging them to have tests, and where appropriate treatment) can be appreciated when put into the context of world epidemiology. Countries differ as to their financial arrangements for their medical services, i.e. whether the patient pays for his treatment at the time or whether payment is covered by some sort of insurance. Because of these differences, Governments are encouraged by the World Health Organization, to finance facilities for investigation and treatment of venereal infection, to any person in need of them. Treatment clinics promise *confidentiality* to their *clients* and this can cause difficulty in the process of contact tracing. In our country regulations came into force in December, 1968, which have made contact tracing more effective. They require that information about a person examined and treated for venereal infection in hospital must be treated as confidential, but permit its disclosure to a doctor, or a person employed under the direction of a doctor, in connection with, and for the purpose of, treatment or prevention of the spread of venereal infection. At a recent Conference two experts told of the diary of a syphilitic prostitute in California, containing the names of 310 men, all of whom she could have put at risk. Furthermore their addresses ranged throughout North America. The 168 contacts traced from those names were all long distance lorry drivers, so you can imagine the potential for spread of infections.

The venereologists find that there is a very definite increase in *casual* sex encounters which makes the tracing of contacts, one of the most important aspects of venereology, very difficult. If a man picks up a girl at a dance hall, he may not remember what she looked like, let alone her name. There is fear that growing unemployment may contribute to a further increase of infection. From experience in the depression of the 1930s this was so. Now that many wives are working, some unemployed men feel the need to prove their virility, their status as men, in a very dangerous way. They go down to the pub in the afternoon, have a few drinks, pick up a woman *casually*, have intercourse with her and pick up an infection too. It seems likely that modern methods of contraception, which provide no element of mechanical protection, such as the condom does, favour the spread of gonorrhoea and non-gonococcal infection particularly.

Our Health Education Council has produced a series of posters which set out simply and clearly the symptoms of these sexually transmitted infections. It has also published a leaflet for persons who suspect they may have contracted one of these diseases, advising them how to get help. Anyone can write or telephone The Health Education Council, Middlesex House, Ealing Road, Wembley, Middlesex HA0 1HH, Tel. 01-998-2731 for the poster, pamphlet or advice on where to go in their own locality. All inquiries are treated in strict confidence. It is thought that we need to redouble our efforts to encourage people who have taken risks of infection to come for examination and treatment, but, to attract these people, enough spacious clinics must be provided with sufficient medical staff, so that patients can be dealt with without long delay, in an unhurried

458

and sympathetic manner in reassuring surroundings, rather than in the crowded grim basements where many of them have to attend now to be seen by a few harassed venereologists. A dial-a-number service for people who may have venereal infection is being launched by several local authorities. A recorded message will give details of symptoms, information about venereal infections and the address of clinics. Some local authorities are arranging for all their venereal disease clinics to be marked on the large-scale maps which are displayed prominently in the city centres. Regular appeals to venereal contacts who failed to return to hospital for treatment are being broadcast by BBC Birmingham. Examples are: An English woman, aged 23, who attended on July 16, number 71/0776; An Englishman, aged 23, who attended on July 1st, number 71/0228; A West Indian man, aged 34, who attended on June 14, number 9729. A hospital official said that about a quarter of the people who registered at the clinic gave false names and addresses and they thought that the radio appeals might help to get them back for treatment.

At present there is campaigning in our Parliament to change the poster laws. Venereal disease notices can only be displayed in public lavatories. This has connotations of shame in the public mind, if only by virtue of their usually squalid location, as though it were necessary to keep them hidden and 'underground'. Shops, post offices and libraries are much healthier suggestions for the placing of venereal disease notices.

Looking into the future—it may be possible to have *immunological control of the venereal infections*. Currently tests of a vaccine to prevent venereal infection are being carried out on volunteer prostitutes in Nevada, where their profession is legal in some areas. The tests are authorized by the Federal Food and Drug Administration in Washington. Nevada has the third highest rate of venereal disease in the country. Most cases occur in the gambling resorts of Las Vegas and Reno, where prostitution is illegal. The trade is legitimate in many rural counties in the state, where the girls are subjected to rigorous weekly health inspections.

SUMMARY OF THE REPRODUCTIVE SYSTEM
The Female External Genitalia
These are called the vulva, and include:

The mons veneris, a pad of fat over the symphysis pubis.

The labia majora, hair-covered skin outside, modified skin inside.

The perineum, the skin between the vagina and the anus.

The labia minora, beginning at the prepuce surrounding the clitoris and meeting posteriorly in the fourchette.

The clitoris, erectile tissue protected by the prepuce.

The vestibule, smooth triangle with the clitoris as its apex, the small lips forming the sides, and the vaginal orifice the posterior base.

The urethral orifice receiving two Skene's glands, in the posterior vestibule.

The orifice of the vagina receiving two Bartholin's glands.

The hymen, a perforated membrane marking the entrance to the vagina.

The fossa navicularis, a depression between the fourchette and hymen.

Bartholin's glands pour their lubricating solution into the fossa navicularis just outside the hymen. Enlargement causes a waddling gait and pain on sitting down.

The Female Internal Genitalia
THE VAGINA

The birth canal connecting the vulva with the uterus. Its external boundary is the hymen, and its internal boundary the deep groove, the fornix, which receives the cervix of the uterus. The groove is deeper at the back, making the posterior vaginal wall 2·5 cm (1 in) longer than the anterior wall. The fornix is divided into the anterior, posterior and two lateral fornices. The modified skin lining is thrown into folds or rugae to allow distension, but this makes it difficult to cleanse. Its walls are kept moist by secretions from the cervix to prevent friction. During reproductive life glycogen is present in the walls and is fermented by Doderlein's bacillus, normally present in the vagina, with the production of an acid medium in which many germs fail to thrive. The bladder and urethra lie in front of the vagina, the perineal body, rectum and anus behind.

THE UTERUS

Lies in the pelvic cavity, separated from the bladder in front by the uterovesical pouch of peritoneum, and from the rectum behind by the recto-uterine pouch of peritoneum. It lies in anteversion, i.e. almost at a right angle to the vagina. The fundus lies above the entrance of the utero-ovarian tubes, the body contains a triangular cavity communicating with the tubes superiorly, and with the cervical canal inferiorly via the internal os. The cervix containing this cervical canal is the lowest portion of the uterus, and it invaginates into the superior vagina helping to form the fornix. The cervical canal communicates with the vagina via the external os.

The uterus is lined by a highly specialized mucous membrane (endometrium) in a continual state of change throughout the reproductive life. The middle layer (myometrium) is composed of plain, involuntary muscle fibres arranged longitudinally, circularly and in figures of eight. The latter arrangement prevents haemorrhage after parturition. The outer layer (perimetrium) is of peritoneum thrown over the uterus and tubes as a sheet is thrown over a line, and the double fold below the tubes and at the sides of the uterus is called the broad ligament.

The uterus is supported on the pelvic diaphragm, and is held in anteversion by the round ligaments attached to the fundus, and coming forward to pass outward through the inguinal canal to be inserted into the pubic bone. The uterosacral ligaments attach it to the sacrum, and two transverse ligaments attach it to the sides of the pelvic cavity.

THE UTERO-OVARIAN TUBES

Two hollow tubes passing from the uterus to the ovaries in the upper fold of the broad ligament. They are narrow at their attachment to the uterus but broaden out into a flute as they near the ovary. This flute has a fimbriated end and the tube is lined with ciliated membrane so that the ova are conducted from the ovary to the uterus.

THE OVARIES

Two small glands lying between the folds of the broad ligament and adherent to its posterior layer. Each is attached to its utero-ovarian tube by a fimbria and to the uterus by an ovarian ligament. There are millions of ovum-containing follicles present in them at birth, and they secrete oestrin to bring about the secondary sex characteristics, and after puberty the oestrin plays a part in the menstrual cycle. The surface of the ovary is smooth in childhood but after successive menstrual periods it is more roughened, as the corpus luteum is converted into fibrous tissue.

THE MENSTRUAL CYCLE

Refer to Figure 225.

THE BREASTS, MAMMARY GLANDS OR MAMMAE

Each breast lies in the pectoral muscle (Fig. 60) on the front of the thorax. They sometimes exhibit premenstrual tension. The central eminence is the nipple surrounded by the areola, a pigmented area, which contains Montgomery's tubercles for lubrication. The milk ducts all converge on the nipple. Lymphatic vessels are numerous and infection and cancer can be quickly spread in breast tissue. Under the influence of the hormone prolactin, from the anterior lobe of the pituitary gland they secrete milk after parturition. Breasts are present in a rudimentary form in the male.

The Male Reproductive Organs
THE TESTES

These glands are contained, one on either side of the septum, in the scrotum. During fetal life they are formed in the abdominal cavity near the kidneys. Shortly before birth they descend the posterior abdominal wall, pass forward on the pelvic brim, and out of the pelvic cavity through the inguinal canal, pushing in front of them a sheath of peritoneum. As they drop into the scrotum the serous membrane around them forms the tunica vaginalis, and the sheath remaining in the inguinal canal becomes fibrous and forms part of the spermatic cord. The testes produce the spermatozoa and the hormone testosterone, that is responsible for the secondary sex characteristics.

461

THE EPIDIDYMIS

A series of convoluted tubules arranged as an oval body on the posterior surface of the testis. It conveys spermatozoa from the testes and deposits them in the vas deferens.

THE VAS DEFERENS

A tube continuing from the epididymis, passing through the inguinal canal where it forms part of the spermatic cord. Once inside the pelvic cavity it passes over the top of the bladder and behind the seminal vesicle to enter this organ at its base.

THE SEMINAL VESICLE

A bladder-like arrangement for the storage of spermatozoa, lying behind the urinary bladder. From it an ejaculatory duct passes into the prostate gland and unites with the urethra.

THE PROSTATE GLAND

Lies at the base of, and encircles the exit from, the urinary bladder. In it the ejaculatory duct joins the urethra. Its secretion renders the spermatozoa more motile for their journey through the penis and along the uterus and utero-ovarian tubes.

THE PENIS

The urethra traverses this tubular organ, at the tip of which is the glans penis covered with the foreskin. It should be possible to draw this foreskin back over the glans; inability to do so constitutes phimosis, the treatment for which is circumcision.

THE SCROTUM

The skin-covered, bag-like structure that hangs from the base of the trunk. It contains the testes as they are thought to function better in a cooler area.

CARE OF THE REPRODUCTIVE SYSTEM

Good diet, fresh air, exercise, meticulous personal cleanliness and a healthy mental attitude to the normal process of reproduction for perpetuation of the human race, and the moral integrity to live by the conventional standards of the society in which we live, all play their part in any consideration of health.

All women over 24 should examine their breasts routinely.

All women over 36 should have a cervical smear test routinely—every five years.

Avoid self-medication during pregnancy.

Avoid smoking during pregnancy. It is undesirable for pregnant women to be infected by toxoplasma from infected cats.

Girls leaving school, who have not had German measles, should be immunized.

462

Female methods. The pill.

Cap or diaphragm and spermicidal jelly.

Intra-uterine device.

Male methods. Condom.

Coitus interruptus.

Rhythm method. Acceptable to those of Roman Catholic faith.

SEXUALLY TRANSMITTED DISEASES

Legally defined in this country. Syphilis

Gonorrhoea

Chancroid (soft sore).

Medically defined. Usually sexually transmitted: non-gonococcal

urethritis

trichomoniasis

genital scabies

lice

warts.

Probably sexually transmitted: genital herpes

candidiasis

(thrush).

Control of sexually transmitted infections. Notification. Statistics.

Those who have been at risk to attend a special clinic.

Case-finding.

Contact tracing.

Arrangements so that infected persons can continue their treatment while travelling.

Use of condoms.

Education of the public, especially young people, so that they know the facts before they make decisions about their sexual behaviour.

Housing in Relation to Health

With the completion of the reproductive system we can now envisage that our imaginary person, Jimmy, will one day consider marriage and will need to think about the provision of a home. The concept of a house or dwelling place is different for the different peoples throughout the world, and can vary from a tent or willow shack, to a mud hut, a bamboo shack, or an igloo. The very latest in houses is an all plastic one, conceived by the Belgians, tried out in Germany, and now there is a show bungalow in the south of England. One great task for the volunteers who go out to help in the developing countries is encouraging house improvement; especially the building of houses with proper ventilation, latrines, pits for refuse, racks for drying crockery, proper food stores and shelters for animals—chickens and cows. You will remember that in the minimum standard of living advocated by the International Labour Organization it merely says—housing of a standard to give protection under healthy conditions. This may seem vague to the people who are used to living in a decent house in the west, but when put into a world context it covers the bare essentials of a dwelling place. There is a world trend towards community nursing, so that health workers need to be able to look at housing from a health point of view.

Even the developed countries have housing problems. Local authorities are encouraged to build houses and let them at an economic rent to those who are unable to buy their own houses. Most local authorities maintain 'temporary accommodation' for those who are rendered homeless; and overnight hostels for those who choose, or resort to the vagrant way of living. A recent Simon Community survey estimated that Britain had 100 000 or so down-and-outs. There are several voluntary organizations that try to help these people and find shelter for them. There is a National Association for Voluntary Hostels. The tinkers are another itinerant group. An estimated figure for tinkers in Scotland is 2 000, half of whom have retained caravans and a standard of living that at least keeps out the rain. The non-caravan tinkers live in tents or shacks. Some of these are made by bending down to the earth the lower branches on a willow tree and holding them down with stones. Tarpaulins are then draped over this framework. The Reverend Sutherland did a survey at one of these 'camps' and found that 25 per cent of the children had spent some of the winter months in hospital compared with 0·45 per cent of children from the local settled population. This surely has relevance for the health workers who care for such patients. Sutherland says that some tinkers have been on the housing list for years. A favourite excuse of local authorities is that tinkers have large families and there are not Council houses big enough for them. Gypsies or travellers form another group that have 'housing' difficulties. A Ministry Circular (No. 26/66) encourages local authorities to make available hygienic sites for gypsy caravans, at which the families can reside during school terms, so that the children become literate. Hampshire tried to wean the gypsies from their traditional pursuits by providing a social worker, special education and housing. The British Gypsy Council expressed disapproval, for nomadism is part of their traditional mode of life. The Social Committee of the Council of Europe took issue with this expression of disapproval, saying that the British Gypsy Council expects too much. From a health point of view, can the world-wide gypsy culture survive in an unmodified form? A stay in hospital holds many extra anxieties for a gypsy used to a free and easy open air type of life.

Housing in Relation to Health

Between the two world wars local councils favoured the pattern of building big new housing estates, to which families were moved from the worst houses in the big towns and cities. Insufficient attention was given to the provision of recreational facilities for the residents, many of whom were used to the close-knit community life of slum-land. Some found life so intolerable that they moved back into bad housing, finding this preferable to lack of friendship and support. Among others who elected to stay, attendance at the doctors' surgeries for minor ailments was increased. Others were able to adapt to the new way of life, and preferred it, to life in a lower amenity area. Some of the growing children on these housing estates experienced increased frustration and boredom in the adolescent period and found their outlet in formation of gangs that became notorious for vandalism and some of the more daring gangs committed crimes against people. Aggressiveness in the defence of territories is well recognized in the animal world. The subjection of human beings to high density living has resulted in antisocial behavioural manifestations. If there is one thing to be learned from the teenage violence in our housing schemes, it is that a social vacuum is poor nurture for good citizens.

After the Second World War, vertical building became more popular. It was hailed as the solution to slum clearance, but it has brought in its wake problems that are as vexing as those created by overflowing tenements. The lifts form a central physical constraint. There can be 400 people in each block, served by two lifts, each of which carries eight people. Housewives organize shopping trips to avoid 'peak hours' when most residents are going to, or coming from work. A half hour wait is not unusual in these rush hours. It is not difficult to imagine the result of this added frustration, after a probable rush hour wait for public transport. Multistorey flats are unpopular with removal men, ambulance men and undertakers, so that the family crises needing such services generate added anxiety. Prams, go-chairs and bicycles take up considerable room and decrease the number of people that can get into the lift and are awkward in a flat. Most Councils have a rule that no pets are to be kept in flats. This denies people the pleasure and responsibility of caring for an animal. Friction can arise about cleanliness and/or misuse of the corridors, lifts and entrance halls.

Multistorey flats are an example of engineering and building techniques running too far in advance of housing techniques such as social planning. One large, grandiose community centre for the residents of six blocks is little used. The sense of isolation experienced by some families is grim. Young mothers do not let their children out to play at ground level because of the difficulty in surveillance and calling them in from such a height. Some Tenants' Associations started campaigning for the Councils to provide play facilities for the pre-school and school children. 'The whole family could be dead and nobody would know a thing about it' and 'You meet people in the lift and you maybe nod to them and talk about the weather or the children but they're not neighbours. As soon as you shut that door you're cut off from the rest of the world' are frequent quotations from research done in these multistorey blocks of flats. To people accustomed to the camaraderie of adversity in the slums, this is perhaps the most upsetting aspect of the new life. In an effort to break down this type of social isolation, one Council installed a 'community room' on the ground floor of *each* of five new blocks. The tenants in each block have formed social clubs, which have organized indoor activities for old people, coffee mornings, beetle drives, bingo sessions, talks from local personalities, and special parties for children. The community spirit engendered by these clubs has enabled them to run outings, and theatre and cinema visits. The tenants go so far as to say, 'We are rediscovering the old ways of making our own entertainment and finding some surprising talent for making other people happy'. However in other places, housing managers say, 'One of the troubles, no matter what facilities are provided is lack of social leaders, people who will take the initiative to start social clubs and other groups. We have been considering the possibility recently of employing community development officers—professional people to bring about the initiative required among the local people to bring them together in a community'. There is increasing awareness of the influence of physical environment on social characteristics, and knowledge of various physical environments, helps health workers to understand the people they are trying to help.

Town and Country Planning oversees the location of industrial and housing areas, trying to decentralize and break these up with green belts to help disperse pollution. Encroachment on agricultural land is a serious matter and such issues as housing density or the balance between blocks of flats and small self-contained houses are still far from any generally agreed solutions. Complex social, economic, aesthetic and sanitary considerations intermingle.

A home involves much more than a mere house and Jimmy and his fiancée will do well to discuss this thoroughly, before they enter into the contract of marriage. This ability to discuss openly and freely is one factor that helps to convert a house into a home.

The over-all planning of housing and redevelopment is the business of the Ministry of Town and Country Planning. Locally the Urban and Rural District Councils make their plans and submit them to the Minister for approval. When an individual desires to build, or reconstruct, a house, he submits plans to the local council, which has the power to approve or reject them. When renovating old property grants are available from the Government, provided the property is brought up to the recommended standard of fitness for habitation. In this country there are 1,800,000 houses that are unfit; 4,500,000 that are below the minimum standard. The 1969 Housing Act is to help do something about them. The recommended standard is as follows:

1. The property must be dry.

2. It must be in a good state of repair.

3. Each room must be properly lighted, heated and ventilated.

4. It must have an adequate supply of wholesome water laid on for all purposes within the dwelling.

5. It must have adequate hot water for all domestic purposes.

6. It must have a good drainage system with an internal water-closet or one otherwise readily accessible, and a fixed bath, preferably in a separate room.

7. It must have adequate facilities for preparing and cooking food, and laundering.

8. It must have a well-ventilated larder.

9. It must have proper facilities for the storage of fuel.

10. It must have a surfaced path to outbuildings, and convenient access to the back door.

The kitchen, living-room and children's rooms should face the sun, whilst the less-used rooms and the larder should be on the shady side of the house.

A nurse or doctor cannot be expected to know the details of house construction, and any person buying or building needs to seek the advice of experts, but a nurse is expected to be familiar with the factors concerning living in a house which contribute to the maintenance of the health of each member of the household.

The first consideration is whether the dwelling is within the money range that can be afforded by the family. The rates levied in the area also have a bearing, because they have to be paid over the years whilst the house is being bought, and it has to be decorated inside and out, and kept in repair throughout this time. The age of the buyer is important, and the number of working years left, over which to make repayment, for one does not want to be still paying for a house on the lessened retirement income. The possibility of lessened income from sickness or injury must be borne in mind, for the basic sickness benefit does not allow for repayment on a house. Some employers make up the sickness benefit to the normal income for a stated period, and this type of employment is worthy of consideration when starting out on a career. Being a house-owner does give one complete jurisdiction over it, e.g. if one wishes to let a bed-sitting-room one is free to do so.

With tenanted property it is important that one reads the agreement carefully before signing it. Rates, repairs, outdoor decoration, inside improvements such as the installation of a new grate, are sometimes undertaken by the landlord and sometimes by the tenant. Arrangements for the giving of notice by either party should be clearly stated in the agreement. The landlord should provide a rent book and the tenant should see that it is signed at each payment.

Locality is important and by the time one is adult and acquiring one's own home certain habits of living are deeply ingrained and therefore not easily changed, e.g. the city dweller seldom settles in the country and vice versa. Councils try to comply with this fundamental of living when procuring accommodation for eventide home, slum clearance, etc.

Where there are children the possibilities of schools, place of worship, ample safe playing space indoors and outdoors, a warm quiet room in which they can do their homework, and ideally individual bedrooms are all-important factors in their development. Consideration should be given to the type of employment offered in the area, if they are not the adventuresome type that want to move away from home in the process of earning a living. Some areas are faced with an unemployment problem amongst the school-leavers, the very group that need to work so that there can be a continuation of the developing and strengthening of their self-regarding sentiment.

Proximity to work may curtail the choice of where to live for if there are several members travelling to work and having to have lunches out, this can make heavy inroads on a limited income. Because of transport difficulties irregular hours of work may further curtail choice, and living out of town is incompatible with many such forms of employment.

The area in which a family chooses to live should provide a sufficient outlet for all their hobbies. If there is not a gardener in the family then a house with a large garden is not for them; on the other hand, there are people who forget the tensions of the day when they go out to tend the garden in the evening. The presence or absence of Clubs, Guilds, Associations, etc. should be ascertained so that family members have the chance of meeting others interested in the same thing. On the new housing estates, community centres are being built in which a great variety of leisure activities are provided.

Shopping arrangements need to be considered for there is nothing worse than a busy housewife having to carry heavy bags. The butcher, the baker, the grocer, the greengrocer and the fishmonger with their vans are still an important part of our national life, and they are welcome visitors at the door of housewives living some distance from the shops. They are especially useful for older people who do not get about so easily.

The older age group also need special consideration in relation to housing, and Councils recognizing this have built pensioners' bungalows, which are small, compact, easily managed and have sunken baths in which the person sits instead of having the legs stretched out in front. Arrangements are made for the health visitor to visit the occupants at regular intervals. Some Councils incorpor-

468

ate these at the side of the semi-detached type of house, so that a young married couple can occupy the semi-detached house, and the parents can live in the bungalow. This maintains family ties and encourages responsibility towards the parents during sickness, it avoids three generations living in intimate contact which can give rise to much stress and strain, and it retains that important feeling of independence for the older people; they continue to feel useful because they can look after the baby when the young housewife goes shopping, and again at least one evening a week to allow the young husband and wife to go out together, a contribution to the mainten-ance of health after marriage.

Lighting, heating, plumbing and ventilation have been dealt with elsewhere in this book, and they each contribute to a healthy, happy home. They are included in the following list of questions which could form the basis of assessment when buying a house. Each person can make additions and subtractions according to his/her personal preferences.

1. When deciding size of dwelling: do we intend to have (any more) children? Ideally each child should have his own bedroom, if not each child should have his own furniture. Is it likely that elderly parents will have to be housed? Do we need a guest room?

2. Do we intend to have a car: is there room for a garage with a safe turn out?

3. Is the house built on dry, well drained land?

4. Has it an adequate damp proof course?

5. Is there insulation against heat loss?—thermolite insulation for outer walls, glass fibre quilting in the roof and double glazing of all windows.

6. Does noise carry throughout the house? Can individual tastes with regard to radio, television, record player and tape recorder be catered for, without upsetting the whole family?

7. Are the walls and stonework in good condition?—Look outside as well as inside.

8. Are there chimneys? Are they in good repair, chimney pots intact, wire balls to prevent birds nesting? If fireplaces have been boarded up, is there a ventilator? A chimney deprived of air flow becomes very damp.

9. Are the window frames of metal or wood? Do they fit properly inside and outside? Apart from the discomfort of draughts, badly fitting windows increase heating bills. Rattling windows are un-pleasant to live with. In a downpour rain can splash between the window and frame on to the sill, then drip down on to the floor. Do the windows open adequately? Have they efficient latches?—especially any windows that can be approached by a drain pipe! What are the window cleaning facilities in the area?

10. Do the door frames fit?—draughts—dearer heating—pools of water after rain. Are there efficient latches on outer doors?

11. Is the roof in good repair?

12. Is there any woodworm?

13. Water supply? Is there a tap straight from the mains for

469

drinking purposes? Is the tank supplying the house clean and covered? Some tanks have an arrangement whereby they can be emptied and swilled out, before refilling. Means of obtaining hot water?

14. Sanitation. The small water closet should be avoided for old people. Is there room to fit gadgets to help old people use the lavatory safely? In a villa-type of house, a downstairs water closet saves wear and tear on the stair carpet. Arrangements for the collection of dry refuse?

15. Heating? Is the house in a smokeless area? If not, what are the Council's plans? If fuel has to be used, is there adequate and safe storage space?

16. Lighting? Natural and artificial.

17. Proximity of kitchen to dining room or area?

18. Place for hanging out washing?

19. Place for sunbathing?

SAFETY OF THE HOME[207, 208, 209, 210, 211, 212, 213]

A home cannot be considered a healthy place in which to live if it is not at the same time a safe place. It should not be a place in which the attitude to accidents is—they happen to other people—and do remember that children take over attitudes very early in life from parents. More accidents occur in the home each year than on the roads, and the numbers are increasing. Home safety is such an important subject for all nations that the World Health Organization has published a booklet[214] about it. An increased accident rate seems to be the price paid for a higher standard of living in the developed countries. In only a few countries are standards of domestic safety legally enforced, and in particular the rapidly developing countries are threatened by the speed of technical advance. The hazards of affluence[215] are—new fuels, very high voltage electricity, very hot water and high speed gadgets which constitute a serious danger in the second half of the twentieth century. A higher standard of living brings with it more and more electrical appliances which are often put on an already outworn or overloaded circuit. Lack of knowledge causes the use of plugs, normally containing 13 amp fuses when bought, for equipment that needs a fuse of lower amperage, e.g. blue 2 amp fuse for standard lamps and television; yellow 10 amp plug for two-bar radiator, etc. There are electric toothbrushes without an earthing wire and electric can openers that can amputate fingers as well as open tins. Electric hedge cutters and powered lawn mowers have cut their cables and have electrocuted the operators. Power saws operated by portable sources of electricity are rarely adequately guarded. Many new dish washers heat the water to a very high temperature; a defective switch can allow the cabinet to be opened so that the operator is sprayed with near-boiling water. Professor Backett draws attention to the little known fact that for bathroom electrical equipment, piped mains water forms a more efficient earth return for electric current than the earth wire of the average domestic circuit. Serious electrical accidents have occurred from the siting of electric kettles where they can be filled from a water tap without first being disconnected. Electric over-blankets are preferred to under-blankets by the fire safety people. The latter should always be switched off before getting into bed. A hot water bottle should not be used with an electric blanket, which should *never* be used for a bed-wetter. It is safest to send electric blankets back to the maker for overhaul after each winter. Electric fires should have a fixed, double guard and should never be used from the same socket as an iron, lamp or electric clock. Electrical equipment should wherever possible be a permanent fixture and trailing flexes should not be allowed. The recessed, shuttered type of plug is the safest. All electrical equipment needs to be overhauled at regular intervals; some firms do carry out this maintenance service. Amateur electrical repairs or improvements should not be done, and skilled men should be procured for all such jobs.

Low gas taps that are easily turned on are a menace to the inquisitive child, and the older type of gas cooker to the older person who is getting forgetful, for they may turn on the supply and forget to ignite it. Do-it-yourself people can inadvertently hammer through pipes. Food can boil over and block the holes in the gas ring, thus pressure builds up in the pipes. Gas pipes become worn with age when they leak, and several accidents occur each year from leaking pipes. Leakage can also occur from perished rubber tubing at the end of flexible tubing on portable equipment. Gas fires should stand on non-combustible material. The light should be applied immediately the gas is turned on, to avoid the risk of explosion. The Gas Board will make a free check of pipes in pensioners' houses. It charges a fee for checking pipes in other houses. In a household with a depressed member, commercial gas (carbon monoxide) offers a way of escape from a life that has become too much for that member, thus ranking as a suicide agent. However, natural gas which is gradually being substituted for commercial gas in this country is not lethal from inhalation, but there is still the risk of explosion with unlit gas. The smell is almost identical to that of commercial gas. To pipe odourless gas to consumers is an offence against the Gas Quality Act of 1961 because a leak might not be detected.

A fine meshed spark guard, of a strength and design that a falling coal will not knock over, should be put in front of an open fire whenever the room is unoccupied. The guard should be fixed in position where there are children and frail old people. Hot ashes in a combustible container can set it alight, so they should be put into a metal container, and set to cool in a safe place. Chimneys should be swept regularly twice a year and smoke issuing from cracks in flues or chimneys should be dealt with immediately by experts. Articles should not be placed on a mantlepiece, nor mirrors above a fireplace, for clothing can easily catch fire whilst stretching up to procure or use such articles. Waste-paper baskets should be made of flame-proof material in case an unextinguished match is thrown in, after lighting a cigarette or pipe.

Any home using oil heating is now (1976) legally required to be equipped with a fire extinguisher, and health visitors and district nursing sisters are asked to bring this to the notice of householders that they visit. Buy only those oil appliances that conform to the British Standards Specification, i.e. have a self-extinguishing device that prevents it from bursting into flames when knocked over. Heaters are safest when placed against a wall, with a guard fixed round them to prevent children knocking them over. Adequate ventilation is essential but heaters should be placed out of draught and on a level surface. Appliances should be filled by daylight, away from flame, and the wicks and burners kept well trimmed and clean. Heaters should not be moved when lighted. Metal cans should be provided for paraffin oil and any other inflammable liquid, and they should be stored in a cool, safe place. Some Councils have banned oil heaters in their multistorey flats.

Beware of the lethal chip pan which causes a high proportion of the fires in the homes in Britain. Boiling fat causes serious burns as fat has a high boiling point and it clings to flesh. Fat must not be overheated and the chip pan never more than one third full of fat. It must not be left unattended on the stove. If the fat does take fire, turn off the source of heat. Put lid on pan, or cover with damp cloth, old tin tray, asbestos sheet, or use an aerosol fire extinguisher if one is easily available. Keep windows and doors closed, for any draught will fan the flame. Current research is attempting to produce a chip pan with a warning device—similar to the whistling kettle.

When ironing, frying and grilling, every person should be trained to switch off the source of heat at any emergency sound, such as the phone or door bell ringing, baby crying, child calling—before attending to these things. Clothes and linen should be aired away from an unguarded source of heat. Aerosol sprays must not be put near a source of heat, and when the container is empty it should not be put on the fire. Most houses have a drawer for odds and ends, and it can cause trouble. If steel wool and a battery come into contact, they can spark furiously and cause a fire. Even the medicine box is a potential fire starter—glycerine mixing with some antiseptics can cause smouldering. There is a danger in never turning out the 'glory hole' as there can be spontaneous combustion of stored paper, oily rags, polish residues and other rubbish.

There should be a fire block between the boiler house of the central heating system and the house itself. Fire resistant doors can be procured. There are some fire retardant paints. Plasterboard is safer than fibreboard. Polystyrene ceiling tiles are dangerous because they melt with heat, and drops of blazing plastic rain down on the room turning a minor outbreak into a potentially serious fire. People should ensure purchase of tiles that are labelled 'flame resistant'. Most adhesives contain a highly inflammable petroleum mixture, therefore a glowing cigarette or pipe, or a naked flame should not come near them. They should be stored in a cool place. A soldering iron should not be put down until it is cool. All windows should be in working order so that they could be used as escape routes if necessary. Smokers should have two golden rules, no smoking after taking sleeping tablets, no smoking in bed. Refraining from smoking when intoxicated will prevent fire. One other cause of domestic fires is listed by the fire prevention officers—the use of naked flame as in blow lamps when thawing out pipes or tanks. Lagging pipes and tanks will help to prevent this temptation. At night before retiring to bed, and when the house is going to be empty, all appliances should be turned off at the wall and any plugs removed. All doors should be kept closed when the family is sleeping and when the house is empty. Closed doors will confine a fire in the room in which it starts for at least 20 minutes.

Visiting patients' homes does not stop at Christmas, and health workers should be aware of the extra risks at this time. The Royal Society for the Prevention of Accidents (RoSPA) have a seven point check list for safety.

1. Do not drape paper chains or holly around lights.

2. Do not buy the tree too early; keep it out of doors or in water until you are ready to decorate.

3. Make sure the tree is embedded in damp earth or sand, and keep it watered regularly.

4. If you use tree lights choose those that come up to British Standard safety requirements, and stop the children from touching them when lit.

5. Place the tree where it will not be knocked over, and away from the fire.

6. If you buy an artificial tree, check that it is made of slow burning or fire-proof material.

7. Provide plenty of deep ash trays for your guests.

The RoSPA also deprecate the use of plastic foam sheeting for fancy dress, as it can catch fire easily and produces molten particles which can cause severe burns. The Society also advises on electrical equipment that might be bought as gifts—items such as toasters, hair dryers, kettles, shavers and percolators should carry the British Electrical Approvals Board symbol, while electric blankets, pressure cookers and saucepans should bear the British Standards Institution kite mark.

Having talked so much about fire it now behoves us to think about elementary fire drill, which it is recommended, families should practice at least annually. Everyone should know how to get out of the house if the fire is downstairs and they are upstairs. There should be a rope and a hook near one upstairs window to which the rope can be attached. All members should gather in the bedroom with this facility leaving their own bedroom windows and doors closed. A rug should be rolled against the bottom of the door and any obvious cracks around the door packed to keep out the smoke. Only open the window if you have to shout out for someone to send for the fire brigade. Even with a serious fire downstairs this bedroom should remain smoke free for 20 minutes by which time the fire brigade will have arrived. Jump only as a last resort, not leaping out, but dropping as it cuts off the height of the fall. Some fire prevention officers talk about 'dreeping'—it means hanging by the fingertips to lessen the height of a fall. There should be a check point outside and all occupants should gather there. Otherwise Dad can think that young Johnny is with Mum, while Mum thinks he is with Dad, yet Johnny might still be terrified inside the burning house. Smoke is more deadly than flame. Some of the resins in simulated leather coverings on furniture give off a dense smoke. Get down on to the floor and crawl—away from the smoke if you know its source, or to the window from which you can be rescued. If there is time and there is something easily available, tie it round the mouth and nose. *Never* go back into a burning building *except for human life*. Panic and ignorance are a fire's best friends.[216] The Fire Brigade will check your house for fire risks. There is no charge, but few people take advantage of this service.

In modern constructions odd steps are avoided and the stairway turns at a rectangular landing, thus the steps do not curve round with one end much narrower than the other. In older buildings where such things cannot be altered, they should be well lit throughout the day and night. A banister rail is a great help to the old and infirm when manipulating a stairway. Adequate switches should be provided so that no person has to turn off the light and then manipulate stairs or corridor before reaching his room. All members traversing them must have well-fitting attire on their feet. Stair carpets must be adequately fixed and free from fraying edges. This applies to all carpets throughout the house; frayed and curled-up edges are a menace. Fitted carpets throughout are safest. The laying of a strip of carpet or matting at an entrance to keep the floor clean and avoid wear and tear can be fraught with danger, as is a door mat on a polished surface.

Baths are an accident-prone area for the elderly. The modern ones are lower and more shallow, and therefore easier and safer to use, though they may be improved by the use of a non-slip mat in the base. Handles can be fixed at strategic points in older baths to decrease the hazard of slipping to the elderly. Scalding accidents are far too numerous amongst small children, and usually occur with an unfixed bath, when the mother pours in a bucket of hot water and turns her back to get a bucket of cold water. Cold water should always be run first as the bath retains heat and can be much too hot to sit on; it decreases the risk of scalding and there is less steam formation to deteriorate decorations.

High on the list of causes of accidents are poisonings that occur from tablets and bottles being left where they are accessible (e.g. in a handbag) to the child who does not understand their danger. Every household should organize a high or locked cupboard in which to place the medicines,[217,218] and all commodities for external use only. Some Councils provide a lock-fast drug cabinet in their houses. Aspro-Nicholas Ltd, the *Daily Mail* and the Royal Society for the Prevention of Accidents have all combined to launch the Aspro Safety Medicine Cabinet as an entirely non-profitmaking operation. It is impossible to rest any medicine or drugs on the cabinet because of its sloping roof. There is also a hidden safety catch method of opening, requiring no keys. Tests were carried out with groups of small children, none of whom were able to open the cabinet on their own. There is also a plastic pill bottle that can only be opened by simultaneous pressure of 12 lb per in^2 (83 kPa) and a twist by the palm of the hand. This pressure is more than most children in the vulnerable age group can produce. Outdoor poisons such as turpentine, paraffin, weedkiller, rat poison, etc., should be in a similar cupboard in an outhouse. They should *never* be transferred to a bottle that has contained a soft drink—as an innocent person may drink the contents expecting it to be a soft drink.

Disposal of unwanted medicines and poisons is a problem and should be done in the safest possible manner. Both liquids and tablets can be put down a lavatory and flushed away. Where there is no water carriage system of sewage disposal it is best to bury them provided animals are not likely to unearth them. Children have been poisoned from bottles put out with rubbish for collection. The problem is now so bad that several Councils have experimented with different types of campaign to get surplus medicines collected safely from houses.

In the bedroom hair lacquer, nail varnish, perfume, hand lotion and liquid makeup have been swallowed by curious children. These things should not be left out on the dressing table.

Sharp tools (including razor blades) should be given a safe place inside and outside the home, so that they are out of reach of the child until he understands the danger attached to their use.

The kitchen abounds with possibilities of burning and scalding, and the sink and cooker should be on the same side to decrease these. The Building Research Centre advocates a working surface on either side of the sink and cooker. Household 'liquids' (e.g. ammonia, turpentine, bleach, silver cleanser) should not be kept in a cupboard under the kitchen sink, for sooner or later, the crawling child will get into the cupboard. When in use pan handles should not project beyond the stove, and kettle spouts should be turned inwards. Scalding can occur on removal of the lid from a boiling kettle. The lidless type avoids this. Asbestos gloves or an oven cloth of durable thick material should always be to hand. Sharp knives should be kept in a wall rack and not in a kitchen drawer. They should never be dropped into a container of hot, soapy water, as later the cutting edge may be grasped with disastrous results. This applies to glassware, each article should be held separately while it is washed.

A good strong stepladder with a broad step and preferably a pole projecting from the top step is a necessity in a house, and climbing up on spindly legged chairs is to be discouraged.

Overhanging tablecloths are a temptation to a child and just need a tug to bring down the teapot over his head. It is probably safest to dispense with a tablecloth whilst the children are small.

Curtains should be dispensed with where they can blow on to the cooker or pilot light of a water heater. Fibre-glass curtains carry a less fire risk than those of conventional materials.

The door of a walk-in larder, storeroom or cupboard should have an inside handle for opening, and should in general not be lockable —in case children lock one another inside in play.

Every child has to learn to use and control such things as matches and fire. It is the child who is not offered the opportunity of learning this control within his own home who goes out and plays with matches, and can thus come to grief setting alight himself, or another child, or someone else's property.

Every child has to learn the habit of tidiness and the earlier it is instituted the better. Encouraging children by giving them a toy cupboard or box, and ensuring that they make use of it after each play period helps to prevent sprained ankles from falling over toys left lying about.

Concern has been shown about dangerous toys. This is a world-wide problem and needs an internationally accepted standard of safety. The American Public Health Service listed toys fifth among items that cause accidents. Regulations under the Consumer Protection Act, 1961, specifying general safety requirements for children's toys are currently in force. They apply to all toys offered for sale regardless of their country of origin. Among other things, the regulations prohibit the sale of toys manufactured from celluloid, require electrical toys to be operated only by low voltage, and prescribe safety requirements for transformers for use with such toys. The amount of lead and other toxic substances which might be present in paint on toys and in crayons is also controlled. Recently there has been concern about large balloons that float easily. They are filled with hydrogen. Contact with a lighted cigarette can cause explosion. When in doubt about the safety of any toy, contact the British Standards Institution, Britain's Safety Council, or the Home Office, Consumer Protection Branch, Whitehall, London.

The foolproof home is a thing of the future, but meantime much can be done by attempting to change the attitude—it happens to others, it cannot happen to me! The Accident Prevention Council advocated a Home Safety Code. Scotland has the worst record in Europe for accidents in the home. Bad living conditions have been blamed, as they cause nervous tension in the families, making them more accident prone. Bad lighting and badly worn tenement closes have also been blamed. It is thought that lower wages contribute to the greater use of oil heaters, which are high on the list of causes of accidents. There is a great need for studies of domestic accidents by social class and type of housing, and the effect of rehousing in relatively safe surroundings. A study is currently proceeding at Bristol. Details of the accident are being collected for every person treated at, and admitted to hospital as the result of a domestic accident.

Finally attempts are being made to legislate for safer homes. Attention is being given to rules about inflammable play suits, electric blankets, domestic electrical equipment, perambulators and pushchairs; the lead content in tinning, and cooking utensils; the toxic content of pencils and crayons, portable cots, oil heaters, inflammable nightwear, childrens' toys and electrical wiring. If you read the daily press you will be informed what the new regulations are, and when they come into force, and what measures you as a citizen can take when you find that the law is contravened.

OVERCROWDING

The following points are taken from the Housing Act, 1936:

1. A house is overcrowded in which two persons of opposite sex and over the age of 10, who are not living together as man and wife, have to sleep in the same room.

2. A room with a floor space of 110 sq ft (10 m²) or over can accommodate two adults. No additional space is stipulated for a baby of under 1 year, and children aged 1 to 10 are counted as half.

3. A room with a floor space of less than 50 sq ft (4·6 m²) is unfit for a bedroom.

4. A room with a floor space of between 50 and 70 sq ft (4·6 m² and 6·4 m²) is suitable for a child under 10.

5. Over the age of 10 years a minimum floor space for a bedroom is 70 sq ft (6·4 m²).

6. Two rooms are considered suitable for three people, three rooms are considered suitable for five people and five rooms for ten people.

It must be understood that these are the absolute minimum conditions which have to prevail before steps can be taken by the authorities to deal with the situation. It is desirable that people should not sleep in the room in which they live, and much less in the room in which they cook. It does not allow sufficient privacy for the occupants to develop their individual interests, and does not supply sufficient safe indoor playing space for the toddler.

Houses are not the only places that can be overcrowded. The possible effects on health of overcrowding are collected in the following list:

1. The atmosphere within the bus, train, room or building does not get changed the minimum number of three times each hour. It thus becomes ennervating and the occupants suffer from apathy.

2. Perspiration has to be removed from too many skin surfaces, the atmosphere becomes laden with water vapour, thus the body is robbed of its cooling mechanism and faintness may ensue.

3. The temperature of the atmosphere rises and together with immobility can cause fainting.

4. Any germs from respiratory mucous membranes, sprayed into the atmosphere during breathing, talking, coughing, sneezing, etc. have the ideal conditions for rapid multiplication, thus infection can easily be spread. The warm, moist, stagnant atmosphere makes the recipients' respiratory mucous membrane lax and atonic and less resistant to infection.

MODIFICATION OF A HOME

Many more people survive bad road and rail crashes and are faced with coming to terms with paralysis of all four limbs, or the lower limbs. Together with these, there are other people who from various disabilities live a 'wheelchair' life. This calls for considerable modification of the home. The trend is towards home dialysis for people with chronic kidney disease, and again this calls for considerable modification of the home. As health workers and district nursing sisters often visit these homes, they need to familiarize themselves with the various types of modification that will help their clients to lead as full a life as possible. Information can be obtained from the Disabled Living Activities Group (DLAG), Central Council for the Disabled, London, SW1.

CARE OF THE HOME

The outside of a home should be kept in good repair. After a high wind the occupier should look at the roof, for a dislodged slate can let in considerable damp and deteriorate ceiling decorations very quickly. Chimney pots if present need regular inspection because a gale may be sufficient to dislodge a portion and cause injury to a passing person. Open gratings over drains need to be kept free from debris, particularly necessary in the autumn with the falling leaves. This also applies to roof gutters and rain-water pipes, because blockage can cause dampness to permeate walls, and damp buildings are never healthy ones in which to live. The outside requires painting every three to five years according to the area, paintwork usually being effective in the country longer than in the city. Flakes of paint have been eaten by children resulting in lead poisoning. This problem is worst in countries that have severe winters. A coat of paint helps to preserve the woodwork and the pipes.

The approach to the home should be kept clean and tidy, and the windows and curtains clean. These are the things that give the first impression to visitors, and it is one way in which we can proclaim to them our basic standards of living. Those who are reasonably (not obsessively) methodical and tidy about the external appearance of the home are very likely to have these characteristics permeating their whole lives. In some countries, e.g. Scandinavia, the householder has a lawful obligation to keep the pavement in front of his house clear of snow. Some Councils in Britain sell rock salt for this purpose.

The inside of the home also needs to be kept safe and clean, with a place for everything and everything in its place. All members of the household should learn to cherish the contents, especially furniture which is now an expensive commodity. The child who learns to respect his parent's furniture is not likely to abuse that belonging to any other person. When grown up one takes into the outside world the standards one has been taught to respect as a child.

Above all else it is the atmosphere created in the home which forms a permanent link in the chain of mental and emotional development. It is here that the child must learn that he cannot expect to have his own way all the time, but is entitled to have his way for half the time, the other half being spent in cheerful co-operation whilst other members have their way. Children are quick to imitate the behaviour enacted before them, and if they witness the parents discussing and managing to reach a solution of the problem being discussed, without rancour or discontent, the children are likely to perpetuate this method of dealing with difficulties in their own lives.

HOUSEHOLD PESTS AND VERMIN

The common flea, the rat flea and the bed bug have already been dealt with on page 131.

Common House Fly (Musca Domestica)

The female produces about 120 eggs at each laying and may produce four such broods. These eggs are white, 0·125 cm long and are usually deposited in manure or in ashpits containing fermenting vegetable matter. They hatch out in eight hours to three days according to the temperature. Active growth occurs and in favourable circumstances this larval stage is complete in four to five days. The ensuing pupal stage takes a similar time for completion, then the case splits and the common fly emerges. It waits until its wings are dry, then flies from this filth to alight perchance on a well-filled dining-table, or on milk in an uncovered bowl. Its life span is six weeks to four months. Flies are most numerous from July to September at which time a fungus disease becomes prevalent amongst them. As they die they may drop into food etc. Only a few remain alive during the winter in warm, secluded places, but these are sufficient to propagate the species. Since the cycle from egg to fly takes eight to ten days the importance of removing manure and emptying dustbins more frequently than this will be appreciated.

Flies have hairs on their legs and sticky pads on their feet to which any of the following can adhere: the germs of typhoid and paratyphoid fever, dysentry, cholera, poliomyelitis and infective hepatitis; decaying animal and vegetable matter capable of causing diarrhoea; the ova of worms. Adherent material may be deposited on kitchen utensils, dish cloths, tea towels, table-cloths, crockery, cutlery and food. Flies liquefy food by vomiting on it and then they suck in the resulting mess into their proboscis. Flies may defaecate while on food. The taste of food, so contaminated, is not changed.

In an attempt to eliminate flies the following measures are important. All dung heaps, refuse tips, etc. to be well away from occupied

479

dwellings. Human excreta in privies and animal excreta in fields to be covered with earth. Water carriage system for the disposal of sewage to be used wherever there is a main water supply, otherwise cesspools. Kitchen and larder windows, and ventilators should be covered with wire gauze. In hot countries all windows and outside doors should be gauze covered. Refuse bins to be away from kitchen and larder windows. Refuse bins must be flyproof, e.g. no holes, lids replaced immediately; bin never over full, emptied regularly at least weekly; the outside must be kept free from 'spills'. All cooking and eating utensils to be kept in a cupboard. Table to be set immediately before meal. Table cloth not to remain in situ between meals. Meals to be cleared away and used crockery and cutlery washed immediately the meal is finished. Tea towels and dishcloths to be kept in flyproof area. Ideally they should be dispensed with and crockery should be 'drip-dried'. Food *never* exposed. Scraps which pets have left should be burnt, wrapped and put in bin or covered in a cool place. Buy fresh food in smaller quantity in summer months. 'Swotting'. Fly papers. Vapona strip exudes the insecticide dichlorvos on slow vaporization, building up until it is lethal. The British Government's Advisory Committee on Pesticides suggests that a person should not come into direct contact with it. In America Vapona strips carry a warning: Do not use in nurseries or rooms where infants, ill or aged persons are confined. Do not use in kitchens, restaurants, or areas where food is prepared or served. Results of research done in the United States of America showed that there was a 15 per cent decrease in red blood cell activity among families exposed to dichlorvos for one year. It also inhibits enzymes which play an important part in the transmission of nerve impulses. DDT papers. Spraying with insecticide. If breeding grounds cannot be destroyed they should be sprayed with borax and cyllin or insecticide.

Moths

These are not known to spread disease, but they are expensive pests when they destroy garments, blankets, carpets and furnishings. Cleanliness is of the first importance and such things must first be clean if they are going to be stored, then wrapped in newspaper for the moth does not like printer's ink, or they can be sprinkled with DDT powder.

Furniture Beetles (Woodworm)

These creatures bore into wood leaving holes as evidence of their presence. Fine sawdust around furniture is often the first warning. There are several effective proprietary preparations which kill the

480

beetles and prevent further holes deteriorating the furniture, but great patience is needed to inject the solution into each hole. It is not wise to introduce new furniture into a room containing some infested furniture, and vice versa. When buying second-hand furniture careful inspection should be carried out before the sale.

Cockroaches

These pests will eat almost anything and are spoilers of food, stores and clothing. They have a peculiar, offensive odour, and they are always found in warm places which accentuates this. There are anti-cockroach powders and sprays available for coping with a small infestation, but large institutions need to call in the Pest Officer to treat the area regularly. If this fails any heated tracks such as along hot water pipes have to be exposed and treated.

Black Beetles

These are often found in association with cockroaches as they favour the same warm places. Some of the DDT preparations are lethal to both pests.

Mice

At three months a mouse is mature and able to have several litters per year, each averaging five young. They can act as host to *Salmonella typhimurium* (*Bacillis aertrycke*) thus excreting large numbers in their faeces, some of which adhere to their fur and paws, and are therefore easily transferred to food, later to produce bacterial food poisoning in the unsuspecting human beings. They can also destroy woodwork and fabrics.

In an attempt to keep down the mice, all food should be kept in metal containers, waste food removed from the premises, mouse holes filled with cement or gas tar, and mouse-traps baited and left at strategic points. In Nature's cycle owls and weasels feed on mice. Cats are usually good mouse catchers.

The anticoagulant warfarin as a lethal agent for mice came on to the market in 1952. By 1958 mice in several countries had developed a resistance to this substance, and by 1966 there were many resistant strains. One great advantage of warfarin is that the mice get back to their nests, curl up and die there in a natural posture, so that the other mice do not become suspicious of the poison. By the end of 1970 a warfarin-based rat bait had been produced which would

481

remain palatable to rats for up to two years. But that was too late to help Australia in 1969 when there was one of the worst plagues of mice in history. Millions of them ravaged crops, grazing land, animal food stores, homes, furniture and gardens, and even attacked people. They nested and fed on the backs of sheep. Officials considered the use of 'Poison 1080' which has no known antidote. Throughout the world in 1969 there were reports of increased numbers of mice doing untold damage. Advertisements for cats abounded in many countries. The strike of dustmen and sewage workers in Britain in recent years resulted in a tremendous increase in the rodent population. The 1949 Pests Act puts the responsibility for eradication and control on the owner or occupier of a building. Many Councils are reluctant to spend money on an effective rodent control service. Some Councils charge for their rodent service.

A new rodenticide called Alphakil has been approved for use by local authorities under the Ministry of Agriculture, Fisheries and Food Notification of Pesticides scheme. The manufacturers claim that one teaspoonful is enough to kill ten mice and that eight mice can be killed for one pennyworth of Alphakil. The majority of mice die within three feet of the bait. This is a disadvantage as mice become wary about death in suspicious circumstances. They quickly become unconscious with a rapid fall in body temperature, leading to death.

Rats

Rats share with mice the ability to excrete *Salmonella typhimurium*, and as they live in filth they can cause bacterial food poisoning. They can do tremendous damage to woodwork and buildings by their strong gnawing action, necessary to prevent their front teeth from getting too long.

Wherever rubbish is dumped there will be rats. In prevention of rat infestation in farm buildings, warehouses, etc., as much food as possible should be stored in metal containers, and all wooden doors fitted with a 30 cm metal strip at the base leaving a maximum space of 0·6 cm between the bottom of the strip and the floor. Pipes need to be fitted with a baffle plate or guard to prevent rats from climbing them, and tie beams of the roof trusses can be similarly treated. A metal base in preference to a wooden base is advocated for haystacks, against which no pole should be left lying as rats can run along same and gain access to the stack where they will breed.

Dealing with infestation on a large scale is the work of experienced Pest Officers who are employed by each local authority. In many instances they only act in an advisory capacity. On a domestic scale a rat-trap is still useful. Rat poison can be used with due thought for pets that may inadvertently pick up the poison. Another disadvantage is that the rat may crawl into an inaccessible place to die, later to give rise to the unpleasant odour of putrefaction. In late 1971 the rat-killer Biotrol came on to the market.

To give some idea of what heavy rat infestation can cost—New York in 1970 spent £8 million on rat control in slum areas, because it was claimed they were doing damage costing £416 million.

Port Health Authorities have special programmes for dealing with rats, as the type found here can transmit bubonic plague. Hydrocyanic acid gas is used to disinfest ships. In November 1971, the World Health Organization issued an international alarm that ship rats in Liverpool docks had developed resistance to poison.

Bacterial food poisoning and plague are not the only diseases spread by rats. They can transmit any of the other bowel infections, rat-bite fever, trichinosis (p. 297) and Weil's disease as well as foot-and-mouth disease amongst cattle.

Housing in Relation to Health

Mosquitoes[219]

There are many species of mosquitoes but all have one point in common—their immature stages take place in stagnant water. They feed on vegetable juices, but the female has a piercing proboscis, as blood sucking is apparently necessary for the development of her eggs. Eradication programmes carried out by the World Health Organization aim at destruction of the breeding-grounds by draining swamp land, and by maintaining an oil film on the surface of stagnant water thereby asphyxiating the larvae. Paraffin cannot be used on water intended for drinking or irrigation. There is a new technique using biodegradable, non-toxic forms of lipid. Spraying swamp lands with DDT preparations is another method, but some of the mosquitoes are gaining resistance to such preparations. Health education programmes in these areas teach people how to avoid being bitten by a mosquito on a domestic scale. If the biting mosquito has already bitten a person suffering from malaria, it will have the malarial parasite in its crop. As it bites successively it will inject the parasite into the next victim and so on. It is therefore an important duty when caring for a patient with malaria to prevent mosquitoes gaining access to that patient. One group of the mosquito family can transmit the virus of yellow fever.

Experts of the World Health Organization have reported on a possible link between the considerable fall in East Africa malaria rates and the occurrence in that area of a new epidemic virus disease, o'nyong-nyong, meaning joint-breaker, fever. Anopheles gambiae and funestus, the two main malaria-transmitting mosquitoes in Africa, have been identified as carriers of the new virus also. O'nyong-nyong fever was first noted in 1959 in Northwest Uganda. It has spread down to Lake Victoria infecting some 750,000 people.

There are a variety of other tiny creatures that inhabit houses in hot climates. They are a nuisance but are not known to transmit disease. People living in these climates usually have adequate knowledge to deal with the pests.

THE RELATIONSHIP OF BAD HOUSING TO DISEASE

1. A poor larder, inefficient arrangements for cooking food, and personal washing, increases the incidence of food poisoning. There may be insufficient protein in the diet as in the main it has to be cooked. Lice flourish on an unclean person giving rise to skin infections such as impetigo.

2. Insanitary arrangements increase the incidence of bowel infections and infestations.

3. Vermin are encouraged and the diseases that they can transmit have just been noted. Can you remember them?

4. Dampness and lack of ventilation encourage the spread of pulmonary tuberculosis and all other respiratory infections including the common cold.

5. Narrow streets of tall houses cut out the sunlight and this was blamed for the high incidence of rickets in this country.

6. Rheumatic fever has been associated with poor housing conditions over the years, and the incidence is lessening as the housing situation is gradually improving.

7. The accident rate is higher in badly kept property.

8. There can be serious interference with the emotional development of children who have to sleep in the same room as their parents.

9. Where there is overcrowding, members of the family tend to spend a lot of time away from the situation, often doing less-desirable things such as drinking and gambling.

Heating

A general discussion on the type of house in which our imaginary person, Jimmy, may choose to live is incomplete without some consideration for the type of heating which may be available. Also all health workers, visiting the homes of the very young and the elderly, are asked to estimate the temperature in the home and to be on the lookout for hypothermia (low body temperature) in their clients. Even if Jimmy is living in a hot climate, he will need some form of heating after sun-down when it can be very chilly. Single-unit heating is still the method most widely used in this country, but there is a move towards greater use of central heating for domestic dwellings. Whichever method is used, in order that fuel shall not be wasted, insulation of all outer walls should be procured, which is best seen to at the construction stage, and includes thermalite insulation for outer walls, glass fibre quilting in the roof space, and double glazing of all windows. Heated air becomes lighter and rises, hence the roof insulation. Where single units are used it is wise to place them on an inner wall, so that the adjacent room will benefit by penetration of warmth along its wall.

Transmission of heat is by conduction, convection and radiation.

Conduction

Heat travels from one molecule to an adjacent molecule, so that it is 'conducted' along. Materials are therefore designated as good or bad conductors of heat, and this property has already been spoken of in relation to clothing, and in the above paragraph air is given as a bad conductor of heat. The heating of a bed either by hot-water bottles or electric blanket is accomplished by conduction. Similarly electric underlays for carpets heat them by conduction, then the air in contact with the carpet is heated and on rising sets off convection currents. Floor heating of any description functions in this way.

Convection

A method of transference of heat only possible in gases and liquids. Those molecules in contact with the warming surface become lighter and rise, cold molecules flowing in to replace them. In this way a constant movement of the molecules is set up and is referred to as convection currents, mentioned in the previous paragraph. This makes heating closely allied to ventilation, as the cold incoming air should be introduced at a low level, should be heated to set up convection currents and should be withdrawn as vitiated air at a high level.

Radiation

Heat rays travel in straight lines from their source to be absorbed by the objects which they strike without heating the intervening atmosphere. Heat rays can be reflected and for this purpose electric fires are fitted with a bright 'reflector'. The sun and the open coal fire are examples of heating by radiation.

SINGLE UNITS USED FOR HEATING
The Open Fire

In spite of all that has been said against the open fire it still conjures up a pleasant picture in most of our minds. Many of the disadvantages have been overcome so that less heat goes up the chimney, and more is radiated into the room. It can be fitted to burn all night so that the room is pleasantly warm the next morning, and constant relighting is avoided, thus economising in matches, paper, sticks, firelighters and gas poker. The drive is toward the use of smokeless fuel, and in several areas the grates have been converted to comply with the prevention of pollution of the outside atmosphere. Eventually the whole country will comply with this prevention. A chimney with or without a fire is a good means of ventilation in a room. A boiler can be placed behind the fire and the heat used to give constant hot water. Domestic rubbish can be burned, and indeed this is one great advantage of the open fire in the sickroom.

On the debit side an open fire entails considerable labour. Coals are heavy to carry, need storage space, and give rise to ash which has to be carried to the ashbin. The chimneys have to be kept in good repair and they need to be swept bi-annually; a fall of soot can prove to be very expensive in cleaning and redecoration. Matches, sticks, paper, firelighters or a gas poker are needed to get the fire alight, and it takes some time before the whole room benefits from the heat. Even then those nearest to the fire are warmest, though their feet may remain cold from floor draught. In a sickroom replenishment may be noisy for the patient.

The Closed Combustion Stove

These are becoming more popular and they are now made in many attractive styles with coloured, heat-resisting glass panels in the

486

doors, which give the appearance of a glow. The doors can be opened to get the direct effect of the fire. Many of them burn smokeless fuel so that the householders do not feel guilty about contaminating the outside atmosphere. Most of them with regular minimal attention burn continuously, and they can be provided with an air duct so that an upper room can be heated at the same time. This constant heat is a great advantage to those who are out for long periods in the wintertime, for it helps to prevent damage which frost might otherwise do. When closed they heat by setting up convection currents, when open there is also radiation of heat. They need to be kept dust free otherwise the smell of charring may be unpleasant. They also assist ventilation in a room, but not to the same extent as an open fire. A boiler can be incorporated behind them, and a constant supply of hot water is appreciated by all members of the family. Domestic rubbish can be burned in them, though this is not recommended in smoke control areas! There is less ash in the slow-combustion process, but it requires to be removed at least twice daily. The coke or anthracite needs storage space, but much less firewood is required than for the open fire. The Aga and Rayburn cookers work on a similar principal, so utilized that they also provide cooking facilities. Chimneys need the same attention as for an open fire, and with both types there is a risk of accident by fire and all that was said about prevention of accidents in the home should be adhered to.

Gas Fires

These consist of jets enclosed within a fireclay candle rising against a fireclay back. The structure may or may not be fixed to a flue, for some types are now portable. By law they must be provided with a fixed, double fireguard, and the new type of safety tap is advisable. The use of gas eliminates storage and the carrying of heavy fuel. Because of a terrible disaster in a multistorey block of flats, gas is not being used in these structures in Britain.

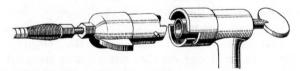

Fig. 229   A modern 'safe' gas fitting.

Where a coin in a meter is used to gain a supply of gas, it is important to turn off the tap when the supply fails, as the next member of the family putting in a coin may not know that the tap to that unit has been left open. Heating is by radiation in the first instance, then each heated object starts off convection currents. Some oxygen is used from the atmosphere and there are some waste products of combustion. Some gas fires emit a hissing noise which can be irritating

487

if it has to be endured for long periods. They must be kept free from dust or the smell of charring can be offensive. For people out at business all day they have many advantages, for such people do not have to worry about the coalman, etc., and gas fires entail much less work in the limited time that is available for domestic activities. The flame can be turned down or off, if the room becomes too warm.

Electric Fires

These consist of an element within a fireclay rod surrounded by a shining reflecting surface. Many types are portable, but it is safest to have them as fixtures where there are young children or old people. They must have a fixed, double guard. There is still an occasional prosecution for the sale of unguarded electric fires. They too heat by radiation in the first instance, then the heated objects set up convection currents, and so the whole room is warmed. They do not use oxygen from the atmosphere and there are no waste products of combustion. The use of electricity eliminates storage and carrying of heavy fuel. There is no labour involved in their use and the provision of heat is noiseless. They can be turned off if a room becomes too warm. They are ideal for dining-rooms which are only occupied for part of each day, and to provide warmth when one member of a family wants quietness for letter writing, studying, etc. They are useful for people with physical impairments such as heart disease, but they do not assist ventilation. There cannot be leakage or taps left on as with gas, but the risk is that of electrical shock should the insulation become worn and the wires exposed.

Electric convector heaters are becoming popular and an interesting project is to go round the shops and acquire literature about all the forms of heating apparatus—it is impossible to mention them all in a book like this.

Paraffin Oil Heaters

These have become popular after a period of being old fashioned, but the new types are considerably improved, and their popularity is due to paraffin being much cheaper than either gas or electricity. Their use involves storage of an inflammable oil, and the heater needs to have the wick attended to, and to be filled with oil at intervals. The Pink Paraffin Service will deliver oil to the house. They can give rise to an unpleasant smell and they heat mainly by convection. They must be put in a safe place free from draughts and must never be carried whilst they are burning. They use a little oxygen from the atmosphere, and they release products of combustion; they do not assist ventilation. One still occasionally reads in the daily press, a coroner's warning about the use of oil heaters in under-ventilated rooms.

488

CENTRAL HEATING

About one in five of the homes in Britain now have central heating. The aim is to provide a temperature throughout the house 30° in excess of the outside temperature. This has many advantages, the main one being that all the rooms in the house can be used to best advantage, whereas with single units of heating, a variety of activities have to be carried out in the one warm room. Central heating can dry the atmosphere and cause people to complain of dry lips, mouth and nose. A bowl of water in each room is a simple remedy. There are many sophisticated humidifiers on the market. The heating panels can be inserted in the walls, ceilings or floors. (Electrically heated murals and pictures are a recent addition to 'wall' units.) They heat the air in immediate contact with them which sets off convection currents and thus the whole area is kept at a constant temperature. The panels (or radiators, more correctly called convectors) can be heated by electricity or by a high or low-pressure water system. The heating of the water can be provided by coal, smokeless fuel, gas, oil or electricity. Many systems can be thermostatically controlled to give the required temperature. Ventilation should be such that the air within each room is changed three times each hour.

Another form of central heating is provided by night-store electrically heated units operating throughout the 'off-peak' period when electricity is cheaper. After installation the units entail no labour. Each unit can be adjusted to a maximum and a minimum heat by turning a knob on the unit. Each unit is very heavy and is therefore a 'permanent' fixture, but one can take the units when removing to another house, i.e. they are not considered part of the 'fabric' of the building. A similar idea is a thermal storage wall. During the day when the wall has stopped charging, air is drawn in by a fan and is passed through the wall. It picks up heat which is ducted throughout the house.

If the individual heating units are set out from the wall and are elaborate in pattern as the old-fashioned radiators (convectors), not only do they act as dust traps, but they discolour the decorations.

A modern form of central heating is by solar radiation. Sunshine is being used to heat several schools in Wallasey, Cheshire. A glass solar wall is incorporated usually in the south side of the building. The intake of heat is regulated by special shutters which work automatically with an electronic device. The shutters shield a black concrete wall which absorbs the solar heat when the shutters are open. During the first January after installation the solar heated building never dropped below 16° C (60° F). In the neighbouring building, the conventional central heating plant had a struggle to keep temperatures in the region of 5° C (40° F). The earth's resources of coal, oil and gas are gradually being exhausted and some scientists think that solar energy will be the chief form used in the future. The French have built a solar furnace in the Pyrenees. A fog signal tower on the south coast is worked by batteries charged by solar cells on the roof.

489

Another form of central heating consists of paint that can be brushed, rolled or sprayed on. It is dark in colour, but can be over-coated with ordinary paint. It conducts a weak electric current which it converts into heat and radiates into the room. Current is led into the paint from small terminals fixed to the wall. It does not exceed 40 volts, so there is no shock or fire hazard. The room warms up quickly because so much of it is heated, and when the circuit is switched off the wall acts like a storage heater.

The Soviet Union has designed a small atomic plant to provide heat and power for remote settlements in the Arctic regions. Arctic settlements are visualized as indoor garden cities where people will live normally through the harsh Polar winter.

CONSERVATION OF HEAT

There is a wall paper that makes use of reflected radiant heat, thus reducing the amount of heat escaping through the walls of a building. It is made from paper backing, a layer of aluminium foil and a plastic surface. The reflective qualities of the foil return to the room a good percentage of the heat which is normally lost. The initial heating comes from a single heating unit—coal, gas, electricity or oil.

DISTRICT HEATING

The development of district heating from a central boiler installa-tion in this country is about 50 years behind the rest of Europe. You will remember that World Health Organization experts recom-mend district heating in the attempt to lessen atmospheric pollution. However a start has been made and there are four schemes in varying stages in Scotland. They are all of the solid fuel type and will pro-vide domestic warm air heating and hot water.

Prevention of Infection

In this story the structure and function of a human being and his personal and environmental hygiene have been discussed, together with their direct relationship to the promotion of health. Around us there are millions of bacteria, many of them doing good and useful work, but a few are capable of producing disease and are therefore called pathogenic organisms; though we can do a great deal to prevent these organisms assaulting the human body, infection is a possible hazard throughout life.

Epidemiology is the study of the distribution and determinants of disease in populations. There are maps of the distribution of *infectious* diseases throughout the world and each country makes its own laws about vaccination and immunization of travellers and immigrants. Each country varies as to the protection offered to its inhabitants, and whether this is paid for under the Health Service or has to be paid for on receipt of service by the individual. Spread of infection in a community is greatly reduced if 60 to 80 per cent of the population is immunized against it—when the herd immunity is spoken of as favourable. Countries help each other by offering massive supplies of vaccine, through the World Health Organization, to those countries in emergency need of them. The mass movement of people into cities in the process of industrialization, of which we spoke at the beginning of this book, is conducive to spread of infection. Measles is a serious problem in Africa at present. The speed of world travel renders every nation at risk to diseases that have been stamped out in the developed countries, e.g. minor outbreaks of cholera in Spain and Britain in the 1970s.

The venereal infections and epidemiology are discussed on page 453.

All workers in this country, exposed to the risk of anthrax, such as those working in glue gelatin, soap and bone meal factories, woollen mills and tanneries, can now be vaccinated against it. Vaccination may be performed by local health authority doctors, general practitioners or factory doctors with vaccine supplied free through the Public Health Laboratory Service.

Several councils are now offering immunization against tetanus to their outdoor employees—those engaged in parks, building, roads, sewers and refuse collection and disposal.

Nature of Infection[220]

The successful invasion, establishment and growth of micro-organisms in the tissues. The invasion of the tissues of the alimentary tract by food poisoning organisms resulting in vomiting, with or without diarrhoea, has been described on page 238, ringworm infection of the skin on page 131, fungal infection of the toes on page 73 and staphylococcal infection of a hair follicle producing a boil on page 241, so that you have already gained considerable knowledge on this subject.

Communicable Disease

Any disease which can be transmitted from one person to another by direct or indirect contact. Hospital sepsis, i.e. sepsis in a wound, is in this category.

Infectious Disease

A disease which can only be caused by its own organism resulting in a particular set of signs and symptoms, and transmissible from one person to another by direct or indirect contact. Any article responsible for the indirect contact is called a fomite.

Incubation Period

The time elapsing from the entry of germs until the appearance of the first sign or symptom. The germs have been developing and multiplying meantime.

Quarantine

Period of separation for those who have been in contact with an infected person. It varies with each disease and is similar to the incubation period.

Epidemic Disease

Reserved for the outbreak of a disease in an area where many people are simultaneously infected.

Endemic Disease

Any disease which keeps recurring in a locality.

Pandemic Disease

An infection spreading over a whole country or the world. Influenza was pandemic after the First World War and again in 1957.

Sporadic Disease

Term applied to scattered or isolated cases; usually the source of infection remains unknown.

Specific Disease

A disease caused by a special organism producing only the symptoms of that particular disease.

SOURCES OF INFECTION

1. **Droplets.** In speaking, coughing, sneezing and breathing, droplets of saliva, sputum or secretion from the nose are sprayed into the

Fig. 230   Sources of infection.

493

atmosphere, and from an infected person or carrier they will contain pathogenic organisms. They travel 6 to 10 yards (5 to 9 m) and can infect any person who happens to be in range, or they can fall to the ground where they dry; thus they can be inhaled or ingested with dust. Diseases spread by droplet infection include tuberculosis, the common cold, influenza, sore throat, scarlet fever, diphtheria, measles, mumps (infectious parotitis), whooping cough and cerebrospinal meningitis (spotted fever). Conditions of overcrowding and inadequate ventilation help infection to spread in this way.

2. **Carriers.** A person who is incubating an infection can spread the germs of that disease. A person who has had an infection and recovered from it can continue to harbour the germs and excrete them. e.g. diphtheria and typhoid fever. Such a person can be called a convalescent carrier, but it is a misleading term because the carrier state can last for years. A person who has never manifested the infection can harbour the germs and excrete them. Such a person can be called a healthy carrier. See page 241 regarding staphylococcal carriers. Anyone identified as a typhoid carrier is excluded from school if a child, and is not allowed to work with food if an adult.

3. **Discharge or secretion from existing lesion.** The diseases in which this can occur are impetigo, ophthalmia neonatorum, chickenpox, smallpox, gonorrhoea and syphilis.

4. **Blood and blood products, e.g. serum.** Syphilis can be spread if the spirochaete in the blood of an infected person enters a skin abrasion on another person. The virus of serum hepatitis can be spread by the communal use of syringes and needles by drug addicts.

5. **Dust and air.** The germs causing tuberculosis and anthrax can live when dried and can therefore be disseminated via dust and air. It has been increasingly realized that the dust in hospital wards is heavily contaminated with streptococci and staphylococci, hence the introduction of cotton cellular blankets and the more extensive use of terry towelling to cut down the production of dust. Suction is considered the safest method for removal of dust; failing that a damping agent should be used. The virus that causes serum hepatitis was found to be airborne in one of the renal dialysis unit outbreaks of that infection.

6. **Soil and road dust.** The spore-forming organisms causing tetanus and gas gangrene live normally in human and animal intestine, from whence they are transferred to the ground.

7. **Fomites.** Any articles that have been sufficiently near an infected person to become contaminated with bacteria, e.g. books, pencils, pens, crockery, cutlery, linen, furniture, etc.

8. **Parasites.** The rat flea can transmit bubonic plague; lice can transmit typhus fever, and their bites can leave a route via which other infections such as impetigo can enter.

9. **Rodents.** The part played by these as vectors of food poisoning has been dealt with (p. 481). Excreta from rats can contain a spirochaete which, if it enters man via an abrasion, can cause a type of jaundice—Weil's disease.

10. **Food.** The part played by food in the dissemination of infection has been discussed on page 238.

11. **Water.** The part played by water in the dissemination of infection has been discussed on page 260.

12. **Milk.** The part played by milk in the dissemination of infection has been discussed on page 255.

13. **Mosquitoes.** In warmer climates these act as vectors for malaria and yellow fever (p. 483).

MODES OF TRANSMISSION OF INFECTION

Some of these have already been mentioned so that revision will be useful at this stage. Athlete's foot and verruca are spread by infected shoes and stockings, baths, bath mats, floors, swimming baths, etc. Ringworm of the finger nails can be spread by direct contact with infected animals, including dogs and cats. The fact that many human beings can carry staphylococci in the skin, nose and mouth gives ample opportunity for the transmission of these germs to cuts, pricks and hangnails. By the same token these and other germs can be transferred to food. A cow suffering from tuberculosis, streptococcal infection of the udders or undulant fever (brucellosis) can pass these germs, via its milk, to man. A staphylococcal carrier can put his germs into milk to produce food poisoning in the consumer. Similarly a carrier of the organisms causing typhoid and paratyphoid fever, and dysentery can transfer these to milk and cause the respective diseases in man. The germs causing typhoid and paratyphoid fever, dysentery and cholera can percolate into water, usually from excreta, and thus transmit these diseases to man. Teats and comforters can transmit thrush and gastro-enteritis to babies. Lavatory seats, cistern and door handles can act as transmitters of the bowel infections, including infective hepatitis. The part played by droplets, dust and air were dealt with in items 1 and 5 under sources of infections. Germs can pass directly along a canal, e.g. the *Bacillus* (*Escherichia*) *coli*, normally present in faeces causes cystitis if it wanders along the short female urethra from a soiled perineum, and little girls can get vulvovaginitis from wriggling along in the sitting position on a dirty floor. The mixing of face flannels and other toilet requisites can spread skin infections and such conditions as pink eye and trachoma. Sexual intercourse can spread the venereal diseases of syphilis, gonorrhoea and non-specific urethritis.

The modes of transmission of infection is an inexhaustible subject, and each time an epidemic occurs the bacteriologists set to work to solve the mystery of where the infection is coming from, and how it is being transmitted.

For infection to occur the infecting organism must be present in large numbers, or be of a virulent variety, the host must be susceptible, and the organism must enter by the appropriate channel. The channels via which infection can enter are the respiratory tract

495

when the process is inhalation, the alimentary tract (ingestion), broken skin or mucous membrane (inoculation), the reproductive tract (direct contact, or ascending infection in the female).

SUSCEPTIBILITY TO INFECTION

This usually refers to a disposition to infection. Susceptibility is increased in conditions of overcrowding, inadequate food and clothing, extremes of atmospheric temperatures, etc.

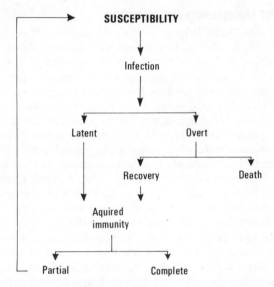

Fig. 231   Susceptibility to infection.

RESISTANCE OF HEALTHY BODY TO DISEASE

Unbroken skin and mucous membrane present a physical barrier to germs as well as pouring out various antiseptic substances (lysozymes) that control the number of organisms on these surfaces; they are assisted in their functions by the movement of cilia where these are present. The hydrochloric acid of gastric juice destroys ingested bacteria. Tears contain another lysozyme capable of dissolving bacteria and thus the eyes are constantly protected. The polymorphonuclear leucocytes possess the power of amoeboid movement and can migrate from the blood vessels to surround and ingest invading bacteria. (phagocytic action). When local infection such as a boil occurs, connective tissue cells multiply to form granulation tissue which shuts off the infected area from surrounding healthy tissue. The lymphatic nodes act as filters to the area that they drain. They also produce lymphocytes which are concerned in the production of antibodies. There is a ring of lymphoid tissue surrounding

496

THE BODY'S DEFENCE AGAINST INFECTION

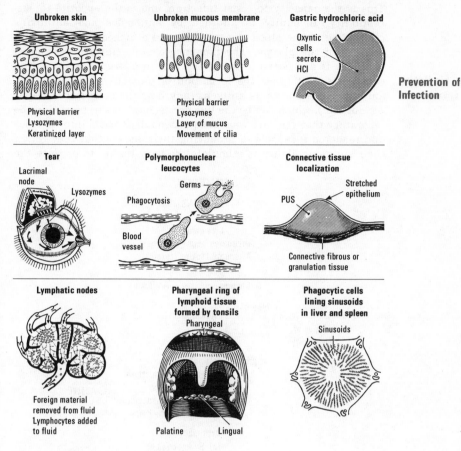

Prevention of Infection

Fig. 232   Resistance of healthy body to disease.

the pharynx, of which the palatine and pharyngeal tonsils are a part, to guard the entrance of organisms in food and air. Phagocytic cells line the sinusoids of the liver and spleen and can deal with any bacteria that gain access to the blood stream.

To keep all these defence tissues at their maximum potential for resisting disease, each person needs to take an adequate diet, to get sufficient fresh air and sunshine, to exercise his body, to avoid over-fatigue and over-exposure to the elements, to keep himself and his surroundings clean, to avoid over-indulgence in those things which are known to harm the tissues, e.g. alcohol, drugs, smoking, etc., and to wear protective clothing wherever this is advisable. Compliance with these laws of living gives a high general resistance, but the body also possesses the ability to manufacture resistance against specific germs, this resistance being spoken of as immunity.

497

IMMUNITY

This is the capacity to resist infection, and is the opposite to susceptibility. Immunity is afforded by antibodies and antitoxins circulating in the blood stream. They are attached to the globulin fraction of the plasma. The immunoglobulins (Ig) are designated G. M. A. D and E the latter interacting in allergic reactions. The substances that cause the formation of protective antibodies and antitoxins are known as antigens (anything the body distinguishes as 'not-self'), which can be the live germs; or germs so treated in the laboratory that they are weakened (attenuated), made into vaccines and rendered incapable of producing their disease, but they retain their antigenic ability to stimulate the body tissues to produce the

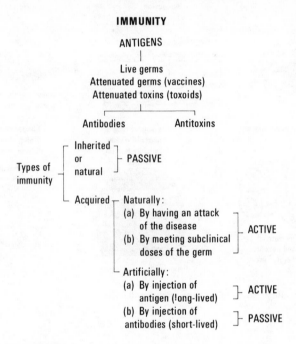

Fig. 233   Immunity.

appropriate antibody; or toxins (poisonous substances made by germs) so treated in the laboratory that they are rendered incapable of poisoning a human being (toxoids), but they retain their antigenic ability to stimulate the body tissues to produce the appropriate antitoxin.

The antibodies and antitoxins circulating in a mother's blood are passed naturally via the placenta to her baby, thus conferring the immunity of the mother on the child. Some Ig G and less Ig M crosses the placental barrier, the smallness of the latter causing the newborn to be susceptible to E. Coli infections. Ig A is preferentially secreted into the colostrum. Ig A is a protein and protein is normally broken down by digestive enzymes but absorption of Ig A is of a special selective nature and is accomplished by special intestinal cells. This inherited natural passive (because the child has not worked to obtain it) immunity to measles and whooping cough appears to last for only three to four months, and in the remainder of the first year of life the baby is vulnerable to these infections. (When planning immunization programmes this fact is borne in mind.[221]) In other instances immunity which appears to be inherited lasts for a lifetime.

Immunity can be acquired *naturally* by one of two ways. First, by having an attack of the infectious disease, whereby the body is stimulated to produce the appropriate antibodies in such quantities that not only does recovery occur, but there are sufficient to remain in the blood throughout life, thus conferring lifelong naturally acquired active immunity. Rubella in the early stages of pregnancy can cause *active* rubella in the baby born at term. Such a baby is infectious and should not come into contact with pregnant women, for as well as the possibility of *active* disease in the newborn, there is danger of congenital abnormality. Secondly, by meeting repeated tiny doses of the germ, insufficient to cause the classical disease, but sufficient to stimulate the body to produce the appropriate antibodies which remain in the blood throughout life, thus conferring lifelong naturally acquired active immunity.

Immunity can be acquired *artificially* by one of two ways. First, by injecting an antigen into the body and allowing time for production of the antibody which then remains in the blood for a specified time. This is also an active type of immunity. Secondly, by injecting ready-made antibodies, but these do not remain in the blood for long. This is a passive type of immunity, and can save life when a person already shows signs and symptoms of the infection; it would otherwise be a race against time as to whether the body could produce sufficient antibody to kill the germs before the germs kill the infected person. The antibodies injected in this passive type of immunity, which is acquired artificially, have been made in the blood of either a human being or a horse.

CONTROL OF EPIDEMICS

1. There is a list of the infectious diseases which have to be notified to the Medical Officer of Health in the area. He thus becomes aware of an epidemic and is able to set the machinery in motion to control spread of the infection.

2. The source of the infection and the mode of transmission has to be found and the bacteriological laboratory will need to help here. Carrier tracing or contact tracing may have to be carried out, and this may entail the use of health visitors, public health inspectors, pest officers, factory inspectors, occupational health nurses and mass radiography. The laboratory staff may need to check on food, milk, water, sewage disposal, etc. Should any of these be the source of infection, immediate steps would be taken to deal with the situation.

3. Isolation of each infected person. If this can be achieved   in the home, then the family can have the advice and support of the health visitor. Hospital beds are available for infectious patients, and trained nurses capable of 'barrier' nursing are available in the hospital and the home.

4. Quarantine of those people known to have been exposed to the infection. These powers are not used so extensively nowadays. When the epidemic is amongst school children, the school nurse visits each morning for inspection. When the epidemic is affecting the adult population, the occupational health nurse may be asked to do a daily inspection of staff. In recent smallpox outbreaks the contacts have been asked to attend an inspection clinic each morning.

5. Susceptibility tests, immunization and vaccination can play a big part in the control of epidemics, as well as in the prevention of infection. The Schick test reveals susceptibility to diphtheria, the Dick test to scarlet fever and the Heaf test to tuberculosis. If the contacts have previously been immunized a booster dose of the antigen will be given, but if intimate contacts have never been immunized then a dose of ready-made antibodies will be considered. Immunization is available for diphtheria, measles, german measles, whooping cough, tetanus, anthrax, smallpox, poliomyelitis, tuberculosis, typhoid fever, yellow fever, rabies and cholera. Immunization to the latter four diseases is not routinely carried out in Britain, but is available to those travelling to endemic areas.

6. Intensified health education, using especially the local radio and television, and newspapers.

7. Provision for disinfection. If the local authority does not possess its own apparatus, then it will probably arrange to use that at the hospital.

8. Overcrowding must not be allowed during epidemics.

9. Strict censure by port authorities, so that notifiable disease is not brought into the country. This includes quarantine of animals to prevent rabies.

Disinfection

An article is rendered non-infectious from the process of disinfection.[222] If at the same time the article can be guaranteed free from germs it is said to be sterile. Any substance or process that kills germs is described as being bactericidal or germicidal. Any substance or process that inhibits the growth of germs without killing them is described as bacteriostatic or antiseptic.

The first division of disinfectants is into a chemical and a physical group.

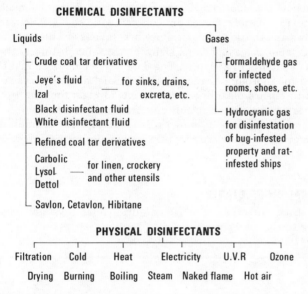

CHEMICAL DISINFECTANTS

Liquids

— Crude coal tar derivatives

Jeye's fluid
Izal — for sinks, drains, excreta, etc.

Black disinfectant fluid
White disinfectant fluid

— Refined coal tar derivatives

Carbolic
Lysol — for linen, crockery and other utensils
Dettol

— Savlon, Cetavlon, Hibitane

Gases

— Formaldehyde gas for infected rooms, shoes, etc.

— Hydrocyanic gas for disinfestation of bug-infested property and rat-infested ships

PHYSICAL DISINFECTANTS

| Filtration | Cold | Heat | Electricity | U.V.R | Ozone |
| Drying | Burning | Boiling | Steam | Naked flame | Hot air |

Fig. 234   Various types of disinfectant.

CHEMICAL DISINFECTANTS

The chemical group is further divided into liquids and gases.

Liquid Disinfectants

The number of such substances now available is legion and the manufacturers issue advice as to the suitability of the liquid for different articles, the strength at which the liquid is to be used and a recommended time of immersion. The strength is inversely proportional to the time, i.e. the stronger the solution the less time it will take to disinfect, and conversely—the weaker the solution the longer time it will take to do its work. The more crude distillations from coal tar such as Jeyes' fluid, Izal, black and white disinfectant fluid, are usually reserved for sinks, drains, excreta, etc. The more refined distillations such as carbolic, lysol and Dettol can be used for linen, crockery and other utensils. Savlon, cetavlon and hibitane have replaced the coal-tar derivates in many hospitals and notes regarding their use are issued from the dispensary.

Gaseous Disinfectants

It is rarely considered necessary to fumigate premises after infectious disease nowadays, but the recommended method is to prepare the room by opening all drawers and cupboards, exposing the bedclothes as much as possible, and tipping the mattress so that as much of its surface as possible comes in contact with the gas, closing the windows and the door on exit. Paraform tablets are then burned in a special lamp, releasing formaldehyde gas which disinfects the room and its contents in 24 hours. The door and windows should then be opened for thorough airing before the usual domestic measures are taken to ensure cleanliness. (See page 73 regarding the treatment of shoes when a person is suffering from athlete's foot.)

Hydrocyanic gas is used by skilled teams for the disinfestation of bug-infested property, and by port authorities for rat-infested ships.

PHYSICAL DISINFECTANTS

Heat, cold, filtration, electricity, ultra-violet rays and ozone all belong to this category.

Heat

Heat can be used in the process of *drying* which, by depriving the bacteria of moisture, kills them. Thus dried foods are sterile.

Burning is the method of choice for all articles of little value.

Boiling for three minutes will kill all but the spore-forming bacteria and probably is the most widely used method of disinfection outside hospital.

Steam in a disinfector or autoclave is widely used for dressings, clothing, mattresses and pillows, etc. In the older type of apparatus the temperature can be raised to 118° C (245° F) to 121° C (250° F), and articles must be subjected to this for 20 to 30 minutes to ensure sterility. In the newer type of apparatus the temperature can be raised to 141° C (285° F) to 148° C (290° F), and this takes $8\frac{1}{2}$ minutes to sterilize the articles concerned.

Dry heat in the form of a *naked flame* is used in laboratories for sterilization of platinum loops, for a needle point used for the first-aid removal of a splinter, for the inside surface of a bowl after

502

application of methylated spirit to same—only used as an emergency method.

Hot air is used for glassware and syringes, but as it is much less penetrating than moist heat, a higher temperature must be achieved, i.e. 160° C (321° F) for one hour, or 180° C (356° F) for half an hour.

Cold

The effect of cold on bacteria can be revised by studying Figure 125. It is mostly used in the refrigeration of food, though it must be remembered that *Salmonella typhi* can live in ice.

Filtration

Into this category comes the filters that can be affixed to water taps, those used at the water works and sewage works, those used for filtering milk, as well as the highly specialized varieties that are used for vital work in the laboratories.

Electricity

This can be used to produce ultra-violet rays and these have been tried with a view to disinfection of the atmosphere in operating theatres and nurseries. The cost has not as yet justified installation on a large scale. Electricity is used in the large-scale pasteurizing plants.

Ultra-violet Rays

Those coming from the sun have a bactericidal property, and over the last decade much greater use has been made of putting articles out in the fresh air and sunshine for disinfection.

Ozone

A gas that sterilizes by liberation of oxygen. It is expensive to produce but is used in one area in France for the sterilization of water and in a recently installed scheme in Scotland.

This merely forms the base on which you can build your knowledge of disinfectants as you use them in the wards.

The last piece of the jig-saw has been put in position to complete the picture of the structure and function of a human being, his personal and environmental hygiene to ensure his healthy sojourn through this life on earth. As germs are no respectors of persons we had to speak of the possibility of infection, and the methods of disinfection.

However well Jimmy has maintained his state of health, he will grow old. People vary as to the age of onset of declining activity, so much so, that it is sometimes necessary to differentiate biological age from physiological age, i.e., a person can be physiologically old at the biological age of 50. Interspersed throughout this book we have spoken of the modifications with regard to bathing, clothing, eating, exercise, recreation, housing and heating, that are necessary for the elderly.

Man is mortal and Jimmy will one day die. Each person needs to come to terms with the *only certain fact about his life*—he will one day die. Many people feel that it is morbid to talk about death. 'Don't talk about it' and 'Goodness, you're not going to die yet' commonly greet an introduction of the subject. Yet it will *certainly happen to each one of us*! Attitudes to death are often culturally determined and beliefs about death religiously determined. Culture and religion can each play a part in what is done to the dead body, and the behaviour of relatives between death and disposal of the dead body, and the ceremony and method of disposal. Each individual has to go through the grieving process of missing the loved one, and this can take up to 18 months.[223]

Having come to terms with the fact that you will die, you need to retain the *sensitivity* whereby you are *appalled by unnecessary death*.

This book shows how societies and each individual in those societies need not take unnecessary risks to his own or anyone else's health and life. Then we can all sing the line from a modern song:

'We can live in a world for which we care'.

References

[1] Crawford, M. (1971). Genetic counselling. *Nursing Times,* 28th October, p. 1351.

[2] Price, W. H. & Whatmore, P. B. (1967). Behaviour disorder and pattern of crime among XYY males identified at a maximum security hospital. *British Medical Journal,* **1,** 533.

[3] de Torok, D. (1971). Alcohol and genes. *Pulse,* 16th October.

[4] Crawford, A. (1964). Is Health Code an essential safeguard for the public? *Medical News,* 3rd April, p. 5.

[5] Wintersgill, P. (1964). Health Code. *Medical News,* 17th April, p. 14.

[6] *Nursing Mirror,* (1971). Different cultures and child care practice. 10th December, p. 21.

[7] Grundy, F. (1968). *New Public Health,* 7th ed. ch. 2. London: Lewis.

[8] *Nursing Times,* (1971). We get the health service we deserve. 12th August, p. 990.

[9] Seiler, H. E. & Mearns, A. G. (1971). *Scottish Public Health Law. A Digest for Nurses.* The Queen's Institute of District Nursing, Scottish Branch, 26 Castle Terrace, Edinburgh EH1 2EH. 2nd ed. 32½p+postage.

[10] *Nursing Times,* (1971). Scottish Health Services due for reorganization. 5th August, p. 946.

[11] Huntly, W. L. (1969). *Aids to Personal and Community Health,* 7th ed. London: Balliere, Tindall and Cassell.

[12] Grundy, F. (1968). *New Public Health,* 7th ed. ch. 1. London: Lewis.

[13] Priest, M. A. (1967). *Modern Hygiene for Nurses,* 3rd ed. London: Heinemann.

[14] Wafinden, R. C. (1971). The fragments that remain or forward the health centres. *Nursing Times,* 2nd September, p. 137.

[15] Swan, W. G. (1967). Port Health. *Nursing Times,* 15th December, p. 1691.

[16] Allen, M. G. (1971. Experiment in teaching. *Nursing Times,* 18th November, p. 1453.

[17] Priest, M. A. (1967). *Modern Hygiene for Nurses.* 3rd ed. Fig. 1. London: Heinemann.

[18] Phillips, W. P. (1968). Unity in tripartite service. *Nursing Mirror,* 12th January, p. 345.

[19] *Nursing Times,* (1971). Will community take over from hospitals? 21st October, p. 1297.

[20] *Nursing Times,* (1971). Out of the 'rut'—the domiciliary midwife. 18th November, p. 1450.

[21] *Nursing Times,* (1971). Midwifery training for part-timers. 25th November, p. 1457.

[22] McCarrick, H. (1971). Key figure of the future—the health visitor. *Nursing Times,* 7th October, p. 1243.

[23] Lamb, J. (1971). Health visitor in Uganda. *Nursing Mirror,* 29th October, p. 33.

[24] Gilbert, J. E. (1970). 'Require no visiting'. *Nursing Times,* 22nd October, p. 1358.

²⁵ Lysistrata, (1971). Another look at health visitors. *Nursing Times,* 4th November, p. 1387.

²⁶ Jones, M. (1971). Letter from Stafford. *Nursing Times,* 2nd December, p. 1514.

²⁷ Altschul, A. (1972). *Psychology for Nurses.* 4th ed. London: Ballière, Tindall and Cassell.

²⁸ Henderson, P. (1969). The changing pattern of disease and disability in schoolchildren in England and Wales. *Nursing Mirror,* 14th November, p. 21; 21st November, p. 48.

²⁹ Coll, A. P. (1962). Health and education hand-in-hand. *Nursing Mirror,* 23rd November, p. 1962.

³⁰ *Nursing Times,* (1971). Careers in nursing. Occupational health nurse. 30th September, p. 1215.

³¹ Thomson, W. (1971). The occupational health team. *Nursing Times,* 2nd December, p. 1499.

³² *Nursing Times,* (1971). OH, oh, oh! 23rd December, p. 1596.

³³ Mellor, V. (1969). The self-inflicted accident. *Nursing Mirror,* 2nd May, p. 27.

³⁴ Howkins, T. P. (1969). British Railways Medical Service. *Nursing Times,* 1st January, p. 18.

³⁵ Bassford, P. A. (1969). Cancer health education in industry. *Nursing Times,* 12th June, p. 748.

³⁶ *Annual Report of the Chief Inspector of Factories.* London: H.M.S.O.

³⁷ *Nursing Times,* (1971). Careers in nursing. The district nurse. 23rd September, p. 1174.

³⁸ Keywood, O. (1971). Home help service. *Nursing Times,* 9th December, p. 1530.

³⁹ Botting, P. M. & Jenkins, T. C. (1967). Portsmouth night attendant service. *Nursing Mirror,* 29th December, p. vi.

⁴⁰ Sprigg, C. (1970). A night nursing service. *Nursing Times,* 21st May, p. 645.

⁴¹ *Nursing Times,* (1969). Working together. 16th January, p. 87.

⁴² Bocock, E. J. & Haines, R. W. (1973). *Applied Anatomy for Nurses.* 4th ed. Edinburgh and London: Churchill Livingstone.

⁴³ Rowe, J. W. & Wheble, V. H. (1972). *A Concise Textbook of Anatomy and Physiology.* 3rd ed. Edinburgh and London: Churchill Livingstone.

⁴⁴ Roper, N. (1969). *Livingstone's Dictionary for Nurses.* 13th ed. pp. 506–515. Edinburgh and London: Churchill Livingstone.

⁴⁵ Roper, N. (1973). *Principles of Nursing.* 2nd ed. Edinburgh and London: Churchill Livingstone.

⁴⁶ Winters, M. C. (1952). *Protective Body Mechanics in Daily Life and Nursing.* Philadelphia: Saunders.

⁴⁷ Filmstrip. *Childrens' Feet.* Camera Talks Ltd., 31 North Row, London W1.

⁴⁸ *Foot Care at all Ages.* (1970). London: British Medical Association.

⁴⁹ Faint, J. (1971). Resting standing up! *Nursing Mirror,* 23rd April, p. 41.

⁵⁰ Rivlin, S. (1971). You and your legs. *Nursing Times,* 25th November, p. 1462.

⁵¹ *Nursing Times,* (1971). You and your legs. 9th December, p. 1539.

⁵² Kline, A. L. & Fegan, W. G. (1971). The effect of footwear on venous return. *Nursing Times,* 30th December, p. 1644.

⁵³ *Nursing Times*, (1972). You and your legs. 6th January, p. 22.

⁵⁴ Cunliffe, W. J. (1970). Warts, verrucae, callosities and corns. *Nursing Times*, 1st January, p. 12.

⁵⁵ *Nursing Mirror*, (1967). Survey shows more boys have bunions. 19th May, p. vi.

⁵⁶ Cohen, A. (1969). Cohen's comment. *Nursing Mirror*, 3rd October, p. 19.

⁵⁷ Sarkany, I. (1971). Detergents and hand eczema. *Nursing Times*, 30th September, p. 1211.

⁵⁸ Barnett, H. (1968). Basic surface anatomy. *Nursing Times*, 11th October, p. 1381.

⁵⁹ Uttley, M. (1971). Sweat glands and perspiration. *Nursing Mirror*, 26th November, p. 35.

⁶⁰ Meacham, F. (1969). Observations on skin surface reactions. *Nursing Mirror*, 30th May, p. 30.

⁶¹ *British Medical Journal*, (1966). Bathing the baby without soap. **2**, 745.

⁶² Gregory, S. & Frick, S. (1967). A new concept in infant bathing. *Nursing Mirror*, 21st April, p. 55.

⁶³ *Medical News*, (1964). Survey of burns in children. 26th June, p. 6.

⁶⁴ *Nursing Times*, (1971). 4th November, p. 1386.

⁶⁵ World Health Organization, (1963). Recommended methods for Vector Control. *Bull. Wld. Hlth. Org.* **29,** Suppl.

⁶⁶ *Nursing Times*, (1968). Swaddled in silver. 20th December, p. 1713.

⁶⁷ *Observer*, (1968). The swaddled babe. Colour supplement. p. 12.

⁶⁸ Keenan, M. (1964). Starting at the bottom. *The Sunday Times*, 23rd Feb.

⁶⁹ Burn, J. L. (1969). Napkin Hygiene. *Nursing Times*, 3rd July, p. 847; 10th July, p. 880.

⁷⁰ Constance, S. D. (1963). Effects of mittens on infants. *Nursing Mirror*, 22nd November, p. x.

⁷¹ Central Council for the Disabled, (1967). *Problems of Clothing for the Sick and Disabled.* London.

⁷² Bliss, M. R., McLaren, R. & Smith, A. N. (1967). Day clothing for geriatric patients. *Nursing Times*, 5th May, p. 598.

⁷³ *Nursing Mirror*, (1964). Reactions of mothers to children's flame-resistant clothing. 2nd October, p. 11.

⁷⁴ Nossal, G. J. V. (1971). *Antibodies and Immunity.* London: Pelican Books.

⁷⁵ *Nursing Times*, (1969). A new light on atherosclerosis. 13th November, p. 1446.

⁷⁶ Malpas, J. S. (1968). The functions of the spleen. *Nursing Times*, 13th September, p. 1224.

⁷⁷ Altschul, A. (1972). *Psychology for Nurses.* 4th ed. Ballière, Tindall and Cassell.

⁷⁸ May, A. R., Kahn, J. H. & Cronholm, B. (1970). *Mental Health of Adolescents and Young Persons. Publ. Hlth. Pap. W.H.O.* No. 41.

⁷⁹ Nuffield Provincial Hospital Trust, (1971). *Stress in Youth.* London: Oxford University Press.

⁸⁰ Evans, C. (1969). Sleep, dreams and computers. *Nursing Times*, 15th May, p. 615.

81 Oswald, I. (1971). The biological clock and shift work. *Nursing Times,* 30th September, p. 1207.

82 Illingworth, R. S. (1969). Sleep problems in children. *Nursing Mirror,* 7th March, p. 38.

83 Glatt, M. M. (1971). What makes an addict? *Nursing Times,* 4th November, p. 1382.

84 Family Doctor Booklet, (1970). *Behind the Drug Scene.* London: B.M.A.

85 Chalke, H. D. (1971). Alcoholism today. *Nursing Times,* 18th March, pp. 313, 331, 332, 333.

86 Freeman, S. (1968). The drinking driver. *Nursing Times,* 13th December, p. 1698.

87 Willis, J. H. (1969). *Drug Dependence.* London: Faber.

88 Dawtry, F. (1969). *Social Problems of Drug Abuse.* London: Butterworth.

89 Nicol, G. (1969). Control of obesity. *Nursing Mirror,* 13th June, p. 30.

90 Gries, A. (1971). Gout. *Pulse,* 23rd October.

91 *Nursing Mirror,* (1969). Undernourished children. 31st January, p. 4.

92 *Nursing Mirror,* (1969). The food we eat. 1st August, p. 37.

93 Lee, C. M. (1968). Feed me with food convenient for me. *Nursing Times,* 27th September, p. 1290; 4th October, p. 1340.

94 *Nursing Times,* (1970). Metrication. 27th August, p. 1109.

95 *Medical News,* (1966). Protein from methane. 17th June, p. 4.

96 *Medical News,* (1966). Fish protein concentrate for humans as well. 25th November, p. 6.

97 *Nursing Mirror,* (1969). Feeding India's children. 14th November, p. 26.

98 Brookes, P. (1971). Malnutrition. A new approach needed. *Nursing Times,* 7th October, p. 1237.

99 Latimer, C. (1971). Nutrition in the Eastern Solomon Islands. *Nursing Mirror,* 8th October, p. 40.

100 Smith, S. E. (1971). Vitamins. *Nursing Times,* 9th September, p. 1109.

101 *British Medical Journal,* (1966). Vitamin E deficiency. 16th April, p. 935.

102 Richards, I. D. G. (1969). Rickets in Britain. *Nursing Mirror,* 17th January, p. 26.

103 Goldberg, A. (1968). Anaemia and iron. *Nursing Times,* 17th May, p. 658.

104 Stanton, B. P. (1971). *Meals for the Elderly.* London: King Edward's Hospital Fund for London.

105 Christie, A. & Christie, M. (1971). *Food Hygiene and Food Hazards.* London: Faber.

106 *Medical News,* (1966). Antibiotics hazardous. 4th November, p. 15.

107 *Medical News,* (1965). French ban drugs in livestock rearing. 27th August, p. 3.

108 *Medical News,* (1964). Difficulties in the control of infective hepatitis. 7th February, p. 19.

109 *British Medical Journal,* (1965). Hand Hygiene. 7th August, p. 315.

110 *Nursing Times,* (1969). Now wash your hands! 6th March, p. 307.

111 *Medical News,* (1966), Ten commandments to ensure cleaner food. 4th November, p. 17.

[112] *British Medical Journal*, (1966). Ministry Drive for Clean Food. 5th November, p. 1149.

[113] Prophylax, (1966). Food hygiene is no laughing matter. *Medical News*, 2nd December, p. 13.

[114] *Hygiene in Catering Establishments*, (1951). Report of the Catering Trade Working Party. London: H.M.S.O.

[115] *Nursing Times*, (1969). Now wash your hands! 6th March, p. 307.

[116] Thrower, W. R. (1964). Towards better milk. *Medical News*, 2nd October, p. 12.

[117] *Medical News*, (1966), Milk standard still too low. 4th November, p. 15.

[118] *Nursing Times*, (1969). Fluoridation in the U.K. The results after 11 years. 21st August, p. 1080.

[119] Working Party on Sewage Disposal, (1970). Report. *Taken for Granted*. London: H.M.S.O.

[120] Jolles, K. E. (1971). An aid for thirsty motorists. *Nursing Times*, 5th August, p. 972.

[121] *Nursing Times*, (1969). Purest water in the world. 21st August, p. 1075.

[122] *Nursing Mirror*, (1971). Cows' teeth and caries study. 14th May, p. 41.

[123] Naylor, M. N. (1971). Prevention of dental caries. *Nursing Mirror*, 18th June, p. 20.

[124] James, P. M. C. (1970). Dental decay: how nurses can help. *Nursing Times*, 5th March, p. 303.

[125] Parkin, S. F. (1968). Teaching children about dental health. *Nursing Mirror*, 15th November, p. 36.

[126] Shotts, N. (1968). The care of childrens' teeth. *Nursing Times*, 16th August, p. 1112.

[127] H.M.S.O. (1969). *The fluoridation studies in the U.K. and the results achieved after 11 years*.

[128] *Nursing Mirror*, (1968). Dental care for infants. 26th April, p. 25.

[129] *Medical News*, (1962). And now the electric toothbrush. 9th November, p. 2.

[130] *Medical News*, (1967). Fluoride scores in tooth test. 17th March, p. 13.

[131] *Medical News*, (1966). Cocoa to counteract caries. 2nd December, p. 5.

[132] *Medical News*, (1964). Caries cut by molybdenum. 17th July, p. 24.

[133] *Nursing Mirror*, (1967). New technique reduces tooth decay. 17th February, p. xvi.

[134] *Nursing Mirror*, (1967). Dentures for children. 27th January, p. xii.

[135] Hopkins, S. J. (1970). Diabetes and insulin resistance. *Nursing Times*, 28th May, p. 677.

[136] Clare, I. (1968). Clean in his habits. *Nursing Mirror*, 24th May, p. 34.

[137] Jobbins, V. (1968). Nappies to pants. *Nursing Mirror*, 26th July, p. 41.

[138] Seaton, D. R. (1968). Common intestinal worms. *Nursing Times*, 5th April, p. 465.

[139] Cole, A. C. E. (1968). Parasitic infection in the U.K. *Nursing Times*, 16th August, p. 1106.

[140] Dawson, J. B. (1963). Taenia in expulsis. *Lancet*, 1, 24.

[141] Silverston, N. A. (1962). Whipworms in rural areas. *Br. med. J.*, 2, 1726.

References

[142] McNicholl, B. (1963). Whipworms in rural areas. *Br. med. J.* **1**, 264.

[143] Cole, A. C. E. (1968). Parasitic infections in the U.K. *Nursing Times*, 23rd August, p. 1147.

[144] Filmstrip. *Refuse Storage and Collection by Paper Sack.* Camera Talks Ltd., 31 North Row, London W1.

[145] *Nursing Mirror*, (1964). Refuse storage and collection by paper sack. 23rd October, p. viii.

[146] Few, E. (1969). Survey in Reading on disposal of soiled dressings. *British Hospital Journal and Social Services Review.* 4th July.

[147] Marett, D. L. & Riley, O. M. (1971). Nursing refuse disposal inquiry. *Nursing Times*, 22nd July, p. 113; 2nd December, p. 191.

[148] *Medical News*, (1963). City's plumbing is criticised by MOH. 19th April, p. 28.

[149] Thrower, W. R. (1963). Pollution and public health: the problem. *Medical News*, 7th June, p. 7.

[150] Wright, L. (1960). *Clean and Decent—History of the Bathroom and the Water Closet and of Sundry Habits, Fashions and Accessories of The Toilet, principally in Great Britain France and America.*

[151] Adam, A. E. (1964). Flushed with success. *Medical News*, 10th July, p. 1.

[152] *Nursing Mirror*, (1966). Appalling standard of hygiene in public lavatories. 26th August, p. 497.

[153] *Nursing Times*, (1966). Public In-conveniences. 19th August, p. 1108.

[154] *Planning for Disabled People in the Urban Environment.* Available from Access for the Disabled, 34 Eccleston Square, London SW1. 75p.

[155] Department of Employment and Productivity. (1968). *Cloakroom Accommodation and Washing Facilities.* London: H.M.S.O.

[156] *British Medical Journal*, (1964). Infection and the water-closet. 13th June, p. 1523.

[157] Dorgu, M. P. (1970). On cleaning a bath. *Nursing Times*, 24th February, p. 283.

[158] H.M.S.O. (1969). *Safety in sewers and at sewage work.* London: H.M.S.O.

[159] Gunn, A. D. G. (1971). Our liquid waste—and the spoilage of the seas. *Nursing Times*, 28th January, p. 110.

[160] Sale, C. (1969). *The Specialist.* London: Putnam.

[161] Baldwin, J. T. (1970). Nursing in a cold climate. *Nursing Mirror*, 25th December, p. 19.

[162] Gunn, A. D. G. (1971). Abuse of the air. *Nursing Times*, 18th February, p. 197.

[163] H.M.S.O. *The Protection of the Environment: the Fight Against Pollution.* London: H.M.S.O.

[164] Royal College of Physicians, (1970). *Air Pollution and Health.* London: Pitman.

[165] H.M.S.O. (1969). *Adults' and Adolescents' Smoking Habits and Attitudes.* London: H.M.S.O.

[166] H.M.S.O. (1969). *Medical Students' Attitudes Towards Smoking.* London: H.M.S.O.

[167] Dunwoody, J. (1971). Smoking: What can we do? *Nursing Times*, 18th March, p. 316.

[168] Fletcher, C. M. & Horn, D. (1970). Smoking and Health. *WHO Chronicle.* August.

[169] Lemin, B. (1971). Here's how to stop smoking! *Nursing Mirror,* 22nd October, p. 32.

Two leaflets: *How to stop smoking* and *Why should I stop smoking?* can be obtained free from local authority clinics.

[170] *Nursing Times,* (1971). Smoking and babies. 18th February, p. 220.

[171] Halliday, N. P. (1970). Smoking and the midwife. *Nursing Mirror,* 25th December, p. 18.

[172] *Nursing Times,* (1971). Smoking and the hospital. 25th November, p. 1481.

[173] *The Psychological Dynamics of Smoking.* (1968). Tavistock Institute, London.

[174] Dixon, M. (1969). Do you eat smoke? *Nursing Times,* 9th January, p. 63.

[175] Four educational leaflets: *Eye Care in the Home; Eye Care on the Road; Eye Care at Work; Eye Care over Forty.* Obtainable from Eye Care Information Bureau, 3 Clements Inn, London WC2. (send a stamped addressed envelope).

[176] *Nursing Times,* (1968). Focus on eyes. 4th October, p. 1339.

[177] *Nursing Mirror,* (1971). For the blind reader. 12th February, p. 19.

[178] *Nursing Mirror,* (1971). New hope for the blind. 14th May, p. 43.

[179] *Nursing Times,* (1970). Sound pictures for the blind. 4th June, p. 719.

[180] Ross, E. (1969). Educating the blind in Jamaica. *Nursing Mirror,* 7th March, p. 40.

[181] *Nursing Times,* (1969). Led by laser. 30th October, p. 1386.

[182] *Nursing Times,* (1971). Wax works trap gnats. 2nd December, p. 1515.

[183] Moody, D. & Chesham, I. (1971). Hearing tests for young children. *Nursing Times,* 17th December, p. 34.

[184] Klein, D. (1970). A hearing clinic. *Nursing Times,* 4th June, p. 712.

[185] *Nursing Times,* (1968). Audiology service in County Durham. 29th November, p. 1625.

[186] Edwards, R. (1968–69). What it means to be deaf. *Nursing Times,* 20th December, p. 1718; 27th December, p. 1759; 2nd January, p. 12.

[187] Department of Education and Science, (1969). *Peripatetic teachers of the deaf.* London: H.M.S.O.

[188] *Nursing Times,* (1971). Telephone for the deaf. 5th August, p. 952.

[189] Keysell, P. (1970). Theatre for the deaf. *Nursing Times,* 14th May, p. 639.

[190] *Nursing Mirror,* (1968). Helping hard-to-hear children to adjust. 12th January, p. v.

[191] *Nursing Mirror,* (1969). Noise and the law, 4th April, p. 39.

[192] *The Law on Noise.* (1969). Public Noise Abatement Society. London.

[193] H.M.S.O. (1968). *Noise and the Worker.* London.

[194] Ward Gardner, A. (1969). Noise at work. *Nursing Mirror,* 15th August, p. 32.

[195] Dicker, K., (1971). The 24-hour bustle. *Nursing Mirror,* 26th November, p. 12.

[196] Myerscough, P. (1972). Health education in pregnancy, *Nursing Mirror,* 7th January, p. 11.

[197] Altschul, A. (1972). *Psychology for Nurses.* 4th ed. London: Ballière, Tindall and Cassell.

[198] Barnard, D. (1968). Health visitors and the Family Planning Act. *Nursing Mirror,* 13th December, p. 34.

References

[199] Lumley, I. B. & Macdonald, R. R. (1968). Immediate postnatal family planning. *Nursing Mirror*, 13th December, p. 35.

[200] Hill, H. (1970). The challenge of family planning. *Nursing Mirror*, 21st August, p. 27.

[201] *Nursing Mirror*, (1971). Barrier methods of contraception. 21st May, p. 21.

[202] Addo, C. (1971). The midwife in family planning. In Ghana. *Nursing Mirror*, 3rd December, p. 34.

[203] Norattejananda, S. (1971). The midwife in family planning. In Thailand. *Nursing Mirror*, 10th December, p. 40.

[204] Dicker, K. D. (1971). The unmentionable diseases. *Nursing Times*, 21st January, p. 94.

[205] Oates, J. K. (1969). The sexually transmitted diseases. *Nursing Mirror*, 31st January, p. 17; 7th February, p. 24; 14th February, p. 24; 21st February, p. 38.

[206] Beanfield, W. K. (1971). Gonorrhoea. *Nursing Times*, 1st April, p. 382.

[207] Allen, P. G. (1964). Good design can help to prevent home accidents. *Medical News*, 13th November, p. 13.

[208] Allen, P. G. (1964). Safety in the home for the elderly. *Medical News*, 20th November, p. 11.

[209] Allen, P. G. (1964). Prevention of children's accidents better than cure. *Medical News*, 27th November, p. 11.

[210] British Medical Association, (1964). *Accidents in the Home.* BMA report including a survey on non-fatal domestic accidents.

[211] *Medical News*, (1965). Fatal facets of a mechanized home sweet home. 27th August, p. 11.

[212] *Prophylax*, (1964). Half of fatal accidents take place at home. *Medical News*, 13th November, p. 29.

[213] Shrand, H. (1964). Accidents in childhood. *Nursing Times*, 4th December, p. 1603.

[214] Backett, E. M. *Domestic Accidents.* Public Health Papers No. 26. Geneva: World Health Organization.

[215] *Nursing Mirror*, (1970). House of hazards. 16th January, p. 12.

[216] Leaflet. *Fire in your Home.* Available free from the Fire Protection Association, Aldermary House, Queen Street, London EC4N 1TJ by sending a stamped addressed envelope.
Fire Prevention, available from same address as 216 on subscription.

[217] *Nursing Mirror*, (1966). Child-proof safety medicine cabinet. 16th December, p. 260.

[218] *Nursing Times*, (1966). Home medicines in safety. 2nd December, p. 1573.

[219] *Nursing Times*, (1970). Mini surgery for mosquitoes. 23rd April, p. 532.

[220] Newsom, S. W. B. (1969). Contaminated water. *Nursing Times*, 11th August, p. 1073.

[221] Stuart-Harris, C. (1972). Immunization. *Nursing Mirror*, 14th June, p. 33.

[222] Maurier, I. M. (1968). Disinfectants in hospital. *Nursing Mirror*, 23rd February, p. 25.

[223] Roper, N. (1973). *Principles of Nursing.* 2nd ed. Edinburgh and London: Churchill Livingstone.

[224] H.M.S.O. (1955). *Report of the Committee on Maladjusted Children.* (The Underwood Report.)

[225] H.M.S.O. (1968). *A Handbook of Health Education.* Department of Education and Science.

Index

A

Abdominal wall, 98
  action of, 101
Abortion, 452
Acarus scabiei, 73
Accidents in the home, 470
Acetabulum, 61
Acne vulgaris, 129
Acromegaly, 385
ACTH, 385
Addiction, 207
Adenoids, 365
Adolescence, 201
Adrenal glands, 393
Adrenaline, 395
Adrenocorticotrophic hormone, 385
Adult, feeding requirements, 231
Air, 333
Albinos, 127
Alcoholism, 209
Aldosterone, 394
Amino acids, 142, 234
Amoeba, 6
Amoebic dysentery, 240
Amoeboid action, 7
Amylase, 282
Anal canal, 283
Androgens, 394
Aneurine, 225
Angiotensin, 347
Angular stomatitis, 272
Animal reservoirs of infection, 242
Ankle joint, 116
Antenatal clinic, 36
Anti-anaemic factor, 277
Antibodies, 145, 498
Anticoagulants, 145
Antidiuretic hormone, 374, 386
Antitoxins, 498
Aorta, 160
Aphthous stomatitis, 272
Aponeuroses, 16
Arachnoid membrane, 182
Arteries, 156
Arterioles, 157
Artificial ventilation, 345
Ascorbic acid, 226
Athlete's foot, 73
Atmosphere, 333
ATP, 7
Autonomic nervous system, 193

B

Baby feeding, 221, 230
Bacillary dysentery, 240
Backbone, 54
Bacteria, 491
  in air, 336
  food poisoning, 243

Bad housing, 484
Balance, sense of, 186, 420
Bartholin's glands, 431
Bath, 323
Bathing, 125
Bathroom, 323
BCG, 38, 366
Bed bug, 131
Bedpan washers, 319
Bidet, 317
Bile, 282
  composition of, 291
  functions of, 291
  pigments, 144
Bilharzia, 298
Black beetles, 481
Blackheads, 129
Bladder, urinary, 375
Blepharitis, 410
Blind people, 411
Blisters, 72
Blood, 141
  clotting, 148
  groups, 149
  pressure, 160
  pressure in kidneys, 374
Body louse, 130
Bone, 17
  flat, 45
  long, 43
Botulism, 239
Bowel, as reservoir of infection, 242
Bracing atmosphere, 337
Breast bone, 75
Breasts, 438
  examination of, 442
Bronchi, 352
Brucellosis, 255
Bunion, 74

C

Caesium, 253, 261
Calciferol, 227
Calcitonin, 387, 391
Calcium, 228
Callosity, 72
Cancellous bone tissue, 17
Cancer, 175, 178, 367
Candida albicans, 456
Candles, 416
Capillaries, blood, 158
  lymphatic, 174
Carbohydrates, 144, 222
  in milk, 252
Carbon dioxide, 336
Cardiovascular system, 141
Care of, cardiovascular system, 167
  digestive system, 293
  ears, 423

513

Care of, endocrine system, 398
  eyes, 406
  face, 128
  feet, 70
  hair, 127
  hands, 88
  home, 478
  lymphatic system, 177
  milk, in the home, 254
    in the hospital, 254
  mouth, 270
  nervous system, 204
  reproductive system, 442
  respiratory system, 362
  skin, 125
  urinary system, 378
Carriers, 241, 444, 499
Cartilage, 16
Cauda equina, 189
Cavernous sinuses, 163
Cellular energy, 7
Central heating, 489
Cerebellum, 185
Cerebrospinal fluid, 182, 189
Cerebrum, 183
Cerumen, 421
Cervical smears, 442
Cervical vertebrae, 56
Cervix, 433, 434
Chancroid, 455
Cheilosis, 272
Chemical closet, 331
Chemical food poisoning, 240
Chemical light, 417
Chilblain, 74
Chimneys, 345
Cholera, 241
Choroid, eye, 401
Choroid plexuses, brain, 182
Chromosomes, 8, 10
Chyle, 176
Cilia, 14
Circle of Willis, 162
Circulatory system, 141
Citizen, individual as, 40
Clavicle, 81
Cleanliness of milk, 253
Clitoris, 430
Closed combustion stove, 486
Closet, chemical, 331
  earth, 330
  railway, 331
  water, 312
Clothing, 133
  choice of, 136
  cleanliness of, 139
  reasons for wearing, 135
Coal fire, 486
Coccyx, 61
Cochlea, 48, 422
Cockroaches, 481
Coeliac artery, 160
Collagen, 16
Collar bone, 81
Common cold, 365
Common flea, 131
Communicable disease, 492
Compact bone tissue, 17

Condensed milk, 250
Conditions interfering with, digestive
  system, 294
  eyes, 409
  feet, 72
  hands, 89
  lymphatic system, 178
  mouth, 271
  respiratory system, 365
  skin, 129
Conduction, 485
Congenital cataract, 410
Conjunctiva, 406
Connective tissue, 105
Conservancy system, 330
Constipation, 294
Contraception, 448
Control of epidemics, 499
Convection, 335, 485
Corn, 72
Cornea, 400
Coronary arteries, 151
Cortisol, 145, 394
Cotton, 136
Coughing, 189
Cranium, 47
Cretin, 389
Cyanocobalamin, 147, 226
Cytoplasm, 5

D

Day nursery, 238
Deaf people, 424
Defaecation, 286
Deglutition, 274
Deltoid, 95
Dental caries, 271
Dependence, 207
  on alcohol, 208
  on amphetamines, 211
  on barbiturates, 211
  on cocaine, 212
  on heroin, 212
  on LSD, 211
  on morphia, 212
  on pot, 210
  on tobacco, 208
Dermatitis, 90
Dermis, 123
Desalination, 259
Designations of milk, 253
Development and maturity, 195
Dextranase, 271
Diabetes, insipidus, 386
  mellitus, 392
Diaphragm, 359
Digestive system, 265
Diseases spread by, milk, 255
  water, 260
Dish-washing machines, 328
Disinfection, 501
Disposable bedpan units, 320
Disposal of refuse, 304
Distillation of water, 263
Divorce, 448
DNA, 5, 6, 9
Dorsal vertebrae, 56

Double helix, 5
Droplets, 242, 493
Drying of milk, 246
Duodenum, 279
Dura mater, 182
Dust, 343, 344, 494
Dustbin, 360
Dysentery, 240

E

Ear, 420
Earth closet, 330
Eczema, 90
Effect on CVS of, emotion, 167
  exercise, 165
  haemorrhage, 167
  posture, 167
  rest, 166
  shock, 167
  sleep, 167
Ejaculatory duct, 441
Elbow joint, 118
Electric cloth, 138
Electric fires, 488
Electric light, 415
Elements, 144, 229
  in milk, 252
Elephantiasis, 178
Embolus, 166
Endemic disease, 492
Endocardium, 153
Endocrine system, 382
Enteric fever, 240
Enterokinase, 283
Enzymes, 145
Eosinophils, 148
Epidemic disease, 492
Epidermis, 122
Epididymis, 440
Epiglottis, 350
Epithelium, 14
Equilibrium, sense of, 186, 420
Erepsin, 283
Ergosterol, 122
Erythrocytes, 145
Ethmoid, 49
Evaporation of milk, 246
Excretal refuse, 311
Exercise, 108, 165
Exophthalmic goitre, 389
Expiration, 361
Eye, 399
Eyebrows, 405
Eyelids, 405

F

Face, 51
Faeces, 286
Fallopian tubes, 434
Fatigue, 110
Fats, 143, 221
  in milk, 252
Feedback mechanism, 383
Femur, 65
Fibula, 68

Filariae, 178
Filum terminale, 189
Finger prints, 122
Fissure of Rolando, 183
Flaccid paralysis, 187
Flannelette, 138
Flat bones, 45
Flat feet, 72
Flies, 479
Fluorine, 230
Folic acid, 226
Follicle stimulating hormone, 385
Fomites, 494
Fontanelle, 47
Food-borne infection, 240
Food poisoning, 238
Foot, 69
Footwear, 71
Foramen magnum, 49
Fractured base of skull, 50
Fresh air, 336
Frontal bones, 48
Fur, 138
Furniture beetles, 480

G

Gall-bladder, 290
Gamete, 13
Gas fires, 487
Gas gangrene, 493
Gas lighting, 416
Gastric artery, 161
Gastric juice, 278, 299
Gastric vein, 278
Gastrin, 277
General practitioner, 36
Genes, 10
Genital herpes, 456
Gigantism, 384
Glucagon, 391
Glucocorticoids, 394
Gluconeogenesis, 223
Gluteal muscles, 102
GNC questions on, air and respiratory system, 368
  blood and circulation, 172
  bone and skeleton, 59, 91
  digestive system, 302
  ear and noise, 427
  endocrine system, 398
  eye, 414
  health and NHS, 242
  housing and household pests, 484
  infection and immunity, 500
  joints, 120
  nervous system, 216
  nutrition, 238
  reproductive system, 463
  respiratory system, 368
  skin and parasites, 133
  urinary system, 381
Gonorrhoea, 454
Goose flesh, 123
Granular leucocytes, 147
Grave's disease, 389
Greenstick fracture, 244
Griseofulvin, 131

Growth hormone, 384
Gum boil, 271

H

Haemoglobin, 146
Haemorrhage, effect on CVS, 167
Haemorrhoids, 294
Hair follicle, 123
Hammer toes, 74
Hand, care of, 88
  description of, 87
  reservoir of infection, 241
Head louse, 129
Heaf test, 38, 366, 493
Health, 19
Health education, 26
Health visitor, 37, 78, 213, 388
Hearing, and balance, 423
  and mental health, 425
Heart, 151
Heating, 485
Hip joint, 114
Hookworm, 297
Hormones, 145, 382
  influencing urine, 374
Hospital lighting, 417
Hospital refuse, 310
Household (dry) refuse, 304
  final disposal of, 309
Housing, 464
Humerus, 82
Humidity, 333
Hyaline cartilage, 17
Hydrochloric acid, 277
Hydrocortisone, 394
Hygiene of foodshops, 233
Hymen, 413
Hypermetropia, 409
Hyperparathyroidism, 391
Hyperpituitarism, 384
Hyperthyroidism, 389
Hypervitaminosis D, 228
Hypoparathyroidism, 391
Hypopituitarism, 384
Hypothyroidism, 388

I

Ileum, 279
Iliopsoas, 101
Ilium, 62
Illegitimacy, 446
Immovable joints, 113
Immunity, 498
Immunization, 37, 499
Incubation period, 492
Infancy and early childhood, 197
Infant and child, care, 444
  welfare clinic, 37
Infant feeding, 230
Infectious disease, 490
  modes of transmission, 495
Inferior turbinate bones, 52
Ingrowing toenail, 74
Innominate bone, 61
Inspiration, 361
Insulin, 391

Intercostal muscles, 78, 360
Internal ear, 422
Intestinal juice, 283
Intestine, large, 283
  small, 279
Intramuscular injection, 95, 102, 103
Intravenous injection, 48, 163
Iodine, 229
Iris, 402
Iron, 229, 252
Ischium, 62
Islets of Langerhans, 391

J

Jail fever, 130
Jejunum, 279
Joints, ankle, 116
  elbow, 118
  freely movable synovial, 113
  hip, 115
  immovable, 113
  knee, 116
  shoulder, 117
  slightly movable, 113
  wrist, 119

K

Kidneys, 370
Kitchen, 326
Knee joint, 116

L

Labia, majora, 429
  minora, 430
Lacrimal apparatus, 406
Lacrimal bones, 51
Lactase, 283
Lactation, diet in, 232
Lacteals, 176, 280
Large intestine, 283
  functions of, 285
Larynx, 350
Latent heat, 123
Later childhood, 200
Latrine, 330
Lavatory pan, 314
Leather, 138
Leisure, 204
Leucocytes, 147
Levator ani, 105
Lice, 129
  genital, 456
Lighting, 414
Linen, 137
Lipase, 283
Liver, 287
  functions of, 289
Long bones, 43
Lorexane shampoo, 130
Lower motor neurone, 93, 94, 96
Lumbar puncture, 189
Lumbar vertebrae, 56
Lungs, 354
Luteinizing hormone, 384
Lymphatic capillaries, 174

Lymphatic ducts, 176
Lymphatic nodes, 174
Lymphatic system, 173
Lymphatic vessels, 174
Lymphocytes, 16, 148, 174

M

Magnesium, 230
Malalignment of teeth, 272
Malar bones, 52
Malnutrition, 218
Maltase, 283
Mandible, 53
Mantoux test, 38
Marriage, 445
  guidance, 447
Mass radiography unit, 39
Maternity hospital, 37
Maxillary sinus, 52
Mediastinum, 178, 355
Medulla oblongata, 188
Meiosis, 9
Meninges, 182
Menstrual cycle, 436
Menstruation, 127
Mental and physical health,
  to the community, 42
  to the family, 41
  to the individual, 41
Mesentery, 282
Mice, 481
Micturition, 378
Midbrain, 187
Middle ear, 422
Midwife, 37
Milk, 246
  from plants, 250
Mineralocorticoids, 394
Mitosis, 8
Mitral valve, 152
Modification of a home, 478
Monocytes, 148
Mons veneris, 429
Mosquitoes, 178, 483
Moths, 480
Mouth, 265
  functions of, 269
  reservoir of infection, 242
Multicellular, 4
Muscle co-ordination, 95, 186
Muscle tissue, cardiac, 106
  functions of, 107
  involuntary, 105
  skeletal, 92
  tone, 190
Muscleworm, 297
Myocardium, 153
Myopia, 409
Myxoedema, 389

N

Nail biting, 89
Nails, 123
Nasal bones, 51
Nerve tissue, 181
Nicotinic acid, 225

Nitrogen, 336
Noise, 426
Non-gonococcal urethritis, 455
Noradrenaline, 395
Normal development, 195
Normality, 196
Nose, 347
  reservoir of infection, 241
Nutrition, 217
Nutritional impulses, 186
Nylon, 137

O

Obesity, 74, 218
Obturator foramen, 61
Occipital bone, 49
Occupational health service, 39
Oedema, 158
Oesophagus, 275
Oil lamps, 416
Omentum, 278
Ophthalmic services, 39
Osmotic pressure, 142
Ossicles, 422
Ossification, 17, 43
Osteitis fibrosa cystica, 391
Osteoblasts, 17
Osteoclasts, 17
Outdoor refuse, 309
Ovaries, 436
Overcrowding, 477
Overweight, 74
Ovum, 13, 385, 436
Oxygen, 145, 335
Oxyhaemoglobin, 146
Oxytocin, 386, 434
Ozone, 503

P

Palatal bones, 53
Palmar arches, 161
Pancreas, 2, 91
Pancreatic juice, 282
Pandemic disease, 492
Paraffin oil heaters, 488
Paralysis, flaccid, 187
Paranasal sinuses, 348
Parasympathetic, 193
Parathormone, 391
Parathyroid glands, 390
Paratyphoid fever, 241
Parietal bones, 48
Paronychia, 90
Parotid glands, 268
Pasteurization, 253
Pectoralis major, 97
Pediculus, capitis, 129
  pubis, 130
  vestimenti, 130
Pelvic floor, 104
Pelvic girdle, 59
Penis, 441
Pepsin, 277
Perception, aural, 420
  visual, 399
Pericardium, 153

Perineum, 105, 430
Periosteum, 17
Peristalsis, 106
Peritoneum, 282
Perleche, 2, 72
Peyer's patches, 282
Phagocytic action, 7
Pharynx, 272, 348
Phosphorus, 229
Phrenic arteries, 162
Pia mater, 182
Piles, 284, 294
Pineal body, 397
Pink eye, 410
Pinna, 421
Pituitary gland, 382
Plague, 131
Plantar arches, 162
Plasma, 141
Plastic, 137
Platelets, 148
Pleura, 356
Pneumoconioses, 366
Pollution of atmosphere, 337
  as it affects health, 344
Polymorphonuclear, 147
Pons varolii, 187
Portable shower, 326
Portal vein, 287
Postural reflexes, 187
Posture, 57, 187
  effect on CVS, 167
Potassium, 229
Pregnancy, 444
  diet in, 232
Prevention, of accidents in home, 470
  of atmospheric pollution, 342
  of infection, 491
Privy, 330
  temporary, 330
Progesterone, 385, 436
Prolactin, 385, 439
Prostate gland, 441
Protein, 220
  in blood, 142
  in milk, 251
Ptyalin, 269
Pubic louse, 130
Pubis, 62
Public health inspector, 250
Pulex irritans, 131
Pulmonary tuberculosis, 366
Pyorrhoea, 271
Pyridoxine, 226

Q

Quadriceps, 103
Quality, of food, 233
  of milk, 248
Quarantine, 492

R

Radiation, 486
Radius, 84
Railway closet, 331
Rat flea, 131

Rats, 482
Rayon, 137
Receptaculum chyli, 176, 281
Recreation, 204
Rectus abdominis, 98
Red blood cells, 145
Red bone marrow, 44
Reflex action, 191
Refuse, excretal, 311
  hospital, 310
  household, 304
  outdoor, 309
  vans, 308
Relapsing fever, 130
Relaxing atmosphere, 337
Releasing factor, 382
Renin, 374
Rennin, 277
Reproductive system, 429
Reservoirs of germs, 241
Resistance to disease, 496
Respiration, 358
Respiratory system, 429
Rest, 205
  effect on CVS, 166
Reticular tissue, 16
Retina, 402
Rib, 77
Riboflavine, 225
Right lymphatic duct, 177
Rights of children, 446
Ringworm, 131
  of the nail, 89
RNA, 6
Roughage, 230
Roundworm, 295
Rubber, 137

S

Sacrum, 60
Safety of the home, 470
Saliva, 269
Salivary glands, 268
Salmonella food poisoning, 239
Sanitation, 303
Sarcoptes scabiei, 131
Scabies, 73, 131
  genital, 456
Scapula, 80
School medical service, 38
School nurse, 178
Sclera, 400
Scrotum, 441
Sebaceous glands, 123
Sebum, 123
Secretin, 281
Self-medication, 442
Semicircular canals, 48, 422
Seminal vesicles, 441
Sense of balance/equilibrium, 86, 420
Sewage works, 329
Sexually transmitted diseases, 453
Shock, effect on CVS, 169
Shoulder, blade, 80
  girdle, 80
  joint, 117

Shower, 125
  portable, 326
Silk, 137
Sinuses, nasal, 348
  venous, 163, 182
Skin, 121, 241
Skull, 46
Sleep, 205
  effect on CVS, 167
Slipped epiphysis, 44
Sluice, 318
Small intestine, 279
Smoking, 178, 363
Sneezing, 189
Social amenities, 36
Socks and stockings, 70
Sodium, 229
Soil and road dust, 493
Solar radiation, 489
Sources of infection, 493
Specific disease, 493
Sperm, 13, 385, 439
Spermatic cord, 440, 441
Sphenoid bone, 49
Spinal column, 54
Spinal cord, 189
Spleen, 179
Sporadic disease, 493
Sputum, 366
Squint, 409
Stale air, 337
Staphylococcal food poisoning, 239
Sterilized milk, 253
Sternocleidomastoid, 96
Sternum, 75
Stomach, 277
Stomatitis, 272
Strontium⁹⁰, 253, 261
Styes, 409
Subarachnoid space, 182
Subclavian artery, 160
Subclavian vein, 163
Sublingual glands, 268
Submaxillary glands, 268
Succus entericus, 283
Sucrase, 283
Suleo, 130
Summary of, carbohydrates, 301
  cardiovascular system, 168
  chemical digestion, 301
  clothing, 139
  connective tissue, 18
  cranium, 50
  digestive system, 298
  epithelial tissue, 15
  eye and lighting, 418
  face, 53
  fats, 300
  food groups, 234
  food poisoning, 245
  joints, 120
  muscular system, 107
  nervous system, 213
  prevention of food poisoning, 245
  protein, 299
  reproductive system, 459
  respiratory system, 367
  skeleton, 90

skin and parasites, 132
urinary system, 379
vertebral column and pelvic girdle,
  63
Superior maxillae, 52
Surfactant, 356
Susceptibility to infection, 453, 496
Sutures of cranium, 47
Swallowing, 189, 274
Sweat, 124
Sweat gland, 123
Sympathetic, 193
Symphysis pubis, 61
Syphilis, 454

T

Tampons, 127
Tapeworm, 296
Tarsus, 69
Teeth, 266
Temperature in relation to bacteria,
  243
Temporal bones, 48
Terylene, 137
Testes, 439
Tetanus, 493
Tetany, 391
Theca, 182
Thiamine, 225
Thoracic cage, 75
Thoracic duct, 176, 222, 283
Threadworms, 295
Throat as reservoir of infection, 242
Thrombocytes, 148
Thrombosis, 14, 148, 167
Thrush, 272, 456
Thumb sucking, 89
Thymus gland, 395
Thyroid gland, 386
Thyrotoxicosis, 389
Thyroid stimulating hormone, 385
Thyroxine, 388
Tibia, 67
Tinea pedis, 73
Tissues, 13
  connective, 15
  epithelial, 14
  muscle, 92
  nerve, 181
Tongue, 268
Tonsils, 267
Toxic goitre, 389
Trachea, 352
Trachoma, 410
Transversus abdominis, 99
Trench fever, 130
Trichomoniasis, 455
Tricuspid valve, 152
Trophic impulses, 186
Trypsin, 283
Tuberculosis, 38, 178, 493
Typhoid fever, 241
Typhus fever, 130

U

Ulna, 86

Upper motor neurone, 93, 94
Ureters, 375
Urethra, 377
Urinary bladder, 375
Urinary system, 370
Urine, composition, 375
Utero-ovarian tubes, 434
Uterus, 432

V

Vaccination, 37, 493
Vagina, 431
Valvulae conniventes, 280
Vas deferens, 440
Vasopressin, 386
Veins, 158
Venereal disease, 454
Ventilation, 345
Ventricles, brain, 182, 184, 187
  heart, 152
Vernix caseosa, 125
Verruca, 73
Vertebral column, 54
Villi, 280
Virus hepatitis, 241, 456
Vitamins, A, 224
  B, 225
  C, 226
  D, 227
  E, 228
  K, 228
  P, 228
  in milk, 252
Vomer, 52

Vomiting, 189

W

Wart, 73, 89
  genital, 456
Wash-basins, 320
Washing machines, 328
Water, carriage system, 311
  closet, 312
  conservation, 264
  cycle, 259
  diseases spread by, 260
  in diet, 223
  pollution, 261
  pollution control works, 329
  purification, 262
  standards, 259
  supply, 255
  uses in body, 256
  uses to a community, 256
  world distribution, 259
  world supply, 258
Welfare of people at work, 39
Whipworm, 297
White blood cells, 147
White fibrocartilage, 17
White fibrous tissue, 16
Whitlow, 90
Wholesale storage of food, 232
Wool, 136
Wrist joint, 119

Z

Zygote, 13

Man, his Health
and Environment

Filmset by Typesetting Services Ltd, Glasgow, Scotland
Printed by T. & A. Constable Ltd, Edinburgh, Scotland